HISTORICAL DICTIONARIES
OF WAR, REVOLUTION, AND CIVIL UNREST
Edited by Jon Woronoff

1. *Afghan Wars, Revolutions, and Insurgencies*, by Ludwig W. Adamec. 1996.
2. *The United States–Mexican War*, by Edward H. Moseley and Paul C. Clark, Jr. 1997.
3. *World War I*, by Ian V. Hogg. 1998.
4. *The United States Navy*, by James M. Morris and Patricia M. Kearns. 1998.
5. *The United States Marine Corps*, by Harry A. Gailey. 1998.
6. *The Wars of the French Revolution*, by Steven T. Ross. 1998.
7. *The American Revolution*, by Terry M. Mays. 1999.
8. *The Spanish-American War*, by Brad K. Berner. 1998.
9. *The Persian Gulf War*, by Clayton R. Newell. 1998.
10. *The Holocaust*, by Jack R. Fischel. 1999.
11. *The United States Air Force and Its Antecedents*, by Michael Robert Terry. 1999.
12. *Civil Wars in Africa*, by Guy Arnold. 1999.
13. *World War II: The War Against Japan*, by Anne Sharp Wells. 1999.
14. *British and Irish Civil Wars*, by Martyn Bennett. 2000.
15. *The Cold War*, by Joseph Smith and Simon Davis. 2000.
16. *Ancient Greek Warfare*, by Iain G. Spence. 2002.
17. *The Vietnam War*, by Edwin E. Moïse. 2001.
18. *The Civil War*, by Terry L. Jones. 2002.
19. *The Crimean War*, by Guy Arnold. 2002.
20. *The United States Army,* by Clayton R. Newell. 2002.

Historical Dictionary of Ancient Greek Warfare

Iain G. Spence

*Historical Dictionaries of War,
Revolution, and Civil Unrest, No. 16*

The Scarecrow Press, Inc.
Lanham, Maryland, and London
2002

SCARECROW PRESS, INC.

Published in the United States of America
by Scarecrow Press, Inc.
A Member of the Rowman & Littlefield Publishing Group
4720 Boston Way, Lanham, Maryland 20706
www.scarecrowpress.com

4 Pleydell Gardens, Folkestone
Kent CT20 2DN, England

British Library Cataloguing in Publication Information Available

Library of Congress Cataloging-in-Publication Data Available

ISBN 0-8108-4099-5 (cloth: alk. paper)

♾™ The paper used in this publication meets the minimum requirements of
American National Standard for Information Sciences—Permanence of
Paper for Printed Library Materials, ANSI/NISO Z39.48-1992.
Manufactured in the United States of America.

To Christine

CONTENTS

EDITOR'S FOREWORD

In some ways the forms of warfare practiced by the Ancient Greeks seem to have almost nothing to do with modern warfare. Weapons have not ceased becoming more powerful and their destructive capability all too easily overwhelms the individual warrior, who has hardly evolved. Yet, it is still the human being who devises the strategies and tactics, flies the planes, navigates the ships, deploys on the ground, and targets and fires the weapons. In this, even the most high-tech forces bear a resemblance to the Greeks. Moreover, alongside the high-tech warfare, there are still countless wars and lesser encounters which are fought between soldiers, some of them using fairly rudimentary weapons and occasionally engaging in hand-to-hand combat in which the individual's ability determines victory or defeat, life or death. This means that, no matter how remote, lessons can be learned from the ancient Greeks and certainly their cunning and courage can be admired and aspired to.

This *Historical Dictionary of Ancient Greek Warfare* is a special sort of primer. The dictionary portion includes entries on many of the aspects that have made the Greeks famous to the present day—their strategies and tactics, their weapons and formations—and which describe their ability not only to defeat but to rule their opponents. There are also entries on famous leaders, mainly military but also civilian, and the decisive battles they fought, on land and at sea, against redoubtable enemies (and among themselves) in a series of wars that never seemed to end. These innumerable battles and wars can be followed more readily over the centuries in the chronology, while the introduction inserts them in the broader context. The bibliography then suggests further reading which should be of interest not only to specialists in Greek warfare or war buffs but to anyone who wants to know more about the history of Ancient Greece.

This latest volume in the series on warfare through the ages was written by Iain Spence, a senior lecturer in the School of Classics, History and Religion at the University of New England in Australia. He is one of the leading authorities on ancient Greek warfare, on which he has lectured and written extensively. His knowledge of the subject and command of the broad outline and many details are immediately obvious. But it is equally heartening to know that he is also familiar with modern warfare as a colonel in the Australian Army Reserve. This combination of the scholarly and practical would have appealed to the ancient Greeks and should do so to modern readers as well.

Jon Woronoff
Series Editor

ACKNOWLEDGMENTS

I would like to thank my family, particularly my wife Christine, for the support they provided during the writing of this book. I owe a particular debt of gratitude to the series editor, Jon Woronoff, for his incredible patience and extremely helpful advice. Anthony Goodman, a previous postgraduate student of mine deserves considerable thanks for the excellent (and free) work he did in preparing and checking large sections of the bibliography. In addition, a large number of students at the University of New England have helped me clarify my thoughts on Greek warfare over the years and I would like to take the opportunity to thank them collectively.

The work of two recent postgraduates, Helen Hartley and Michael Leichsenring, was particularly timely in helping me with the entries in the areas of naval warfare and non-hoplite infantry, respectively. Another student, Christopher Webber, provided me with an early version of his book *The Thracians, 700 B.C.-46 A.D.*, which has also enhanced the quality of several entries. My (unfortunately) ex-colleagues Doctors Lea Beness and Tom Hillard also provided me with some useful, and interesting, material—particularly concerning sieges. Mrs. Gina Butler provided me with last-minute assistance with some urgent formatting problems—in her usual cheerful and efficient manner.

Plates 1 and 2b are reproduced by permission of the Trustees of the British Museum and plate 2a by permission of the Ashmolean Museum, Oxford. Plates 3 and 4 are by the author and figures 1-7 are by Christopher Spence.

READER'S NOTES

Entries. The dictionary entries are constructed with different levels of detail. Broad, or first level, entries (such as "Sicily" or "Macedon") provide a general overview of a topic. Further details can then be gained from the second level of entry referred to within these (such as "Syracuse" or "Philip II.") Even more detail can then be located by using the cross-references within these (such as the names of individual Syracusan tyrants ("Gelon," "Hieron," etc.) or specific wars and battles fought by Philip II). Entries for topics such as the "Peace of Callias" are listed as "Callias, Peace of."

Cross-references. Cross-references to other entries are indicated by boldface type. These are indicated only the first time they appear within an entry (although if an entry has sections, cross-references are indicated in each section). For cross-references involving names of peoples, see the entry for the city or region. A cross-reference in the plural may refer to an entry in the singular form (and vice versa). So, for "Athenians," "Macedonians," and "mercenaries," see "Athens," "Macedon," and "mercenary;" for *psilos* see *psiloi.* Where there are several entries under a heading (e.g., where several people or places share a name), the cross-reference indicates the appropriate section of the entry by a number in parentheses. For example, "Pausanias (2)" refers to the king of Sparta, 409-395—the second entry under "Pausanias." Wars are a special case. They are referred to in the text of entries by their most common name, although this may not necessarily appear as a main entry in the dictionary. If not, it will be found as a subentry under the main entry. For example, the First Peloponnesian War does not appear as a separate entry in the dictionary but is a subentry under "Peloponnesian Wars" as "(1) First Peloponnesian War (459-445)." Similarly, the Third Syrian War is a subentry under "Syrian Wars" as "(4) Third Syrian War (246-241)."

Ancient References. The ancient authors are frequently cited so the dictionary can also serve as a handy reference for the best (or best readily available) sources of evidence for the entries. The system used is the ancient referencing system. The advantage of using this system is that it is more precise than a page reference and the reader can consult any edition of that author and quickly find the reference. Most modern translations now provide the ancient numbering system.

Prose writings were subdivided into books, chapters, and sections, and references are traditionally given by these—for example Herodotus 1.23 and Thucydides 9.83.2, meaning Herodotus, book 1, chapter 23, and Thucydides book 9, chapter 83, section 2. A "book" in an ancient author's work is roughly equivalent to a chapter in a modern work (Herodotus is divided into nine books, Thucydides into eight); "chapters" roughly equate to a paragraph or two and "sections" are usually a couple of sentences long. References to ancient verse works are usually by line. References to speeches are by the speech number, title, and section, but note that for some orators different numbering systems exist and speech numbers are not always consistent. So if in doubt focus on the title.

Transliteration of Greek Names and Words. In the past, Greek names were traditionally rendered into English in a Latinized form. For example, Greek has no letter "c" and uses "ai" where Latin has "ae," and many Greek names end in "os" whereas the Latin ending is "us." However, the modern tendency is to use a straight transliteration of the Greek. This system has "Perikles" and "Granikos" for "Pericles" and "Granicus," "Aischylos" (or "Aiskhylos") and "Hipparchos" (or "Hipparkhos") for "Aeschylus" and "Hipparchus," "Peisistratos" for "Pisistratus," and "Kounaxa" for "Cunaxa." I have followed the traditional system in this dictionary as this is the most familiar to the general reader but have given the transliterated name in brackets at the start of the entry. In the text, transliterated Greek words (except for proper names) are denoted by italics.

Hellenistic Monarchs. The same names are often repeated in Hellenistic royal families, and in antiquity an epithet or nickname was often used to distinguish between rulers of the same name. Ancient historians refer to the monarchs of Hellenistic Greece in a variety of ways, including by the original epithet or by number in the dynasty. For example, "Antigonus I" and "Antigonus Monophthalmus" ("Antigonus One-Eye") denote the same man. For the sake of clarity this dictionary uses both number and epithet for all such monarchs, so the Antigonus in the example appears as "Antigonus I (Monophthalmus)." The one exception to this is Alexander III (the Great) who throughout is referred to simply as "Alexander the Great." It was some time after Alexander's death that the *diadochoi* assumed royal titles and prior to this they were known just by name and/or nickname. For ease of reference the dictionary consistently refers to members of this first generation of kings by their royal title—even when the period concerned was prior to their assumption of the monarchy.

Dates. All dates are B.C. unless otherwise indicated. Many dates in ancient Greek history are approximate and a variety of dating systems operated in different cities. In Athens the year was dated by the name of the chief *archon* or magistrate and ran from the month Hekatombaion, which usually fell in high summer, to the same month the following year. When an ancient source simply dates an event by the Athenian *archon* year (e.g., "in the year of Themistocles"), it could have occurred in either of two years under our modern dating system. Such dates are traditionally rendered like the following: 493/2, 321/0. Greek calendars were based on the lunar cycle, which meant adding extra dates or omitting dates in order to bring it into line with the solar cycle. Precise dates are therefore often very difficult to establish.

Periods. For ease of reference to broad periods of Greek history, I have used several terms which are in general use (although not always with agreement on their exact dates). These are: "Mycenaean" = circa 1600-1100 B.C.; "Dark Age" = circa 1100-700 B.C.; "Archaic" = circa 700-500 B.C.; "Classical" = circa 500-323 B.C.; and "Hellenistic" = circa 323-30 B.C. Note that the divisions between these periods are approximate.

Translations. A wide variety of ancient Greek sources are readily available to the modern reader in excellent translations such as the Penguin and Oxford Classics series. However, all translations in this book are my own, unless otherwise indicated.

Measurements. All measurements are given in both English and metric (the conversions are generally rounded to the nearest whole number). Individual Greek states used different measurement systems, which sometimes makes accurate conversions difficult.

Illustrations. All the drawings are by Christopher Spence and all the maps were produced by the author.

ABBREVIATIONS

Aen. Tact.	Aeneas Tacticus
Anab.	*Anabasis* (of either Arrian or Xenophon)
Arr.	Arrian
b.	born
c.	circa (around the time of)
CAH	*Cambridge Ancient History*
cf.	compare
Curt. Ruf.	Curtius Rufus
d.	died
Diod. Sic.	Diodorus Siculus
ed.	edited/edition
FGrH	F. Jacoby, *Fragmente der griechischen Historiker*. Berlin, 1923-
fl.	floruit (flourished, was active around the time of)
fr.	fragment
Hdt.	Herodotus
Hell.	Xenophon, *Hellenica* (*Historia Graeca* [*HG*] or *A History of My Times*)
Hipparch.	Xenophon, *Hipparchikos* (*de Equitum Magistro* [*Eq. Mag.*] or *Cavalry Commander*)
IG	*Inscriptiones Graecae*
Lak. Pol.	Xenophon, *Lakedaimonion Politeia* (*Respublica Lacedaemoniorum* [*Resp. Lac.*] or *Constitution of the Lacedaemonians*)
lit.	literally
Mor.	Plutarch, *Moralia*
OG	V. D. Hanson, *The Other Greeks*
Paus.	Pausanias
PH	Xenophon, *Peri Hippikes* (*On Horsemanship*)
pl.	plural
Plut.	Plutarch
Polyb.	Polybius
sing.	singular
Thuc.	Thucydides
trans.	translated
Xen.	Xenophon
*WACG*²	V. D. Hanson, *Warfare and Agriculture in Classical Greece*, rev. ed.
WWW	V. D. Hanson, *The Western Way of War*

xvii

ILLUSTRATIONS

Maps

Plates

Figures

CHRONOLOGY

This chronology outlines the major events of warfare in the Greek world from circa 720-30 B.C. As wars do not occur in isolation from political and social events and movements, these are listed where relevant. The chronology also includes those events outside the Greek world (e.g., in Persia) which impinged on Greek security or led to hostilities.

The same names often ran in the families of Hellenistic rulers, so they were distinguished at various times by an epithet and/or number, for example, Antiochus III (the Great). As modern writers can refer to them either by name and epithet or by name and number, both are listed here. Where a ruler had more than one title, for example, Ptolemy VIII (Euergetes Physcon), only the more common one is used.

Many of the dates, especially those prior to the middle of the fifth century B.C. and those in the early and mid-third century B.C., are uncertain. For general problems with precise dating in the ancient Greek world, see the Reader's Notes. Where possible, entries for a specific year are in chronological order within that year. A sentence which starts with a question mark means that the date is uncertain but that the event probably happened in or around that year. Question marks modify only the sentence which they start. All dates are B.C. and all battles are land battles unless otherwise specified.

c. 1270   Destruction of Troy (level VI; level VIIa destroyed c. 1190).

c. 1250   Mycenae damaged by attack (or perhaps natural disaster).

c. 1200   Mycenae attacked and severely damaged.

c. 745   Greek migrations overseas; a major wave of colonization begins.

c. 740-720   First Messenian War. Sparta, under King Theopompus, conquers Messenia and reduces its population to serf-like status as helots.

c. 700   Lelantine War on Euboea, between Chalcis and Eretria and their respective allies for possession of the Lelantine Plain. Eretria is defeated and Chalcis becomes dominant in the north, especially the Chalcidice.

c. 685-668   ?Second Messenian War. The Messenians, under Aristomenes, and with Arcadian support, revolt against Sparta and are initially successful but then lose the battle of the Great Foss and are besieged at Eira (traditionally for 11 years) before surrendering (possibly starts as late as 630).

c. 669   Battle of Hysiae; Argos defeats Sparta.

c. 660   First recorded major Greek sea battle, Corinth against Corcyra.

c. 620   War between Corcyra and Ambracia.

c. 595-590   First Sacred War.

c. 570   Egypt gains control of Cyprus.

c. 560-550   War between Sparta and Tegea. Tegea is made a dependent state (but not reduced to the same status as Messenia).

c. 559-556   Miltiades the Elder establishes Athenian control in the Thracian Chersonese (Gallipoli Peninsula).

c. 550   Syracuse destroys Camarina.

c. 547   Battle of the Champions; Sparta secures the disputed territory of Thyreatis from Argos.

546   Cyrus the Great of Persia conquers the Ionian Greeks.

c. 540   Croton, Metapontum, and Sybaris together destroy Siris.

c. 535   Etruscan and Carthaginian fleet defeat Phocaea and Massilia at the battle of Alalia (Corsica).

c. 525   Persia gains control of Cyprus.

524   Cumae defeats Etruscan invaders. ?Miltiades the Younger becomes ruler in the Chersonese.

c. 519   Pisistratid Athens assists Plataea and defeats Thebes in battle. The territories of Plataea and Hysiae are expanded at Theban expense.

c. 512   Darius I of Persia conquers Thrace and the Hellespont.

511-510   Spartan expeditions against the Pisistratids of Athens. The first is defeated but the next, under King Cleomenes I, expels the Pisistratids.

c. 510   Croton destroys Sybaris.

508/7   Athenian power struggle between Cleisthenes and Isagoras, who is supported by Sparta. Cleomenes I is besieged on the Acropolis and forced to withdraw under terms, abandoning many of Isagoras' supporters.

c. 506   Reforms of Cleisthenes at Athens form the basis for a new army structure. The Spartans organize a coalition attack on Attica; simultaneously a Peloponnesian force ravages as far as Eleusis, Boeotia seizes Oenoe and Hysiae in northwest Attica, and Chalcis attacks northeast Attica. On the same day, in two separate battles in Boeotia and Euboea, Athens defeats the

Boeotian League and Chalcis. Athens annexes some of Chalcis' territory and establishes a 4,000-strong cleruchy. Thebes allies with Aegina, which raids the Attic coast, beginning hostilities with Athens which continue intermittently until 481.

c. 500 Persian expedition fails to take Naxos.

499 Ionian Greeks revolt against Persia.

499/8 Aristagoras, a leader of the Ionian Revolt, visits Greece, seeking support against Persia. He is rebuffed by Sparta but Athens and Eretria send ships. The rebel Greek force burns Sardis.

494 The Persians defeat the Ionians at the naval battle of Lade; Miletus falls to the Persians. Sparta decisively defeats Argos at the battle of Sepia.

493 End of the Ionian Revolt. Miltiades the Younger returns to Athens from the Chersonese. Themistocles becomes *archon* at Athens; Piraeus is fortified.

492 Mardonius establishes Persian control over Thrace, Thasos, and Macedon. A Persian fleet invading Greece is sunk in a storm off Athos.

491 Persian embassies demand the submission of Greek states. Many states, especially the islands (including Aegina), comply. Athens appeals to Sparta, which takes hostages on Aegina to prevent it assisting the Persians.

490 Persia invades Greece (via the Cycladic Islands). Siege and destruction of Eretria. Athens and Plataea defeat the Persians at the battle of Marathon.

c. 489 Miltiades leads an Athenian naval expedition to the Cyclades to free them from Persia. Miltiades dies from wounds received at Paros.

487 War between Athens and Aegina.

486/5 Egypt and Babylon revolt against Persia. Darius I of Persia dies and is succeeded by Xerxes I.

485 Gelon becomes tyrant of Syracuse.

c. 483 Themistocles' naval bill leads to a major expansion of Athens' navy. The Persians make preparations to invade Greece.

481 Congress of Greek states meets at the Isthmus of Corinth to discuss resistance to the impending Persian invasion. End of hostilities between warring Greek states (including Athens and Aegina).

480   Second Persian War. Persia invades, the Greeks abandon their position in Tempe (Thessaly), and Thessaly medizes. A Greek army is destroyed at Thermopylae; the accompanying fleet withdraws from Artemisium. Most of Boeotia medizes. The Athenians evacuate their city, which the Persians then sack. The Greek fleet defeats the Persians at Salamis. Xerxes I returns to Persia with a large part of his army, leaving Mardonius in command in Greece. Carthaginian invasion of Sicily is defeated at Himera.

479   Mardonius is defeated and killed at Plataea by the Greeks under the Spartan Pausanias. Persia suffers a naval and land defeat at Mycale (Ionia).

479/8   Rebuilding of Athens' city walls.

478   Pausanias leads a combined Greek expedition against the Persians in Cyprus and Byzantium.

478/7   Foundation of the Delian League (Confederacy of Delos), a naval alliance led by Athens and designed to prosecute hostilities against Persia in Ionia and the Aegean.

476/5   Disgrace of Pausanias and his recall to Sparta. Cimon of Athens now leads the Greek campaign against Persian-occupied territory in Thrace.

c. 471   Ostracism of Themistocles. He is subsequently accused by Sparta of stirring up trouble against it in the Peloponnese and flees to Persia.

c. 468/7   The Delian League crushes the revolt of Naxos, which is made a subject member.

c. 467/6   Athenian land/naval victory over Persia at Eurymedon (Ionia).

c. 465   Thasos revolts against the Delian League and is besieged. Local tribes destroy an Athenian colonizing expedition at Ennea Hodoi (Nine Ways) near Drabescus in Thrace.

464   Earthquake at Sparta is followed by a helot revolt and a Spartan siege of the rebel helots on Mount Ithôme in Messenia (Third Messenian War).

c. 463/2   Thasos surrenders to Delian League forces and is reduced to tributary status and stripped of its mainland possessions.

462   Athens answers a Spartan appeal for help against the helots on Mount Ithôme, dispatching a hoplite force under Cimon. Constitutional reforms of Ephialtes reduce Cimon's power. Cimon is dismissed by the Spartans from Mount Ithôme. Egypt revolts against Persia.

c. 462/1   Athens allies with Argos and Thessaly (aimed at Sparta).

461   Ostracism of Cimon and murder of Ephialtes (Athens).

c. 460   Death of Themistocles (Persia).

460   A fleet of 200 Athenian and Delian League ships is sent to aid the Egyptian revolt against Persia. End of the Third Messenian War. Rebels on Mount Ithôme evacuate under terms and are settled in Naupactus by Athens.

459   The Athenian/Delian League fleet captures Memphis.

459/8   Athens allies with Megara and builds long walls there to control the coast road into Attica from the Isthmus of Corinth. Outbreak of the First Peloponnesian War between Athens and the Peloponnesian League. Corinth and Epidaurus defeat Athens at Halieis. Athens defeats a Peloponnesian fleet at Cecryphalea; Aegina joins in the war against Athens.

458   Athens defeats an Aginetan and allied fleet off Aegina, lands troops, and blockades Aegina by land and sea. Peloponnesian troops are sent to Aegina. Simultaneously a Corinthian army advances against Athens but is heavily defeated in the Megarid by an ad hoc army of Athenians above and below the normal military age.

458/7   Athenian alliance with Segesta (Sicily).

457   Building of the Long Walls of Athens, connecting the city to its port, the Piraeus. A Spartan expedition to northern Greece forces the Phocians to restore territory to Doris and reestablishes the Boeotian League under Thebes as a counterweight to Athens. On its return, this expedition defeats a combined Athenian-Argive army at the close-run battle of Tanagra (summer). Athens defeats the Boeotians at Oenophyta (autumn), gaining control of most of Boeotia and Locris and securing the alliance of Phocis.

457/6   Aegina surrenders to Athens and is enrolled in the Delian League. ?Athens acquires Troezen.

456   Persians blockade the Greek expeditionary force to Egypt on Prosopitis, an island in the Nile. An Athenian naval expedition to the Corinthian Gulf under Tolmides ravages the Peloponnesian coast, burns the Spartan dockyards at Gytheum, and secures the alliance of Zacynthus and Cephallenia. Tolmides also captures Chalcis (Aetolia) and settles it with anti-Spartan Messenians and then defeats Sicyon in a land battle.

455   Athens unsuccessfully attempts to extend its influence into Thessaly. ?Pericles' naval expedition to the Corinthian Gulf demonstrates Athenian naval domination of the area; Achaea allies with Athens.

454   Greek expeditionary force to Egypt surrenders and the Treasury of the Delian League is permanently moved to Athens for safety.

454/3   Athens assists Phocis to gain control of neighboring Delphi.

451   Five Years' Truce between Athens and the Peloponnesians. Thirty Years' Peace between Sparta and Argos.

c. 450   Athenian cleruchies established on Naxos, Andros, and possibly Carystus.

450-449   Cimon leads a naval expedition to Cyprus, defeating a Persian fleet. Cimon dies (from illness) during the siege of Citium. Siege of Citium is raised, but Athens defeats the Persians on land and sea near Salamis (Cyprus).

449   Sacred War between Sparta and Phocis over the occupation of Delphi.

449/8   Peace between Athens and Persia (Peace of Callias or de facto cessation of hostilities?).

448   Spartan expedition to Delphi restores its autonomy and ends Phocian control. ?Athenian alliances with Leontini and Rhegium (Sicily/Italy).

447   Athens restores Phocian control over Delphi. Athens' defeat at Coronea and loss of Boeotia (?early 446). Phocis ends its alliance with Athens. Athenian cleruchies are sent to the Chersonese to secure it from the Thracians and assist in controlling the Black Sea grain route. ?Athenian cleruchies are established on Imbros and Lemnos.

446   Peloponnesians invade Attica under King Pleistoanax of Sparta. Euboea and Megara revolt against Athens; Euboea is regained by Pericles but Megara stays free.

446/5   Thirty Years' Peace between Athens and the Peloponnesians begins. Athens loses Achaea, Troezen, and the Megarid. Aegina remains a member of the Delian League but Athens guarantees its autonomy.

445   Athenian cleruchies established on Euboea and at Brea (Thrace).

441/0   War between Samos and Miletus over Priene (Ionia). The Milesians are beaten and appeal to Athens. Contrary to the Delian League terms, Athens orders Samos to stop fighting and submit to arbitration. Samos revolts.

440   Pericles' expedition against Samos. He expels the opponents of Athens, takes hostages, establishes a pro-Athenian democracy, and withdraws. Samian exiles, with Persian support, return to Samos and seize control, handing the Athenian garrison over to the Persians. Byzantium revolts against Athens. Naval defeat of the Samians by Pericles; siege of Samos.

439 Surrender of Samos, which becomes a subject member of the Athenian empire and is forced to hand over its fleet, destroy its walls, and pay an indemnity to Athens. Byzantium surrenders to Athens.

437/6 Athens defeats the Edonians at Ennea Hodoi and founds Amphipolis and a naval base at nearby Eion. ?Phormio leads an expedition to the Corinthian Gulf and Ambracia; capture of Ambracia. Pericles' expedition to the Black Sea establishes several settlements in the area.

435 Conflict between Corinth and Corcyra over Epidamnus. The Corinthians lose a naval battle off Leucimme and the Corcyraean fleet ravages as far down as Elis (Peloponnese).

433 Athens allies with Corcyra. The Corinthian fleet defeats the Corcyraeans at Sybota but withdraws under threat from an Athenian force.

433/2 Athens renews treaties with Leontini and Rhegium (Sicily/Italy).

432 General revolt against Athens (Chalcidice), sparked by Potidaea. Athens besieges Potidaea and passes the "Megarian decree," restricting Megarian access to Athens (and possibly the whole empire).

432/1 Peloponnesian League conferences at Sparta decide against arbitration and in favor of declaring war against Athens.

431 A Theban surprise attack on Plataea fails. The Peloponnesian War (Archidamian War) begins between the Spartan and Athenian alliance systems. Peloponnesian invasion of Attica.

430 Peloponnesian invasion of Attica. Plague at Athens. Failure of Pericles' attack on Epidaurus; the Athenians fine Pericles. Phormio is sent to Naupactus to secure the Gulf of Corinth. Potidaea surrenders to Athens.

429 Peloponnesians besiege Plataea. Pericles dies. A rebel army defeats an Athenian force outside Spartolus (Chalcidice). Phormio defeats Peloponnesian fleets off Patrae and later off Naupactus (Corinthian Gulf).

428 Peloponnesian invasion of Attica. Revolt of Mytilene against Athens.

427 Peloponnesian invasion of Attica. Fall of Mytilene to Athens. Plataea surrenders to the Peloponnesians—the city is razed and its defenders executed. Civil war at Corcyra. Athenian expedition to Sicily under Laches.

426 Demosthenes campaigns in northwest Greece, winning a battle at Olpae, but losing to the Aetolians at Aegitium.

425 Peloponnesian invasion of Attica. The Athenians fortify Pylos. The Spartans attack them, occupying the nearby island of Sphacteria. Cleon secures Athens' refusal of Spartan peace terms. Athens captures 292 Spartiates on Sphacteria and threatens to execute them if Sparta invades Attica again. Messenians ravage Spartan territory from Pylos. Delian League tribute rates are reassessed. Corcyraean oligarchs are exterminated by democrats.

424 Athenian defeat at the battle of Delium (Boeotia). Brasidas captures Amphipolis and Torone (Chalcidice). Thucydides (the historian) is exiled. Congress at Gela removes the Athenian excuse for action in Sicily.

423 One-year armistice between Athens and Sparta.

422 Cleon's expedition to the Chalcidice recaptures Torone and several other cities. Athens is defeated at Amphipolis; Cleon and Brasidas (the Spartan general) are killed. Peace negotiations between Athens and Sparta.

421 Peace of Nicias signed between Athens and Sparta; Corinth and Thebes refuse to sign. Fifty-year alliance between Athens and Sparta begins.

421/0 Sparta and Boeotia ally.

420 Alliance between Athens, Argos, Mantinea, and Elis.

418 Sparta defeats Argos at Mantinea. An oligarchy is established at Argos. Fifty-year alliance between Sparta and Argos begins.

417 Alliance between Athens and Argos is renewed.

416 Athens besieges Melos, which surrenders unconditionally. The Athenians massacre the adult male Melians, sell the rest of the inhabitants into slavery, and send out a cleruchy of 500 to garrison the island.

415 Athenian expedition to Sicily under Alcibiades, Nicias, and Lamachus. Recalled to Athens to face charges, Alcibiades flees to Sparta.

414 Athenians besiege Syracuse; Sparta sends Gylippus to assist.

413 Peloponnesian War breaks out again (Decelean War). Athenian expeditionary force to Sicily is destroyed. On Alcibiades' advice, the Spartans invade Attica and seize and fortify Decelea.

413/12 Euboea, Lesbos, Chios, and Erythrae prepare to revolt against Athens; Sparta directs its assistance to Chios and Ionia.

412 Revolt of Chios, Erythrae, and Clazomenae and forced defection of Teos. Miletus, Lebedus, and Erae revolt. A democratic coup in Samos is backed by Athens. Cities on Lesbos revolt but Mytilene is soon recaptured.

412/11 Sparta secures Persian assistance against Athens. Cyme and Phocaea defect to the Peloponnesian side. Clazomenae refuses to accept Spartan orders. Cnidus revolts against Athens. Athens besieges Chios.

411 Boeotians capture Oropus from Athens. Minor Peloponnesian naval success against Athens at Syme is followed by Rhodes' revolt against Athens. Oligarchic rule begins at Athens (the Four Hundred); the Athenian army and fleet at Samos remain democratic. Abydus revolts against Athens, followed by Lampsacus, which is immediately recaptured. Byzantium revolts against Athens. Alcibiades is recalled by the Athenians on Samos and elected general. Miletus drives out its Persian garrison. A broader oligarchic government (the Five Thousand) takes control in Athens. Euboea revolts against Athens. Athens achieves naval victories at Cynossema and Abydus (Ionia).

411/10 Evagoras regains Cyprus from the Phoenicians.

410 Athenian naval victory over the Peloponnesians at Cyzicus (Ionia). Full democracy is restored at Athens. Athens refuses Spartan offers of peace.

409 War between Segesta and Selinus (Sicily); Segesta appeals to Carthage. Athens wins the battle of Abydus and defeats Megara at Cerata.

408 Carthaginian expedition to Sicily destroys Selinus and Himera. Byzantium surrenders to Athens.

407 Alcibiades returns to Athens and is elected general.

406 Athenian naval defeat at Notium (Ionia); Alcibiades is exiled. Athenian naval victory at Arginusae (Ionia) results in the trial and execution of Athenian generals for failing to recover Athenian sailors from the water. Athens refuses Spartan peace offers. Carthaginians capture Acragas (Sicily).

406/5 Dionysius I seizes Syracuse and is beaten by the Carthaginians (Gela).

405 Lysander is reappointed to Spartan naval command in Ionia. The Athenian fleet is destroyed at Aegospotami. Carthaginians raze Acragas and besiege Gela. Plague casualties cause Carthage to make peace, but the terms isolate Syracuse.

405-404 Siege of Athens.

404 Athens surrenders to Sparta, ending the Peloponnesian War. Athens' Long Walls are demolished. The Thirty Tyrants take power at Athens. Alcibiades murdered. Athenian democratic exiles seize Phyle and Piraeus;

civil war erupts in Athens. Dionysius I, aided by Sparta, defeats an uprising. Athenian democrats defeat the oligarchs at Munychia.

403   Fall of the Thirty Tyrants at Athens and restoration of democracy.

402   War between Sparta and Elis.

401   Cyrus the Younger, aided by Greek mercenaries—the "Ten Thousand"—attempts to oust his brother as king of Persia. Cyrus is killed at the
battle of Cunaxa but the Ten Thousand successfully withdraw across the
Persian Empire.

400   Sparta declares war on the Persian satrap Tissaphernes and sends Thibron to command forces against him in Asia Minor.

400/399   Sparta, under King Agis II, ravages Elis and imposes severe
terms.

399   Sparta expels the Messenians who had been settled by Athens at
Naupactus and Cephallenia. Agis II dies and is succeeded by Agesipolis II.

398   Thibron's replacement, Dercylidas, continues the Spartan campaign in
Asia Minor. Dionysius I of Syracuse attacks and captures Motya, starting a
war between Syracuse and Carthage.

397   The Carthaginians recapture Motya and defeat a Syracusan fleet.

396   King Agesilaus II takes command of Spartan forces in Asia Minor
and wins some successes. The Persians send Timocrates to Greece with
money to stir up anti-Spartan feeling. Carthaginians besiege Syracuse.
Pharax is sent from Sparta with 30 ships to aid Dionysius I. Plague devas-
tates the Carthaginian army and Dionysius I wins a naval and land victory.

395   Outbreak of the Corinthian War between Sparta and a coalition of
Thebes, Athens, Argos, and Corinth. The Spartan general Lysander is defeated and killed at Haliartus and King Pausanias is forced to withdraw
from Boeotia. Sparta loses control of northern Greece. Euboea, Acarnania,
and the Chalcidice join the anti-Spartan alliance. Agesilaus II attacks Sardis
and Phrygia. Dionysius I establishes strategic settlements at Entella, Adranum, Aetna, and Tyndaris and reinforces other Greek cities, especially
Leontini.

394   Agesilaus II is recalled to Sparta. The Athenian Conon and the Per-
sians defeat the Spartan fleet at Cnidus; Spartan naval dominance ends, and
Persian control of Ionia is restored. Sparta, assisted by Elis, Sicyon, Tegea,
Pellene, and Epidaurus, defeats the anti-Spartan coalition at Nemea and
Coronea.

394/3 Long walls built at Corinth by the anti-Spartan forces help keep Sparta pinned in the Peloponnese. Dionysius I campaigns against the Sicels in inland Sicily but fails to capture Tauromenium.

393 Assisted by Persia, Athens begins to rebuild the Long Walls from the city to the Piraeus. Dionysius I fails to take Rhegium.

c. 393 Athens recovers Lemnos, Imbros, and Scyrus and allies with Chios.

393/2 Dardanian invasion of Macedon.

392 Union of Corinth and Argos. Sparta gains temporary control of Corinth's long walls. Conon dies (Cyprus). Dionysius I and his Sicel allies defeat the Carthaginians under Mago, who makes peace on terms much less favorable than in 405. Sicel states and Tauromenium are abandoned to Dionysius I. Athens rejects Spartan peace terms.

391 ?Athenian Long Walls are completed. Thibron wins over Magnesia, Ephesus, and Priene (Ionia), but is defeated by the Persians under Struthas. Teleutias' naval successes in Ionia lead to a Samos-Sparta alliance. Agesilaus II captures Lechaeum and part of Corinth's long walls. He then invades Acarnania, forcing it into the Peloponnesian League. He captures Piraeon at the north of the Isthmus of Corinth, blockading Corinth and controlling the isthmus. Naval defeat of Dionysius I by the Italiote coalition off Rhegium.

390 Iphicrates' light troops defeat a *mora* of Spartan hoplites at Lechaeum (near Corinth) and the north of the isthmus is retaken by coalition forces. Lucanian allies of Dionysius I defeat Thurii in battle.

389 Athens assists Evagoras of Cyprus' revolt against the Persians. Thrasybulus' naval campaign in the northern Aegean brings over to Athens Byzantium, Chalcedon, Thasos, Samothrace, and the Thracian Chersonese and captures Lesbos. Athens regains control of the Dardanelles and the Propontis (Sea of Marmara) and the route to the Black Sea. Dionysius I defeats the Italiote coalition at Elleporus; Rhegium surrenders and Caulonia is captured and depopulated.

388 Thrasybulus is killed at Aspendus by locals upset by his financial exactions to maintain his fleet. Iphicrates ambushes and defeats a Spartan army near Abydus. Sparta gains the support of Syracuse. Dionysius I attacks and depopulates Hipponium and forces Rhegium into subjection; he now controls both sides of the Straits of Messina and begins a program of settlement and the creation of dependencies in the Adriatic.

387 The Athenian fleet is cut off in the Dardanelles by Antalcidas and Athens' grain supply is interdicted. Dionysius I captures Rhegium.

386 The King's Peace (Peace of Antalcidas), a common peace between the Greek states, backed by Persia, officially ends the Corinthian War. The peace guarantees the autonomy of Greek states (except Lemnos, Imbros, and Scyrus, which go to Athens, and Cyprus and Clazomenae, which go to Persia). Sparta uses the autonomy clause to reduce Theban power in Boeotia and to undo the Corinthian-Argive union.

385-384 Sparta captures Mantinea and breaks it up into its original four villages. The Persians defeat Evagoras at sea (Citium) and besiege Salamis (Cyprus). Dardanian invasion of Molossia (Epirus).

384 Dionysius I raids Etruria.

383 War between Syracuse and Carthage. Dionysius I wins a battle at Cabala (an unidentified location in Sicily), killing the Carthaginian commander, Mago. Dardanian invasion of Macedon.

382 Spartan expedition is sent against the Chalcidian Confederacy after Acanthus and Apollonia complain that Olynthus is infringing their autonomy. A Spartan force under Phoebidas seizes the Cadmea (citadel) at Thebes; Sparta repudiates this action but retains control of the Cadmea. Plataea is rebuilt by Sparta as a counterweight to Thebes. Persians attack Cyprus.

381 Sparta besieges Phlius to force the restoration of oligarchic exiles. The Spartans are defeated at Olynthus and Teleutias is killed.

380 ?Evagoras makes peace with Persia, retaining his rule but paying tribute.

379 Phlius surrenders to Sparta. Olynthus surrenders to Sparta; the Chalcidian Confederacy is dissolved. With Athenian help, Pelopidas expels the Spartans from Thebes. Dionysius I captures Croton but suffers a major defeat at Cronium (Sicily) and makes peace with the Carthaginians, surrendering territory and influence in the west of Sicily.

378 Sphodrias leads an abortive raid on Athens. Sphodrias' actions are repudiated by Sparta but he is acquitted on his return. An enraged Athens allies with Thebes. A Spartan army under Agesilaus invades Boeotia but achieves little.

378/7 The Second Athenian Confederacy is formed with the stated aim of forcing the Spartans to respect Greek freedom. The Spartans invade Boeotia, achieving little.

376 An Athenian fleet under Chabrias defeats a Spartan fleet near Naxos; 17 additional cities join the Second Athenian Confederacy.

375 An Athenian fleet under Timotheus sails around the Peloponnese, enrolling Corycra, Cephallenia, Molossia, and parts of Acarnania in the Second Athenian Confederacy and defeating a Peloponnesian fleet off Acarnania. Chabrias commands an Athenian force in Thrace and northern Aegean. Spartan hoplites are defeated at Tegyra by a smaller Theban force.

374 Athens and Sparta make peace, but shortly afterward war breaks out again when Timotheus interferes in an internal dispute on Zacynthus. Thebes invades Phocis but a Spartan army under Cleombrotus I prevents its conquest. Jason of Pherae forces the submission of Pharsalus and now controls all of Thessaly. Evagoras is assassinated (Cyprus).

374/3 Sparta dispatches a fleet and army to Corcyra, capturing the island and besieging the city. Athens and Jason of Pherae ally.

373 Athenian fleet under Timotheus, sent to relieve Corcyra, is delayed by lack of money. Timotheus is replaced by Iphicrates. The Corcyraeans rout the Peloponnesian force, which withdraws. Thebes seizes and destroys Plataea. Jason of Pherae and King Amyntas III of Macedon ally.

371 Restatement of the King's Peace, a common peace backed by Persia, guarantees the autonomy of Greek cities, the withdrawal of garrisons, and a general disarmament. The day after signing, Thebes withdraws from the treaty. The Spartans invade Boeotia, demanding the dissolution of the Boeotian League, and are heavily defeated at the battle of Leuctra. Jason of Pherae captures Heraclea and attacks Phocis.

371/0 General unrest in the Peloponnese, including anti-Spartan risings in Tegea, Mantinea, and probably Megara, Sicyon, Corinth, and Phlius. The Second Athenian Confederacy forms a defensive alliance with many states in the Peloponnese. Euboea and Acarnania join the Boeotian coalition.

370 Assassination of Jason of Pherae and collapse of unity in Thessaly. Mantinea is rebuilt as a city. The Arcadian League is formed and appeals to Thebes for assistance against Sparta.

370/69 Epaminondas' first Peloponnesian campaign (begun in winter 370 and lasting three months) helps to consolidate the Arcadian League, liberates Messenia from Sparta, and ravages Laconia.

369 Athens sends troops under Iphicrates to the Isthmus of Corinth to assist Sparta. Athens allies with Sparta. Alexander II invades Thessaly but internal conflict in Macedon forces his return; intervention of Thebes. Iphicrates fails to take Amphipolis. Epaminondas' second Peloponnesian campaign (summer). Sicyon and Pellene are forced into the Boeotian League; Corinthian and Epidaurian land is ravaged. The Boeotians withdraw when Dionysius I of Syracuse sends troops to help Sparta. ?Megalopolis is founded as Arcadian League capital (?368).

368 Pelopidas and Boeotian troops free Larissa (Thessaly) from Macedon and check Alexander of Pherae. The future Philip II of Macedon is among the hostages given to Thebes. Alexander II of Macedon is murdered; intervention of Athens. Pelopidas is seized during negotiations with Alexander of Pherae; a Theban expedition fails to recover him, partly because of Athenian help to Alexander. Negotiations at Delphi fail to establish a common peace (Sparta refuses to recognize Messenian independence). Syracuse and Carthage wage war. Dionysius I captures Selinus, Entella, Eryx, and Drepanon but fails at Lilybaeum, losing his fleet there. Sparta, helped by troops from Dionysius I, wins the Tearless Battle against the Arcadians.

367 Athens allies with Dionysius I of Syracuse, who shortly afterward dies (of natural causes); Dionysius II accedes. A Boeotian expedition against Alexander of Pherae secures the release of Pelopidas. Peace is established between Syracuse and Carthage.

366 Epaminondas' third Peloponnesian campaign; Achaea joins the Boeotian coalition. Thebes recovers Oropus from Athens. Athens allies with Arcadia. Corinth and Phlius make peace with Thebes and recognize Messenian independence (partial acceptance of the Peace of Pelopidas). Dionysius II sends troops to Greece to help Sparta and Athens against Thebes. ?Satraps' revolt (Persia). Elis and the Arcadian League go to war over Triphylia.

365 Timotheus secures Samos and settles Athenian cleruchs there. Perdiccas III succeeds to the Macedonian throne and assists Timotheus' activities in the Chalcidice. Dionysius II sends troops to assist Sparta against Thebes. ?The future Philip II of Macedon returns home from Thebes.

364 Thebes destroys Orchomenus (Boeotia). Thebes defeats Alexander of Pherae at Cynoscephalae but Pelopidas is killed. As a result of this battle Alexander of Pherae is subordinated to the Boeotian League, now dominant in central Greece. Timotheus is active in the Chalcidice. ?Athenian cleruchs are sent to Potidaea. A Theban fleet under Epaminondas detaches Byzantium and Rhodes from Athens. The Arcadian League splits after quarrels between Mantinea and Tegea and Theban interference to support Tegea.

362 Boeotian League defeats a Greek coalition (Sparta, Athens, Mantinea, Achaea, and Elis) at the battle of Mantinea; Epaminondas is killed.

362/1 General peace is declared in Greece (Sparta refuses to sign).

361 Thessalian League allies with Athens against Alexander of Pherae. Athens allies with Arcadia, Achaea, Elis, and Phlius. A Boeotian army is sent to Arcadia to maintain control of Megalopolis. Agesilaus II of Sparta campaigns in Egypt, assisting the satraps' revolt.

360 Satraps' revolt ends, restoring Persian royal authority in most of Asia Minor. Agesilaus II dies in Egypt (of natural causes). Timotheus fails to take Amphipolis.

359 Illyrians defeat Macedon in battle, killing Perdiccas III; accession of Philip II (initially as regent).

358 Alexander of Pherae is assassinated. Philip II defeats the Paeonians, then decisively defeats the Illyrians under Bardylis and expands the Macedonian frontier. Philip makes a formal peace with Athens.

357 Athens recovers Euboea (after anti-Theban uprisings) and the Chersonese. Philip II besieges and captures Amphipolis. Chios, Rhodes, Byzantium, and Cos revolt against Athens (start of the Social War). Chabrias dies. Philip besieges and captures Pydna. Dion invades Sicily to overthrow Dionysius II and captures Syracuse.

356 Philip II captures Potidaea, sells the inhabitants into slavery, and hands it over to the Chalcidians. Alexander the Great is born. Rebel coalition defeats an Athenian fleet at the battle of Embatum (or Embata), near Chios. Timotheus and Iphicrates are tried at Athens for failing to vigorously prosecute the war. Iphicrates is acquitted; Timotheus is fined, goes into exile, and dies soon afterward. Chares helps Artabazus' revolt against Persia. Philip defeats the Illyrians and forces the submission of the Paeonians. Dionysius II is defeated at sea and blockaded in Ortygia (the island citadel of Syracuse), later escaping. Internal strife develops in Syracuse between Dion and Heraclides. An Italiote coalition attacks Greek cities in southern Italy.

355 End of the Social War; Athens recognizes the independence of Chios, Cos, Byzantium, and Rhodes. Phocis, under Philomelus, seizes Delphi. Amphictionic League declares war on Phocis (Third Sacred War).

354 Phocians defeat the Thessalians in Opuntian Locris. The Boeotians and Locrians defeat and kill Philomelus at the battle of Neon. Dion is assassinated (Syracuse).

353 Onomarchus of Phocis forces Locrian Amphissa to surrender, secures Doris and Thermopylae, rebuilds Orchomenus (Boeotia), and allies with

Pherae. Athens allies with Cersobleptes of Thrace and refuses a Megalopolitan request for help against Sparta. Philip II captures Methone. Onomarchus defeats Philip in two battles in Thessaly.

352   Athenians under Chares reoccupy Sestus. Philip II takes Pagasae and marches south, defeating the Phocians and killing Onomarchus at the Crocus Plain. Philip captures Pherae, gaining control of Thessaly and the Thessalian League, and is probably elected as *tagus* (although possibly not until 344 or 342). He marches to Thermopylae but withdraws when faced by Phocis and a coalition of allies, including Athens, Sparta, and Achaea. Sparta, aided by Phocis, Elis, Achaea, Mantinea, and Phlius, unsuccessfully attacks Megalopolis (supported by Arcadia, Argos, and Boeotia). Philip helps Perinthus and Byzantium against Cersobleptes of Thrace.

352-351   Chares and Charidemus lead an Athenian expedition to the Hellespont.

351   Philip II defeats Cersobleptes of Thrace, who makes peace. Cyprus revolts against Persia. Sparta defeats Megalopolis in battle and forces a truce. The opposing Phocian and Boeotian forces in the Peloponnese return home.

350   Thebes obtains help from Artaxerxes II of Persia.

349   Athenian alliance with Olynthus. Philip II campaigns against the Chalcidian Confederacy, forcing most of it to submit. Athens prepares an expedition to assist Olynthus and the rest of the Chalcidian Confederacy. Cersobleptes rebels against Macedon. Euboea revolts against Athens (perhaps instigated by Philip).

348   Phocion lands in Euboea and wins the battle of Tamynae, but fails to recover any of the rebelling cities. Athens recognizes the independence of Euboea. Philip II captures and destroys Olynthus. Philip and Athens make preliminary peace moves.

347   Dionysius II recovers Syracuse. Boeotia appeals to Philip II for help in the Sacred War; his troops defeat the Phocians in a minor clash at Abae.

346   Spartan and Athenian expedition to Thermopylae to block Philip II fails to secure the pass because of internal Phocian dissension. Peace of Philocrates is established between Philip and Athens and her allies (from which Phocis and Cersobleptes are excluded). Cersobleptes submits to Philip. Athens declines to send troops to help Philip against Phocis; he occupies Thermopylae anyway and then subdues the rest of Phocis, ending its power and the Third Sacred War. Philip joins the Amphictionic League.

345 Sicilian Greeks elect Hicetas as general and appeal to Corinth for help against tyrants and the Carthaginians.

344 Demosthenes' mission in the Peloponnese to raise support against Philip II. Philip campaigns successfully in Illyria and reorganizes Thessaly. Dionysius II is defeated by Hicetas and penned up in Ortygia. Timoleon sails to Sicily, defeats Hicetas at Hadranum, receives the surrender of Ortygia from Dionysius, and liberates Syracuse.

343 Timoleon resettles Syracuse. Tarentum (Taras), a Spartan colony in southern Italy, appeals for help against the Italiotes; King Archidamus leads a mercenary force to assist.

343/2 Persia reconquers Egypt. Megara and Chalcis (Euboea) ally with Athens. Philip II expels the regent of Epirus, places Alexander I on the throne of Molossia, and enlarges its territory. Corinth, which has colonies in the area, appeals to Athens. Athens allies with the Achaean League, Corinth, Corcyra, Leucas, and Ambracia and establishes friendly relations with Cephallenia, Messenia, Arcadia, and Argos.

342 Athens allies with Messenia and sends troops to Ambracia; Philip II withdraws. Philip dethrones Cersobleptes; Thrace becomes a tribute-paying province of Macedon. Timoleon leads an unsuccessful campaign against Hicetas in Sicily.

341 Demosthenes' mission to the north detaches Byzantium and Perinthus from Philip II. Athenian troops under Phocion install a pro-Athenian democracy at Megara. The Euboean League is formed and allies with Athens. A Carthaginian expeditionary force to Sicily is decisively defeated by Timoleon at the Crimisus River (?339).

340 Philip II besieges Perinthus and Byzantium; Athens declares war. He seizes the Athenian grain fleet in the Hellespont. Byzantium initially refuses to admit Athenian troops but lets in a second expeditionary force under Phocion. Persia allies with Athens and provides financial assistance against Philip. Amphictionic Council (dominated by Philip) declares war against Locrian Amphissa (Fourth Sacred War or Amphissaean War).

339 Sieges of Perinthus and Byzantium are raised. Philip II's Thracian expedition defeats the Scythians under Atheas. Philip is given command of Amphictionic League forces and occupies Elatea, avoiding the Theban garrison at Thermopylae and opening the route to Athens. Athens and Boeotia ally against Macedon and send a joint force to block Philip's route south. Timoleon overthrows tyrannies in Sicily and makes peace with Carthage.

338 Philip II destroys Amphissa, opening a route to Attica through Delphi and Boeotia and ending the Fourth Sacred War. Athens and Boeotia reject a peace offer from Philip and are decisively defeated at the battle of Chaeronea. The Boeotian Confederacy is disbanded and Athens makes peace, dissolving the Second Athenian Confederacy in order to escape occupation. Philip ravages Sparta and imposes control over the Peloponnese. He then forms a league of Greek states at Corinth (ratified in early 337), unifying all Greek states south of Olympus (except Sparta) as a "Hellenic League" (League of Corinth). Lucanians defeat and kill Archidamus at Mandonion (Italy), ending Spartan assistance to Tarentum (Taras).

337 The League of Corinth under Philip II declares war on Persia.

336 A vanguard of around 10,000 men under Parmenio crosses over to Persia. Philip II is assassinated; he is succeeded by Alexander the Great. Widespread unrest and revolts occur, including in Thrace, Thessaly, Ambracia, and Thebes. A bloodless invasion of Thessaly leads to Alexander's acceptance as *archon* of the Thessalian League and then his recognition by the Delphic Amphictiony and election as general of the League of Corinth.

335 Alexander campaigns in Thrace, heavily defeating the Triballi and securing the region. He then allies with the Celts and defeats the Illyrians outside Pelium. Thebes revolts and is captured by Alexander. After a vote by the League of Corinth, Thebes is razed.

334 Alexander's Persian campaign starts. He defeats a Persian army at Granicus. Lydia submits and Alexander sets up anti-Persian democracies in Ionia. Alexander besieges and captures Miletus, disbands his fleet, and attacks the Persian ports by land. He takes Halicarnassus, securing Caria. Alexander I of Molossia lands in Italy to assist Tarentum (Taras) against natives and wins considerable success.

334/3 Alexander conquers Lycia, Pamphylia, and western Pisidia and winters in Gordium.

333 Alexander conquers Cilicia and defeats Darius III at Issus.

332 Alexander besieges and captures Tyre. He rejects Darius III's offer to share the Persian empire. Alexander takes Gaza, securing Syria and Palestine. Egypt surrenders. Agis III of Sparta works against Macedon in Greece. A nonaggression pact is established between Rome and Alexander I of Molossia (who now heads the Italiote League and controls much of southern Italy).

331 Foundation of Alexandria, Egypt. Cyrene surrenders. Alexander defeats Darius III at Gaugamela and occupies Babylon and Susa. Agis III organizes a Peloponnesian revolt against Macedon, but Megalopolis, Argos, and Messenia remain loyal. Agis besieges Megalopolis but is killed by Antipater, ending the revolt. A split in the Italiote League results in war in southern Italy between Tarentum (Taras) and allied cities and Alexander I of Molossia, who is assassinated at the battle of Pandosia.

330 Alexander invades Persia and defeats Ariobarzanes at the "Persian Gates" (the pass into Persia). Persepolis and other royal palaces are occupied. Alexander reaches Ecbatana (the capital of Media) and pays off the Thessalian and Hellenic League troops who wish to return. Darius III is assassinated by Bessus. Alexander follows the assassins and captures Hyrcania and Areia. He executes Philotas and his father Parmenio.

329 Alexander enters the Caucasus and Bactria.

328 Alexander conquers Bactria, capturing Bessus, crosses the Oxus River, and takes Sogdiana. Sogdian revolt is suppressed. Alexander kills Cleitus. The Pages' Conspiracy against Alexander results in the execution of Callisthenes and others.

327 Alexander invades India via Afghanistan.

326 Alexander crosses the Indus and defeats Porus at the battle of the Hydaspes River; Porus is left as a client king but Alexander establishes garrison cities at Bucephala and Nicaea. At the River Hyphasis Alexander's army refuses to proceed. Alexander withdraws to the Hydaspes, abandoning hopes of further eastward conquest. Nearchus voyages down the Jhelum. Siege of Sangala; conquest of the Malli. Alexander is wounded.

325 Alexander crosses the Gedrosian desert, suffering many casualties from the harsh conditions. Nearchus returns to the Persian Gulf.

324 Alexander reaches Susa. Macedon enforces the restoration of Greek exiles (opposed by Athens and Aetolia). Macedonian army mutiny at Opis fails. Macedonian veterans are discharged and sent home under Craterus, who is appointed regent in Greece. Alexander reaches Ecbatana. Harpalus, fleeing from Alexander, arrives in Athens with a large sum of money to incite a revolt; he fails and is arrested, escapes, and is murdered at Taenarum. Demosthenes is tried and exiled over the Harpalus affair.

323 Death of Alexander at Babylon (10 June). Army elects Philip Arrhidaeus and (the unborn) Alexander IV kings of Macedon. Perdiccas is made regent and commander in Asia and Antipater the commander in Macedon and Greece. Craterus is appointed to exercise the ceremonial and ritual aspects of royal power in Macedon. Senior members of Alexander's court are appointed as satraps of various regions, including Lysimachus in Thrace,

Antigonus in Phrygia, Ptolemy in Egypt, and Eumenes of Cardia in Cappadocia (not yet under Macedonian control). Greeks revolt against Antipater (Lamian War). Athens and Aetolia ally. Demosthenes returns to Athens. Antipater is besieged in Lamia.

322 A Greek fleet is defeated by the Macedonians off Abydus. Perdiccas invades Cappadocia and Pisidia, bringing them under Macedonian control. A Macedonian fleet defeats the Greeks off Amorgos. Antigonus Monophthalmus, accused by Perdiccas of treason for not supplying troops to assist in Cappadocia, flees to Greece; Ptolemy is acquitted on the same charge. Lysimachus defeats Seuthes III of Thrace in battle. A Greek fleet is defeated in the Malian Gulf. Ophellas conquers Cyrene for Ptolemy. Greeks are defeated at Crannon. Macedon garrisons Athens. Suicide of Demosthenes.

321 Aetolia comes to terms with Macedon; end of the Lamian War. Philip Arrhidaeus and Eurydice marry. Prompted by Antigonus Monophthalmus and supported by Ptolemy, Antipater and Craterus attack Perdiccas.

320 Craterus is defeated and killed in Cappadocia by Eumenes, serving as Perdiccas' commander in chief. Perdiccas invades Egypt but his troops mutiny and kill him. Antipater is appointed regent; he makes Seleucus satrap of Babylonia and Antigonus, with command against Eumenes, satrap of Phrygia, Lycia, and Pamphylia. Ptolemy annexes Palestine and Syria.

319 Antigonus Monophthalmus defeats Eumenes and besieges him in Nora, then defeats Attalus and Alcetas. Antipater dies (of natural causes). Polyperchon is elected regent but is opposed by Antipater's son, Cassander. Polyperchon declares the Greek cities free.

318 Eumenes submits to Antigonus Monophthalmus. Polyperchon appoints Eumenes to command against Antigonus. Phocion is executed in Athens. Antigonus defeats Polyperchon's fleet, under Cleitus, at anchor near Byzantium, severing communications between Polyperchon and Eumenes.

317 Agathocles seizes power at Syracuse. Eumenes captures Babylon. Inconclusive battle of Paraetacene between Antigonus Monophthalmus and Eumenes. Eurydice appoints Cassander regent for Philip Arrhidaeus. Olympias (mother of Alexander the Great) seizes Pydna and executes Philip Arrhidaeus and Eurydice.

316 Antigonus Monophthalmus defeats and captures Eumenes at Gabiene, subsequently executing him. Antigonus is proclaimed "King of all Asia" and consolidates his power in Media and Persia. Cassander takes over Macedon, capturing Pydna and executing Olympias. Cassander rebuilds Thebes. Seleucus flees Babylon and joins Ptolemy in Egypt.

315   Antigonus Monophthalmus occupies Syria. His rejection of an ultimatum from Ptolemy, Cassander, Lysimachus, and Seleucus results in war.

314   Antigonus Monophthalmus occupies Phoenicia, besieges Tyre, builds a navy, and declares the Greek cities free. He sends Aristodemus to the Peloponnese. Cassander defeats a coalition of Glaucias of Taulantia, Epidamnus, and Apollonia, supported by Corcyra.

313   Tyre falls to Antigonus Monophthalmus. Antigonus' armies suffer reverses in Thrace.

312   Demetrius, son of Antigonus Monophthalmus, is defeated at Gaza by Ptolemy. Seleucus, supported by Ptolemy, recaptures Babylon and Media from Antigonus. Carthage invades Sicily.

311   Ptolemy, Cassander, and Lysimachus make peace with Antigonus Monophthalmus; Seleucus carries the war on alone. Agathocles is defeated at the Himeras River; Carthaginians besiege Syracuse.

310   Agathocles of Syracuse invades Africa to attack the Carthaginians. War breaks out again between the Macedonian leaders in Greece and Asia. ?Cassander murders Alexander IV and his mother Roxane (possibly in 308).

309   Antigonus Monophthalmus makes peace with Seleucus. Polyperchon marches against Cassander, espousing the cause of Barsine and Heracles, Alexander the Great's illegitimate son. Cassander buys off Polyperchon and persuades him to murder Barsine and Heracles.

309/8   Agathocles allies with Ophellas of Cyrene and then murders him.

308   Magas, Ptolemy's half-brother, becomes governor of Cyrene. Antigonus Monophthalmus murders Cleopatra, Alexander the Great's sister. ?Seleucus defeats Antigonus' general, Nicanor, in the east. Antigonus and Seleucus sign a nonaggression pact. Ptolemy's invasion of the Peloponnese fails.

307   Agathocles is defeated in battle against the Carthaginians in Africa. Demetrius, son of Antigonus Monophthalmus, frees Athens from Cassander.

306   Peace is declared between Syracuse and Carthage. Demetrius returns to Asia, decisively defeats Ptolemy in a naval battle off Cyprus, and takes the island. Antigonus Monophthalmus and Demetrius are proclaimed kings.

305   Demetrius I besieges Rhodes, earning the epithet "Poliorcetes" (the Besieger). Agathocles defeats the Sicilian coalition and controls most of Sicily.

305/4   Ptolemy, Seleucus, Lysimachus, and Cassander, in coalition against Antigonus I (Monophthalmus), follow his example and have themselves elected kings.

304   Rhodes makes peace with Antigonus I (Monophthalmus), remaining independent. Cassander besieges Athens but withdraws when Demetrius I (Poliorcetes) arrives with an army. Agathocles is proclaimed king of Syracuse.

304/3   Demetrius I (Poliorcetes) captures the Isthmus of Corinth, securing both Corinth and Sicyon. He also conquers Achaea and most of Arcadia.

302   Demetrius I (Poliorcetes) restores the Hellenic League. He is appointed commander and embarks on a war against Cassander, who allies with Ptolemy I (Soter), Seleucus I (Nicator), and Lysimachus. Demetrius is recalled to Asia by Antigonus I (Monophthalmus) to meet this threat. Lysimachus campaigns in Asia Minor. Ptolemy attacks Jerusalem. Mithridates I (Ctistes) founds the kingdom of Pontus.

301   Ptolemy I (Soter) captures southern Syria and Phoenicia and regains Judaea, but then returns to Egypt. Antigonus I (Monophthalmus) is killed at Ipsus fighting Lysimachus, Cassander, and Seleucus I (Nicator). Demetrius I (Poliorcetes) escapes and holds Cyprus, Tyre, and Sidon. Antigonus' kingdom is divided; Seleucus gains Syria, Lysimachus gets most of Asia Minor. Ptolemy gets nothing because of his failure to help, but retains Phoenicia and the southern parts of Syria he had captured. Cassander gets Macedon and Greece; the Hellenic League is dissolved.

301/0   Demetrius I (Poliorcetes) returns to Greece, basing himself in Corinth.

c. 300   Agathocles of Syracuse marries Theoxene, Ptolemy I (Soter)'s (?)stepdaughter (?Magas' sister), and captures Corcyra (possibly in 298).

299   Ptolemy I (Soter) and Lysimachus ally; Lysimachus marries Ptolemy's daughter Arsinöe. Cassander's son Alexander marries another of Ptolemy's daughters. Seleucus I (Nicator) and Demetrius I (Poliorcetes) ally (possibly in 298). ?Agathocles besieges Croton.

298   Seleucus I (Nicator) and Demetrius I (Poliorcetes) occupy southern Anatolia.

298/7   Cassander attacks Corcyra but is driven off with the help of Agathocles of Sicily. ?Lachares seizes power in Athens (possibly in 296/5).

297   Cassander and his son Philip IV die (of natural causes); Macedon is divided between his younger sons, proclaimed joint kings by the army as Antipater I and Alexander V. Ptolemy I (Soter) restores Pyrrhus I as king in Epirus.

296   Demetrius I (Poliorcetes) besieges Athens.

295   Ptolemy I (Soter)'s relief fleet fails to break the blockade of Athens. While Demetrius I (Poliorcetes) is busy in Greece, Ptolemy gains Cyprus, Seleucus I (Nicator) Cilicia, and Lysimachus the Ionian ports (295-294).

294   Athens surrenders. Demetrius I (Poliorcetes) garrisons the Piraeus. Alexander V appeals to Pyrrhus I of Epirus and Demetrius for help against Antipater I. Pyrrhus restores him but occupies Ambracia, Acarnania, Amphilocia, and parts of western Macedon. Demetrius marches north but Alexander tells him he is no longer needed. Demetrius murders Alexander and becomes king of Macedon. Antipater flees to his father-in-law, Lysimachus.

293   Demetrius I (Poliorcetes) conquers Thessaly.

292   Aetolia and Boeotia revolt against Demetrius I (Poliorcetes), who quickly takes Thebes.

291   Demetrius I (Poliorcetes) wages desultory war against the Aetolian League and Pyrrhus I. Second revolt of Boeotia against Demetrius fails; Thebes is retaken.

290   Demetrius I (Poliorcetes) garrisons Corcyra. The Aetolian League captures Phocis and bans the supporters of Demetrius from the Pythian games. Agathocles of Syracuse allies with Demetrius.

289   Death of Agathocles of Syracuse. Demetrius I (Poliorcetes) invades Aetolia and Epirus. Pyrrhus I of Epirus defeats the Macedonian army under Pantauchus in Aetolia.

288   Pyrrhus I invades Macedon, but is driven back.

287   Lysimachus invades eastern Macedon, capturing Amphipolis; Pyrrhus I simultaneously invades from the west. Temporary truce is declared between Demetrius I (Poliorcetes) and Pyrrhus. Demetrius marches against Pyrrhus but is deserted by his army and flees. Pyrrhus and Lysimachus are proclaimed joint kings and divide Macedon between them. Demetrius reestablishes himself in central Greece and the Peloponnese. Athens revolts against Demetrius. Ptolemy I (Soter) sends a fleet to Greece to assist against Demetrius. Demetrius comes to terms with Ptolemy and Athens, keeping Corinth, Chalcis, and Attica (including Eleusis and the Piraeus), but recognizing Athenian freedom. A truce is negotiated between Pyrrhus and Demetrius, who crosses to Asia to fight Lysimachus, leaving his son, Antigonus Gonatas, in charge of his Greek possessions. Pyrrhus seizes Thessaly.

286 Demetrius I (Poliorcetes) campaigns in Ionia and captures Sardis but is chased into Cilicia by Lysimachus' son, Agathocles. Athens fails to recover the Piraeus from Antigonus Gonatas.

285 Demetrius I (Poliorcetes) surrenders to Seleucus I (Nicator). Lysimachus drives Pyrrhus I from Macedon and Thessaly and becomes king of all of it, executing Antipater I. Ptolemy I (Soter) makes Ptolemy II (Philadelphus) joint ruler; Ptolemy Ceraunus (Ptolemy I's son by a previous marriage) flees Egypt and joins Lysimachus.

284 Lysimachus conquers Paeonia. ?Refounding of the Achaean League.

283 Demetrius I (Poliorcetes) dies of natural causes while in captivity. His son succeeds to the throne as Antigonus II (Gonatas). Demetrius' admiral in Miletus surrenders his fleet, Tyre, and Sidon to Ptolemy I (Soter), who now has undisputed naval superiority.

283/2 Death of Ptolemy I (Soter); Ptolemy II (Philadelphus) becomes sole king of Egypt. Lysimachus executes his son, Agathocles, whose wife, children, and supporters flee to Seleucus I (Nicator); serious unrest develops in Macedon.

282 Seleucus I (Nicator) attacks Lysimachus, who is deserted by most of Anatolia.

281 Seleucus I (Nicator) defeats and kills Lysimachus and captures Ptolemy Ceraunus in the battle of Corupedium. Antigonus II (Gonatas) takes Athens. Areus I reconstitutes the Peloponnesian League.

281/0 Seleucus I (Nicator) invades Thrace but is murdered by Ptolemy Ceraunus, who is accepted by the armies of Lysimachus and Seleucus as king in Lysimachus' stead. Seleucus' son succeeds to the Syrian throne as Antiochus I (Soter). Ptolemy Ceraunus beats off a naval attack by Antigonus II (Gonatas), marches to Macedon, defeats Antigonus again, and clashes with Pyrrhus I. Pyrrhus is invited to Italy to help the Greek cities in southern Italy against Rome.

280 Ptolemy Ceraunus becomes king of Macedon. An uprising led by Areus I of Sparta is defeated by the Aetolian allies of Antigonus II (Gonatas). ?Achaean League reconstitutes itself. War breaks out between Antiochus I (Soter) and Antigonus. Pyrrhus I invades Italy to help the Greek cities and defeats the Romans at the Siris River. Ptolemy Ceraunus murders Lysimachus' youngest sons.

279   Gauls invade Thrace and then Macedon. Ptolemy Ceraunus is defeated, captured, and killed. Gauls invade again, beating the Macedonians under Sosthenes. The Aetolians and allies defeat the Gallic invasion at Delphi; the Gauls suffer heavy casualties on their retreat north. Antigonus II (Gonatas) and Antiochus I (Soter) make peace. Pyrrhus I defeats the Romans at the battle of Asculum but suffers heavy casualties.

278   Carthage blockades Syracuse. Pyrrhus I crosses to Sicily, eventually clearing the Carthaginians from everywhere except Lilybaeum. Gallic bands ravage Thrace and Byzantium; some then cross into Asia.

277   Gauls ravage Thrace and the Chersonese but Antigonus II (Gonatas) defeats them at Lysimacheia. He invades Macedon, fighting three other claimants to the throne.

276   Antigonus II (Gonatas) becomes king of Macedon, secures control of Thessaly, and allies with Antiochus I (Soter). Pyrrhus I requests help from the other Greek monarchs. Despite threats, Antigonus refuses. Pyrrhus returns to Italy and winters in Tarentum (Taras).

275   Pyrrhus I is defeated at Beneventum by the Romans, withdraws to Epirus, and prepares to attack Macedon. ?Antiochus I (Soter) defeats the Gauls at the "Elephant Battle" and drives them into Phrygia (perhaps in 270).

274   Pyrrhus I overruns Upper Macedon and defeats Antigonus II (Gonatas) in battle at Aous. Antigonus flees to Thessalonica. ?Magas, governor of Cyrene, allies with Antiochus I (Soter), revolts against Ptolemy II (Philadelphus), and overruns part of Egypt. A Libyan revolt causes him to withdraw.

274/3   Pyrrhus I's son defeats Antigonus II (Gonatas) in battle and secures control of Thessaly. First Syrian War begins between Antiochus I (Soter) and Ptolemy II (Philadelphus).

272   Pyrrhus I invades the Peloponnese and besieges Sparta. During this siege, Antigonus II (Gonatas) recovers much of Macedon and moves against Argos. Pyrrhus is killed in street fighting in Argos. Antigonus now dominates much of Greece, except Epirus and Aetolia. The Romans capture Tarentum.

271 First Syrian War ends.

270 Romans capture Rhegium.

268/7 Athens, Sparta, and Ptolemy II (Philadelphus) form an alliance against Antigonus II (Gonatas) and begin the Chremonidean War.

266 Antigonus II (Gonatas) invades Attica and reaches the Isthmus of Corinth. His Gallic troops mutiny and he returns home. Areus I of Sparta marches north but is held by the fortifications at Corinth and withdraws.

265 Areus I is killed fighting Antigonus II (Gonatas) at Corinth. The Peloponnesian alliance collapses, with the Achaean League seceding and Mantinea joining the Arcadian League. Antigonus besieges Athens.

c. 264-262 Alexander II of Epirus' invasion of Macedon fails.

263 Hieron of Syracuse allies with Rome against Carthage (during the First Punic War). Pergamum, under Eumenes I, declares independence from Seleucid rule.

262 Romans campaign in Sicily and take Agrigentum. Eumenes I of Pergamum defeats Antiochus I (Soter) in battle near Sardis.

262/1 Antigonus II (Gonatas) captures Athens; end of the Chremonidean War. Ephesus falls under the control of Ptolemy II (Philadelphus).

261 Death of Antiochus I (Soter); accession of Antiochus II (Theos), who allies with Antigonus II (Gonatas). Ptolemy II (Philadelphus) and Antigonus make peace.

260 Second Syrian War pits Ptolemy II (Philadelphus) against Antiochus II (Theos) and Antigonus II (Gonatas). Acrotatus of Sparta is defeated and killed at Megalopolis.

259 Death of Magas of Cyrene. ?Antiochus II (Theos) recovers Ephesus and Miletus.

258 ?Antigonus II (Gonatas) defeats the fleet of Ptolemy II (Philadelphus) off Cos (?261).

255 Peace between Ptolemy II (Philadelphus) and Antigonus II (Gonatas).

253 Peace between Ptolemy II (Philadelphus) and Antiochus II (Theos), ends the Second Syrian War.

251 Aratus frees Sicyon from Macedon; it joins the Achaean League.

250/49 Antigonus II (Gonatas) loses Corinth when its governor, Alexander, rebels and allies with the Achaean League (?or 252).

249 ?Alexander is proclaimed king in Corinth.

248 Rome renews its naval alliance with Hieron of Syracuse.

247 ?Arsaces, chief of the Parnoi, captures Parthia from Seleucid empire.

246 Death of Ptolemy II (Philadelphus); accession of Ptolemy III (Euergetes). Death of Antiochus II (Theos); accession of Seleucus II (Callinicus). Civil war breaks out between Seleucus and a rival claimant to the Syrian throne, a half-brother by his father's second wife Berenice, the sister of Ptolemy III (Euergetes) and leads to the Third Syrian (Laodicean) War between Ptolemy and Seleucus. Ptolemy overruns Syria and Mesopotamia but finds Berenice and son dead and returns to Egypt to put down a revolt.

246/5 Death, from natural causes, of Alexander in Corinth.

245 Aratus is elected general of Achaean League. Antigonus II (Gonatas) regains Corinth. Ptolemy III (Euergetes) launches a naval offensive against Seleucus II (Callinicus) in the Aegean. The Aetolian League defeats Boeotia. ?Antigonus attacks Ptolemy, defeating him in a naval battle off Andros, and recovers Delos and most of the Cyclades.

244 Seleucus II (Callinicus) recaptures Mesopotamia. Ptolemy III (Euergetes) captures Ephesus and Samos. ?Diodotus detaches Bactria from Seleucid rule (a gradual process, completed around this date). Agis IV becomes king of Sparta.

243 Aratus captures Corinth from Antigonus II (Gonatas) and brings Epidaurus and Troezen into the Achaean League, sparking the revolt of Megara and war with Antigonus. ?Antiochus Hierax is made king of Asia Minor in return for support from his mother Laodice to his elder brother Seleucus II (Callinicus) in the Third Syrian War. ?The Aetolian League and Epirus ally and divide Acarnania between them.

242 Achaean League allies with Sparta and elects Ptolemy III (Euergetes) admiral.

241 Rome occupies all of Sicily after Carthage surrenders and ends the First Punic War. Ptolemy III (Euergetes) makes peace with Seleucus II (Callinicus), ending the Third Syrian War and giving Ptolemy much of Syria. An Aetolian League attack on the Achaean League, encouraged by Antigonus II (Gonatas), is beaten off. Eumenes I of Pergamum dies and power is assumed by his nephew Attalus (c. 224 proclaimed king as Attalus I [Soter]). Agis IV of Sparta is executed by political opponents.

241/0   Antigonus II (Gonatas) makes peace with the Achaean League.

240/39   Antigonus II (Gonatas) and the Aetolian League ally against the Achaean League. Antigonus dies of natural causes and is succeeded by Demetrius II. Alexander II of Epirus dies (also of natural causes), leaving dynastic problems. Laodice declares her son Antiochus Hierax, and not his brother Seleucus II (Callinicus), the rightful heir to the Seleucid throne.

239   Seleucus II (Callinicus) invades Lydia, starting the War of the Brothers. Antiochus Hierax, aided by Mithridates II of Pontus and Gauls, defeats Seleucus at Ancyra. The Aetolian and Achaean Leagues, supported by Elis and Boeotia, ally against Demetrius II.

238   War erupts between Antiochus Hierax, with his Gallic allies, and Attalus of Pergamum. Rome occupies Sardinia and Corsica. ?Attalus defeats the Gauls. Aratus leads the Achaean League against Argos, an ally of Demetrius II.

236   ?Peace between Seleucus II (Callinicus) and Antiochus Hierax ends the War of the Brothers. Antiochus gets Asia Minor north of the Taurus River.

235   ?Cleomenes III becomes king of Sparta. Aratus defeats and kills Aristippus of Argos in battle at Cleonae. Megalopolis joins the Achaean League. Attalus I (Soter) campaigns against Antiochus Hierax.

234   Orchomenus (Arcadia) and Mantinea join the Achaean League.

233   The army of Demetrius II, led by Bithys, inflicts a heavy defeat on Aratus and the Achaean League at Phylacia, causing Orchomenus (Arcadia) and Mantinea to leave the Achaean League and join the Aetolian League.

232   Dardanians under Longarus heavily defeat Demetrius II in battle. ?Epirus becomes a republic and allies with Aetolia. Acarnania becomes independent but is attacked by the Aetolians, who besiege Medeon.

231   Demetrius II pays Agron of Illyria to relieve the siege of Medeon in Acarnania. Agron defeats the Aetolians but dies and his widow Teuta becomes regent. Attalus I (Soter) campaigns against Antiochus Hierax.

230   The Illyrians under Teuta ravage Elis and Messenia and seize Phoenice (Acarnania). An Epirote army sent to help is defeated outside Phoenice and the Illyrians ravage northern Epirus and raid Italian shipping. Epirus appeals for help to the Aetolian and Achaean Leagues, which send troops. Teuta recalls her army for action against the Dardanians. Epirus ransoms Phoenice and allies with Teuta.

229 First Illyrian War between Rome and Teuta of Illyria. Corcyra, Dyr-rhachium (Epidamnus), Apollonia, and Issa ally with Rome. Demetrius II is defeated and killed in battle with the Dardanians. His cousin Antigonus is made regent, becoming (?227) king of Macedon as Antigonus III (Doson). Argos joins the Achaean League. Athens regains its independence from Macedon. The Aetolian League hands Tegea, Mantinea, Orchomenus (Arcadia), and Caphyae over to Sparta. Antigonus drives the Dardanians out of Paeonia. Attalus I (Soter) defeats Antiochus Hierax at Coele and again (229-228) at Harpasus in Caria.

228 Teuta of Illyria surrenders to Rome, which establishes a protectorate over the Illyrian coast; Demetrius is made ruler of Pharos. Aegina joins the Achaean League. Antigonus III (Doson) secures Thessaly and allies with Epirus. He hands over territory in Thessaly and Achaea to ensure the Aetolian League's neutrality. Cleomenes III fortifies part of Megalopolitan territory, sparking the Cleomenic War between Sparta and the Achaean League.

228/7 Seleucus II (Callinicus) fails to regain Parthia from Arsaces II when his aunt Stratonice and his brother Antiochus Hierax revolt. He defeats Antiochus (who flees to Thrace), then captures and kills Stratonice.

227 Roman provinces of Sardinia-Corsica and Sicily are created. Cleomenes III defeats Aratus at Lycaeum near Megalopolis. Aratus captures Mantinea. Cleomenes defeats Aratus at Ladocaea near Megalopolis. Cleomenes stages a coup and reforms the Spartan constitution along "Lycurgan" lines, restoring the land allotment system and common messes. ?Antigonus III (Doson)'s anti-Egyptian expedition to Caria makes some temporary gains.

226 Antiochus Hierax is murdered by Gauls in Thrace. Cleomenes III retakes Mantinea, invades Achaea, and defeats the Achaean League army in battle at Hecatombaeum (near Dyme in the northwest Peloponnese). Aratus secures the rejection of a proposal that Cleomenes becomes head of the Achaean League. Aratus and Antigonus III (Doson) negotiate. Seleucus II (Callinicus) dies in a riding accident; Seleucus III (Soter) accedes.

225 Argos and Corinth ally with Cleomenes III of Sparta, who secures numerous defections from the Achaean League.

224 Aratus becomes dictator of the Achaean League. Cleomenes III besieges Sicyon but leaves when Aratus allies with Antigonus III (Doson). Argos is betrayed to Antigonus, who restores Macedon's influence in Greece by forming the Hellenic League, comprising the Macedonian, Thessalian, and Boeotian Leagues and other smaller groupings.

223 Antigonus III (Doson) invades Arcadia, takes Tegea, and lets Aratus destroy Mantinea. Cleomenes III captures and razes Megalopolis. Seleucus III (Soter) dies and is succeeded by Antiochus III (the Great).

223-20 Achaeus recovers most of Seleucid Asia Minor from Attalus I of Pergamum.

222 Cleomenes III ravages Argive territory. Antigonus III (Doson) defeats Cleomenes at Sellasia (July) and captures Sparta. Cleomenes and a few supporters escape to Egypt. The Dardanians attack Macedon and Antigonus returns home and defeats them, but falls ill. Molon, satrap of Media, revolts against Antiochus III (the Great) and defeats his army.

221 Death of Antigonus III (Doson); accession of Philip V. Death of Ptolemy III (Euergetes); accession of Ptolemy IV (Philopator). Aetolian League attacks Epirus and Messenia. Antiochus III (the Great) defeats the rebel Molon, who commits suicide.

220 Second Social War begins in Greece, pitting the Hellenic League under Philip V against Aetolia, Sparta, and Elis. Aetolian invasion of Messenia leads to the defeat of Aratus and the Achaean League at Caphyae. Achaeus revolts against Antiochus III (the Great). Antiochus suppresses revolts in Media and Babylon. Rhodes declares war on Byzantium.

219 Second Illyrian War is started by Demetrius of Pharos' piracy. A Roman naval force swiftly defeats Demetrius, who takes refuge with Philip V. The Fourth Syrian War, between Antiochus III (the Great) and Ptolemy IV (Philopator), begins with Antiochus seizing control of southern Syria and northern Palestine. Lycurgus of Sparta invades Argos and declares war on the Achaean League. Philip defeats the Dardanians and then invades Aetolia and Acarnania. Cleomenes III and his supporters rebel in Egypt and commit suicide or are killed. Rhodes wins the war with Byzantium.

218 Philip V invades Aetolia and Sparta. Attalus I (Soter) campaigns against Achaeus.

217 Ptolemy IV (Philopator) decisively defeats Antiochus III (the Great) at Raphia (June 22), recapturing Palestine and southern Syria and ending the Fourth Syrian War. Peace of Naupactus between Philip V and Aetolia ends the Second Social War.

216 Antiochus III (the Great) and Attalus I (Soter) campaign against the rebel Achaeus. Native Egyptian uprisings occur in Upper Egypt. Philip V campaigns in Illyria but withdraws when a Roman naval force arrives.

215 Death of Hieron; Syracuse allies with Carthage against Rome (during the Second Punic War). Philip V allies with Hannibal against Rome, starting the First Macedonian War. Antiochus III (the Great) captures Sardis and besieges Achaeus in the citadel.

214 Marcus Valerius Laevinius and his fleet off the Illyrian coast cause Philip V to withdraw. Philip campaigns in the Peloponnese, ravaging Messenia. Demetrius of Pharos is killed in an assault on Mount Ithôme.

213 Marcus Claudius Marcellus besieges Syracuse. Hannibal occupies Tarentum but fails to dislodge the Romans from its citadel. Aratus dies of natural causes. Philip V campaigns in Illyria. Sardis' citadel falls to Antiochus III (the Great); Achaeus is executed.

212 Antiochus III (the Great) recovers the rebellious province of Armenia. Philip V campaigns in Illyria and captures Lissus (on the coast).

211 Rome allies with the Aetolian League and Attalus I (Soter) against Philip V. The Romans capture and sack Syracuse (death of Archimedes). ?Lycurgus of Sparta dies; Machanidas becomes regent for his son, Pelops.

210 Rome recovers Agrigentum and the rest of Sicily. Machanidas of Sparta allies with the Aetolians. Philip V captures Echinus. Ptolemy V (Epiphanes) becomes joint ruler of Egypt with Ptolemy IV (Philopator). Publius Sulpicius Galba campaigns in Greece against Philip and captures Aegina. Antiochus III (the Great) campaigns to recover Parthia and Bactria.

209 Rome recaptures Tarentum. Peace negotiations between Philip V and the Aetolians fail. Antiochus III (the Great) campaigns in Parthia, capturing most of the cities and forcing Arsaces II to acknowledge him as overlord.

208 Naval campaign of Attalus I (Soter) and Publius Sulpicius Galba against Philip V. Antiochus III (the Great) attacks Bactria and besieges its ruler, Euthydemus, in the capital, Bactra.

207 Philopoemen, general of the Achaean League, defeats Sparta in battle at Mantinea and kills Machanidas. Philip V raids Aetolia and sacks Thermum. Nabis becomes regent, and then king, of Sparta.

206 The Aetolian League makes peace with Philip V. Antiochus III (the Great) makes peace with Euthydemus of Bactria, who acknowledges him as overlord. Antiochus advances into north India and subdues it.

205 Rome and Philip V negotiate the Peace of Phoenice. Ptolemy IV (Philopator) dies. Upper Egypt achieves independence from the Ptolemies (until c. 185).

204   Antiochus III (the Great) returns to Syria. Ptolemy V (Epiphanes) becomes sole ruler of Egypt, under the regent Agathocles. Nabis of Sparta raids Megalopolis. Cretan pirates, backed by Philip V, campaign against the Cyclades and the Rhodians.

203   Philip V uses his fleet to capture Thasos and raid the Hellespont.

203/2   Philip V and Antiochus III (the Great) ally against Ptolemy V (Epiphanes), planning to partition Egypt's overseas possessions.

202   Antiochus III (the Great) invades southern Syria (Fifth Syrian War). Philip V campaigns against Egyptian possessions in the Aegean and as a result Rhodes declares war on Philip.

202/1   Attalus I (Soter) joins Rhodes against Philip V.

201   Attalus I (Soter) and Rhodes appeal for Roman assistance against Philip V. Nabis of Sparta attacks Messenia but is beaten back; the Achaean League declares war on Nabis. Antiochus III (the Great) captures Gaza.

201/0   Philip V campaigns in Asia Minor, taking Samos from Ptolemy V (Epiphanes), defeating the Rhodian fleet at Lade, and ravaging Pergamum. Off Chios, Attalus I (Soter), Chios, Pergamum, Byzantium, and Rhodes decisively defeat Philip in a naval battle and blockade him at Bargylia.

200   Philip V attacks Athens; Rome warns him off but Philip ignores this and campaigns in the Hellespont, capturing Abydus. Rome declares war (Second Macedonian War). The Romans establish a base at Apollonia (Illyria). Philopoemen defeats Nabis of Sparta. Antiochus III (the Great) defeats the Egyptians at Panion (southern Syria) and occupies all their external territories. ?Death of Euthydemus of Bactria; accession of Demetrius I, who over the next few years extends Greek rule over northern India.

199   Abortive Roman attack on Macedon. The Aetolian League joins Rome against Philip V.

198   Flamininus outflanks Philip V at Aous and forces him to withdraw to Macedon. The Achaean League joins Rome against Philip V.

198/7   Peace between the Achaean League and Nabis of Sparta.

197   The Romans defeat Philip V at Cynoscephalae; he makes peace. Antiochus III (the Great) occupies southern Asia Minor, including Ephesus. Death of Attalus I (Soter); Eumenes II (Soter) becomes king of Pergamum.

196   Smyrna, Lampsacus, and Eumenes II (Soter) appeal to Rome against Antiochus III (the Great). Flamininus proclaims Greek freedom. Conference between Antiochus III (the Great) and the Romans at Lysimacheia. Antiochus ignores Roman demands he withdraw from Europe but they do not interfere in his subsequent occupation of the Gallipoli Peninsula.

195   Rome defeats Nabis of Sparta. Antiochus III (the Great) campaigns in the Gallipoli Peninsula. Hannibal joins Antiochus. A treaty between Antiochus and Ptolemy V (Epiphanes), giving Antiochus possession of southern Syria, ends the Fifth Syrian War.

194   The Romans withdraw all their troops from Greece. Antiochus III (the Great) completes his occupation of the Gallipoli Peninsula.

193   Nabis breaks his treaty with Rome and attacks the Achaean League. Philopoemen and Flamininus defeat Nabis. The Aetolians ally with Antiochus III (the Great). Relations between Rome and Antiochus break down.

192   The Aetolians assassinate Nabis; Sparta joins Achaean League. Aetolian League seizes Demetrias for Antiochus III (the Great), who lands there and declares Greece free. Rome, joined by Philip V and the Achaean League, declares war on Antiochus.

191   Antiochus III (the Great) captures Chalcis; defeated by the Romans and Philip V at Thermopylae, he withdraws to Ephesus. The Romans march against the Aetolian League. Eumenes II (Soter) and Rhodes join a Roman naval expedition which defeats Antiochus' fleet off Corycus.

190   Armistice with the Aetolian League frees Roman troops to fight against Antiochus III (the Great). Antiochus' fleet is defeated at Side and Myonessus. Aided by Philip V, a Roman army crosses to Asia Minor and, helped by Eumenes II of Pergamum, defeats Antiochus at Magnesia.

189   The Aetolian League surrenders to Rome, effectively ending its power in Greece. Sparta secedes from the Achaean League. Peace negotiations between Rome and Antiochus III (the Great). Rome subdues Galatia.

188   The treaty of Apamea between Rome and allies and Antiochus III (the Great), seriously reduces the power of the Seleucid empire. The Achaean League forces Sparta to submit and destroys its constitution and walls, essentially ending its role as an independent state.

187   Antiochus III (the Great) of Syria is killed while plundering a temple in Elam; accession of Seleucus IV (Philopator).

186   Prusias I of Bithynia and Eumenes II (Soter) of Pergamum at war.

183   Messenia revolts against the Achaean League; Philopoemen captured and killed. End of war between Bithynia and Pergamum (Rome decides in favor of Pergamum). War between Pontus and Pergamum.

182   Hannibal commits suicide. Messenia surrenders to the Aetolians.

180   Death of Ptolemy V (Epiphanes) of Egypt; accession of Ptolemy VI (Philometor).

179   Death of Philip V of Macedon; accession of Perseus. Pergamum wins war against Pontus.

175   Death of Seleucus IV (Philopator) of Syria; accession of Antiochus IV (Epiphanes).

172   Eumenes II (Soter) complains to Rome about Perseus; Roman mission to Greece.

171   Outbreak of Third Macedonian War between Rome and Perseus, who is supported by Genthius of Illyria. Perseus' early successes in Thessaly.

170   Perseus stops the Romans entering Macedon and repels them in Illyria.

169   The Romans penetrate the Macedonian frontier but are unable to advance further. War between Syria and Egypt (Sixth Syrian War). Antiochus IV (Epiphanes) occupies most of Egypt except Alexandria. Civil strife between Ptolemy VI (Philometor) and Ptolemy VIII (Physcon).

168   The Romans under Aemilius Paullus decisively defeat the Macedonians under Perseus at Pydna. The Romans take Illyria and capture Genthius. Ptolemy VI (Philometor) and Ptolemy VIII (Physcon) reconcile. The Romans force Antiochus IV (Epiphanes) to end the siege of Alexandria.

167   Romans seize 150,000 people in Epirus and sell them as slaves; end of Third Macedonian War. Rome deposes Perseus, splits Macedon into four republics and Illyria into three. Rome removes Rhodes' Asia Minor possessions and establishes Delos as a free port, effectively breaking Rhodian power. Jewish revolt against Antiochus IV (Epiphanes) causes his destruction of Jerusalem, the massacre of its inhabitants, and the attempted suppression of the Jewish faith. Outbreak of the Maccabaean revolt.

166   Rome declares Galatia free from Pergamum.

165   Antiochus IV (Epiphanes) prepares an expedition to the east.

164   Re-dedication of the Temple of Jerusalem by Judas Maccabaeus. Death of Antiochus IV (Epiphanes) from natural causes in Media, leaving Lysias as regent for Antiochus V. Ptolemy VI (Philometor) is forced to flee Egypt by Ptolemy VIII (Physcon) and seeks refuge in Rome.

163 Roman commission further weakens Syrian military power. Demetrius, son of Antiochus IV (Epiphanes), flees Rome for Syria. Egypt partitioned—Ptolemy VIII (Physcon) rules Cyrene and Ptolemy VI (Philometor) rules Egypt proper.

163/2 Judas Maccabaeus attacks the Seleucid fort of Akra in Jerusalem.

162 Lysias and Antiochus V defeat Judas Maccabaeus' brother, capture Beth-Zur, and briefly besiege Jerusalem.

162/1 Demetrius lands in Syria, executes Lysias and Antiochus V, and is crowned as Demetrius I (Soter).

161 Rome ends its alliance with Ptolemy VI (Philometor) and supports Ptolemy VIII (Physcon)'s attempts to take Cyprus. This fails but Ptolemy Philometor spares his brother, restoring him to rule in Cyrene. Judas Maccabaeus defeats a Seleucid army. Roman defense pact with Judaea.

160 A Seleucid army, led by Bacchides, defeats and kills Judas Maccabaeus. His brother, Jonathan, continues guerilla war against the Seleucids. Death of Eumenes II (Soter) of Pergamum; accession of Attalus II.

159 Prusias II of Bithynia invades Pergamum with Galatian assistance but the Romans force him to withdraw and pay reparations.

156-154 Frontier problems between Bithynia and Pergamum.

154 Massilia requests Roman help against the Oxybian Ligures and receives most of their land when the Romans defeat them.

153/2 Attalus II assists the pretender Alexander Balas to get Rome to recognize his claim to the throne of Syria.

152/1 Alexander Balas invades Syria and secures the assistance of the Jewish leader Jonathan against Demetrius I (Soter).

151/0 Death of Demetrius I (Soter) in battle against the pretender Alexander Balas (supported by Pergamum, Cappadocia, and Egypt). Alexander Balas succeeds to the throne of Syria and Jonathan is made governor of Judaea. Sparta attempts to secede from the Achaean League. Greek embassies request Rome to adjudicate. ?Death of Mithridates IV (Philopator Philadelphus); accession of Mithridates V (Euergetes).

149 Andriscus, a pretender to the throne of Macedon, takes over the Macedonian republics, starting the Fourth Macedonian War with Rome. Andriscus overruns Thessaly.

148 Rome defeats and captures Andriscus at Pydna, ending the Fourth Macedonian War. Rome creates a province from Macedon, Thessaly, and Epirus. ?Parthia annexes Media from Syria.

147 Demetrius (son of Demetrius I) invades Syria to regain his father's throne from Alexander Balas. Revolts break out against Alexander. Egypt intervenes in Syria, securing Phoenicia on the way. A Roman commission determines Sparta, Corinth, and Argos should be removed from the Achaean League. Because of the violent Achaean reaction to this the Romans do not enforce this decision, but prohibit further attacks on Sparta.

146 Alexander Balas flees to Cilicia. Demetrius crowned as Demetrius II (Nicator). Achaean League declares war on Sparta. Rome declares war on Achaean League. Rome defeats an Achaean League army at Scarphaea (near Thermopylae). Boeotia, Euboea, and Phocis join the Achaean League against Rome. The Romans sack Corinth, killing the males and selling the women and children into slavery; Achaean League dissolved.

145 Alexander Balas returns to Syria but is defeated at the battle of Oneoparas and murdered by his own officers. Ptolemy VI (Philometor) dies of wounds received in this battle. Syrian Greeks under Diodotus revolt against Demetrius II (Nicator), espousing the cause of the infant son of Alexander Balas, Antiochus VI (Epiphanes). Ptolemy VII (Neos Philometor) king of Egypt under the regency of his mother, Cleopatra II.

145/4 Civil unrest in Egypt; Ptolemy VIII (Physcon) returns, marries Cleopatra II, and assassinates Ptolemy VII (Neos Philometor). Ptolemy VIII (Physcon) becomes king and purges Alexandria of opponents.

144 Syria is effectively partitioned between Diodotus (who has eliminated the Jewish leader Jonathan) and Demetrius II (Nicator). Diodotus holds coastal Syria, Demetrius has Cilicia and the eastern satrapies.

142/1 Diodotus murders Antiochus VI (Epiphanes) and is proclaimed king of Syria as Diodotus Tryphon. Judaea, under Simon, becomes essentially autonomous in return for Jewish support to Demetrius II (Nicator).

141 Seleucid garrison of Akra in Jerusalem surrenders. Parthians in possession of Babylonia.

140 Simon becomes priest-king of Judaea. Demetrius II (Nicator) campaigns with some success against the Parthians in the east.

139 Parthians defeat and capture Demetrius II (Nicator). Demetrius' brother Antiochus returns to Syria and becomes King Antiochus VII (Sidetes). Attalus II of Pergamum dies and is succeeded by Attalus III.

138   Antiochus VII (Sidetes) captures and kills Diodotus Tryphon and becomes sole king in Syria.

135-134   Antiochus VII (Sidetes) recovers Palestine.

133   Death of Attalus III, who bequeaths Pergamum to Rome.

132   Pergamene revolt under Aristonicus.

132/1   Civil war in Egypt. Ptolemy VIII (Physcon) is expelled and Cleopatra II becomes sole ruler.

130   Rome suppresses the Pergamene revolt and creates the province of Asia (c. 129-123). Antiochus VII (Sidetes) recaptures Media and Babylonia from Parthia. Ptolemy VIII (Physcon) returns to Egypt.

129   Roman Senate refuses to ratify Perperna and Manlius Aquilius' Asian settlement. Parthians free Demetrius II (Nicator) and send him back to Syria. Antiochus VII (Sidetes) is defeated and killed in battle against Parthia. Demetrius becomes king of Syria again.

129/8   Cleopatra II invites Demetrius II (Nicator) to advance into Egypt but he withdraws to Syria to quell a revolt instigated by Cleopatra Thea.

128/7   Cleopatra II flees to Syria. Ptolemy VIII (Physcon) returns to Alexandria but civil war continues in Egypt.

126/5   Demetrius II (Nicator) is defeated by the Egyptian-backed pretender Alexander Zabinas in Lebanon and subsequently murdered (probably by his wife Cleopatra Thea). Seleucus V briefly secures the throne but is quickly murdered (probably by his mother Cleopatra Thea).

125   Antiochus VIII (Grypus) becomes co-ruler of Syria with his mother, Cleopatra Thea. War continues between them and Alexander Zabinas.

124   Cleopatra II returns to Alexandria and reconciles with Ptolemy VIII (Physcon), although some troubles continue.

123   Antiochus VIII (Grypus) captures and executes Alexander Zabinas.

121/0   Cleopatra Thea poisoned; Antiochus VIII (Grypus) sole ruler in Syria.

120   Mithridates V (Euergetes) of Pontus is assassinated at Sinope and succeeded by his queen and two minors. The future Mithridates VI (Eupator), Mithridates' son by an earlier wife, flees.

118   Formal end of civil war in Egypt.

116   Death of Ptolemy VIII (Physcon); dynastic problems in Egypt.

112   Mithridates VI (Eupator) seizes power in Pontus and embarks on a policy of territorial expansion.

104   Mithridates VI (Eupator) occupies Galatia and Cappadocia, with the support of Nicomedes III (Euergetes) of Bithynia.

103   ?Mithridates VI (Eupator) and Nicomedes III (Euergetes) of Bithynia divide Paphlagonia (sometime between 108 and 103).

96   Ptolemy Apion bequeaths Cyrene to Rome. Murder of Antiochus VIII (Grypus) leads to continual civil war in Syria or foreign occupation.

96/5   Rome installs Ariobarzanes I as king of Cappadocia. Rome orders Mithridates VI (Eupator) out of Paphlagonia and Cappadocia.

95   Tigranes II (the Great) becomes king of Armenia.

94   ?Death of Nicomedes III (Euergetes) of Bithynia; accession of Nicomedes IV (Philopator).

91/90   Mithridates VI (Eupator) expels Nicomedes IV (Philopator) from Bithynia.

89   Rome sends Manlius Aquilius to Asia Minor to help Nicomedes IV (Philopator) and Ariobarzanes I of Cappadocia. Mithridates VI (Eupator) withdraws but is provoked to fight by Nicomedes' toll on shipping in the Bosphorus and raid on Pontus (done on Aquilius' instructions).

88   Mithridates VI (Eupator) overruns Asia Minor, instigating a massacre of Romans and Italians (the "Asiatic Vespers"), starting the First Mithridatic War. Athens joins Mithridates, who unsuccessfully besieges Rhodes.

87   Lucius Cornelius Sulla lands in Greece and besieges Athens.

86   Athens captured. Chios sacked by Mithridates VI (Eupator).

86-85   Lucius Cornelius Sulla defeats Archelaus, Mithridates VI (Eupator)'s general in Greece, at Chaeronea and then Orchomenus (Boeotia), ending the war in Greece. Unrest in Asia Minor against Mithridates.

85   A Roman army, under Fimbria, defeats Mithridates VI (Eupator) in battle at Rhyndacus (Pergamum). Mithridates surrenders to Sulla, ending the First Mithridatic War. He gives up his conquests, surrenders his Aegean fleet, and pays an indemnity. Huge fines are imposed on those cities involved in the "Asiatic Vespers" (massacre of Romans and Italians). Rome restores Nicomedes IV (Philopator) and Ariobarzanes I and rewards Rhodes.

83 Sulla's legate, Murena, invades Cappadocia and Pontus claiming that Mithridates VI (Eupator) is rearming; Second Mithridatic War. Tigranes II (the Great) of Armenia occupies Syria, Cilicia, and Phoenicia, essentially ending Seleucid rule.

83/2 Mithridates VI (Eupator) drives Murena out of Cappadocia; end of Second Mithridatic War on the previous terms (Treaty of Dardanus) when Sulla disowns Murena's actions.

80 Civil disorder in Egypt. Lucius Cornelius Sulla proclaims Ptolemy XI (Alexander II) king. Alexandrians murder Ptolemy XI; Ptolemy XII (Auletes) seizes the throne.

c. 76/5 Mithridates VI (Eupator) makes an agreement with Sertorius, the rebel Roman leader in Spain.

74 Rome makes Cyrene a province. Nicomedes IV (Philopator) dies and bequeaths Bithynia to Rome. Mithridates VI (Eupator) declares war on Rome (Third Mithridatic War) and successfully invades Bithynia. He then invades the province of Asia and besieges Cyzicus.

74/3 Lucius Licinius Lucullus forces Mithridates VI (Eupator) to withdraw from Cyzicus, with the loss of much of his force (mid-winter).

73 Lucius Licinius Lucullus defeats a Mithridatic fleet off Lemnos.

72 Lucius Licinius Lucullus defeats Mithridates VI (Eupator) at Cabira, destroying his army. Pirates defeat Marcus Antonius at sea off Crete.

71-70 Lucius Licinius Lucullus mops up against Mithridates VI (Eupator), who flees to Tigranes II (the Great) in Armenia.

70 Quintus Metellus campaigns against Cretan pirates.

69 Lucius Licinius Lucullus invades Armenia, defeats Tigranes II (the Great) outside Tigranocerta, captures the city and pursues Tigranes. Antiochus XIII (Asiaticus) proclaimed king of Syria (approved by Lucullus).

68 Lucius Licinius Lucullus' army mutinies and he is forced to retire to Nisibis. Mithridates VI (Eupator) returns to Pontus and resumes hostilities.

67 Quintus Metellus captures Crete, which becomes a Roman province. Mithridates VI (Eupator) defeats a Roman force under Triarius at Zela and regains his kingdom. Tigranes II (the Great) recovers Armenia.

66 Pompey new Roman commander in the east. Phraates III of Parthia invades Armenia, preventing its support to Mithridates VI (Eupator). Pompey destroys Mithridates' army at Nicopolis.

65 Mithridates VI (Eupator) regains the Crimea and raises a new army.

64 Pompey annexes Syria; end of Seleucid monarchy.

63 Pompey in Damascus and Jerusalem. Suicide of Mithridates VI (Eupator) after his son Pharnaces and army mutiny; end of Third Mithridatic War.

62 Pompey's Asian settlement joins western Pontus to Bithynia, and parts of Judaea to Syria to create new Roman provinces. It also establishes several client kings and kingdoms.

59 Rome recognizes Ptolemy XII (Auletes) as king of Egypt.

58 Cyprus a Roman province. Alexandrians expel Ptolemy XII (Auletes).

57 Ptolemy XII (Auletes) arrives in Rome.

54 Rome restores Ptolemy XII (Auletes) to the throne of Egypt.

51 Ptolemy XII (Auletes) dies and is succeeded by Ptolemy XIII and Cleopatra VII (joint rulers).

49 Civil war in Egypt between Ptolemy XIII and Cleopatra VII.

48 Alexandrine War between Cleopatra VII, supported by Caesar, and Ptolemy XIII and his supporters. Caesar is besieged in Alexandria.

47 Mithridates of Pergamum rescues Caesar. Ptolemy XIII dies, ending the Alexandrine War. Cleopatra VII and Ptolemy XIV become joint rulers of Egypt.

44 Murder of Ptolemy XIV. Cleopatra VII associates her three-year-old son with her on the throne as Ptolemy XV (Caesarion).

41 Marcus Antonius arrives in Egypt.

34 "Donations of Alexandria." Antonius declares Ptolemy XV (Caesarion) the legitimate son of Caesar and "King of Kings," and Cleopatra VII "Queen of Kings." Together they are given rule of Egypt and Cyprus, while the three infant children of Antonius and Cleopatra are made rulers of most of the rest of the east.

31 Octavian defeats Antonius and Cleopatra VII at Actium.

30 Octavian enters Alexandria. Suicide of Antonius and Cleopatra VII and execution of Ptolemy XV (Caesarion). Egypt is made a Roman province.

INTRODUCTION

The ancient Greeks were hardy travelers and colonizers. From the mainland of Greece they settled as far afield as the Crimea and the shores of the Black Sea, the coastlines of Asia Minor, Egypt, Italy and Sicily, southern France, and parts of Spain and Africa. Greek civilization and influence were later spread as far east as northern India with the conquests of Alexander of Macedon and his successors. The main emphasis of this dictionary is on mainland Greece in the sixth to the fourth centuries B.C. However, it also covers other parts of the ancient Greek world, including the military activities of the Greeks in Italy and Sicily and the later Hellenistic Greek and Macedonian monarchies in the east. Important aspects of earlier Greek and Homeric warfare are also included, but full coverage starts with the eighth century B.C.—although the surviving information is rather patchy prior to the sixth century B.C. The available material starts to reduce around the mid-second century, when mainland Greece was annexed by the Romans, and dictionary entries cease at 30 B.C. This was when Egypt, the last truly Hellenistic monarchy, was formally incorporated into the Roman empire—although it had not been entirely independent for many years prior to this.

Warfare was an important part of ancient Greek life and did much to shape Greek society. It also exerted a profound influence on the history of the Western world—an influence which is still felt today. At a fundamental level, Greek force of arms must have played a major role in the establishment of Greek colonies in the Black Sea and Mediterranean coastlines and the consequent spread of Greek culture throughout these regions. This did much to shape the societies and history of the area. There are also very specific examples of Greek warfare determining the course of Western (and Near Eastern) history.

In 480-479, military success against a Persian invasion allowed Greek society to evolve into its mature form and heavily influenced its direction. Prior to the Persian Wars Athens was really a second-rank Greek state, but her transformation into a naval power and her performance during the wars placed her on an equal footing with Sparta—previously secure as the dominant military power in mainland Greece. If the Greeks had failed in 480-479, the course of Greek, and ultimately Western, history would have been quite different. It is difficult to imagine the fifth-century flowering of Athenian democracy, with all its rich political, literary, artistic, and cultural legacy, occurring under Persian rule. There would have been no Theban hegemony and probably no Alexander the Great. Instead of Greek influence being spread deep into Asia Minor by Alexander's military triumphs, Persian influence would have been spread across Greece and perhaps as far as the western Mediterranean. Conceivably, under these circumstances Rome would never have developed from a small town on the Tiber into one of the most successful imperial powers the world has seen.

The Social and Geographical Context of Greek Warfare

The term "Greece," although used above, is rather misleading, as there was no unified Greek nation in antiquity. To borrow Metternich's famous phrase, Greece proper at that time was no more than "a geographical expression" (in fact the word "Greece" is derived from the Latin term; the inhabitants referred to their country as Hellas and themselves as Hellenes). About 80 percent of its area is mountainous and movement consequently is restricted, particularly in winter. Except in areas like Thessaly, where wide plains aided communication, Greek settlements often developed semi-independently from their neighbors. Greek society therefore evolved not as a unified state but as a collection of autonomous city-states or *poleis* (singular: *polis*). Each *polis* had its own urban center, surrounded by its agricultural hinterland, the *chora,* and was often separated from its neighbors by mountain ranges or broken country which hindered easy communication. The foundation of wealth was agriculture and it seems likely that in most periods and places upwards of 85-90 percent of the population earned their living from agriculture—most as small farmers. City-states guarded their independence jealously and did not normally grant citizenship to foreigners (including Greeks from other city-states), often preventing them from owning land locally or even intermarrying with locals.

Despite the existence of regional dialects and the proud independence of the individual city-states, the ancient Greeks did share a common language and cultural heritage. Together, Greeks celebrated panhellenic festivals at Delphi and Olympia and had a common pantheon of gods and a common body of myth and legend—with appropriate local variation. While they regarded themselves as Athenians, Spartans, Thebans, or Argives first, and only second as Greeks (or more accurately Hellenes), there was a general sense of "Greekness"—especially compared with foreigners, who were collectively, and contemptuously, referred to as *barbaroi.*

However, despite occasional attempts by philosophers to identify foreigners as the only legitimate opponents in war, the ancient Greeks spent a considerable proportion of their time fighting each other. This was such a feature of Greek life that there was a fairly lively debate among the Greeks over whether their natural state of affairs was peace, punctuated by unnatural periods of war—or war, punctuated by unnatural periods of peace. Greek warfare probably originated in very early times as cross-border raids against neighboring settlements to acquire booty, especially cattle, sheep, and other movables—sometimes motivated by profit, sometimes by the need to survive, and sometimes by revenge for an insult or quarrel. Another element, which remained a feature throughout most ancient Greek warfare, was an agonal or competitive one. This saw warfare as a rite of passage for the young men, an opportunity to prove themselves as full members of society—and for a city to prove its worth against another city.

But despite some ritual elements, warfare was not a ritual or game and had the potential to profoundly affect Greek societies and individuals in several ways. The loss of a battle and the consequent inability to protect agricultural areas could lead to starvation. Few Greek cities could afford to lose two crops in a row without resorting to imports. The loss of a war could lead to slavery or death for the citizens of the defeated *polis*. The relatively constant inter-Greek warfare also helped prevent the Greeks from becoming united—which ultimately allowed their conquest, first by the Macedonians and then by the Romans.

The motives for war became more complex as Greek society evolved. As populations grew and exceeded the relatively small areas of good agricultural land, one emerging motive for war was the need to acquire more territory. This need was partially met in the eighth and seventh centuries by overseas colonization (which often involved fighting the local inhabitants and exposed Greeks to other methods of warfare). However, this was not always possible, or was sometimes seen as a less desirable option. One good example of this is Sparta. Instead of engaging in large-scale overseas settlement, the Spartans chose to subjugate neighboring Messenia and turn its inhabitants into serfs who produced food for their Spartan masters.

The development of *poleis* as autonomous states with often insular or parochial outlooks also led to opportunities for conflict and to an apparently natural desire for Greek states (or at least the larger ones) to dominate others. As Thucydides records in a speech he ascribes to an Athenian delegation to Sparta in 432:

> We have not done anything out of the ordinary, nor anything contrary to human nature in accepting an empire when it was offered to us and then refusing to give it up. Three very powerful motives prevent us from doing so—security, honor, and self-interest. And, again, we were not the first to act in this way. It has always been a rule that the weak should be subject to the strong. (Thuc. 1.76.2)

In the fifth and fourth centuries in particular, a variety of states, including Athens, Sparta, and Thebes, and dynasts such as Jason and Alexander of Pherae and Philip II of Macedon attempted to gain hegemony, or rule, over the whole of Greece. In some instances, notably involving Athens, commercial motives such as a wish to dominate overseas markets and trade routes may have played some part in instigating military action. In most cases, though, the simple desire to dominate others and to preserve one's own freedom of action was quite sufficient as a motivation for war.

Specific wars of course arose from individual causes, but broad changes in motivation can be identified over time. Warfare initially took the form of raids and minor clashes but became increasingly sophisticated and more complicated. As warfare progressed from the simple clash between neighboring cities, more and more effort and resources were put into it to ensure success. By the fourth century it had taken on a much more Clausewitzian character to involve the continuation of political objectives by other means. As might be expected, the methods of waging war also developed over time and its effects became increasingly serious.

Early Greek Warfare and the Development of the Hoplite and the Trireme

Prior to the seventh century, Homeric or aristocratic warfare seems to have emphasized the individual. The literature, at any rate, gives the impression of aristocratic champions fighting each other in single combat under the approving gaze of the general mass of soldiery. But this is probably an exaggeration, as Homer provides hints of formations which must have required more than the relatively small number of aristocrats to man. In addition, not all Greek soldiers at Troy were aristocrats—Thersites, an Achaean soldier beaten by Odysseus for daring to speak out at an assembly, was clearly not of noble birth (*Iliad* 2.211ff.). However, warfare must have been largely restricted to those who could afford the armor, weapons, and other equipment (including, in the very early period, chariots, and later on, cavalry horses).

The late eighth century saw the introduction of hoplite equipment (large round shield, helmet, breastplate, greaves, spear, and sword) and the first half of the seventh century the introduction of hoplite tactics. This was a revolution in warfare—instead of loose formations involving considerable individual combat, battles now essentially consisted of massed formations of heavy infantry pushing against and hacking at each other until one side broke and ran.

The hoplite, or heavy infantryman, remained the dominant military arm in most Greek states south of Thessaly until the late fourth century, although light infantry and cavalry had increasing successes (and always predominated in areas such as Aetolia, Thessaly, and Thrace). This hoplite domination meant that warfare was essentially conducted at a fairly slow pace and was restricted to the relatively flat and unbroken terrain that allowed the hoplites to maintain their formation. Outside his formation, the hoplite was vulnerable to faster-moving adversaries and one of the main aims of the hoplite general was to prevent his phalanx from losing cohesion.

The primacy of the hoplite also had important social consequences. Considerably more citizens could afford to serve as hoplites than could afford service in the cavalry or mounted infantry. The hoplite reform therefore to some extent democratized warfare (although it still required the possession of moderate wealth). It also helped to democratize Greek society by extending the numbers of those actively involved in the defense of the state. The earliest Greek political assemblies were probably gatherings of the citizens of military age who debated such fundamental issues as declaring war or making peace, and the more who served as soldiers, the broader the political participation. The hoplite also made an important contribution to Greek social and philosophical concepts of the ideal citizen—for example, heavily influencing the form of Plato's ideal state. Because the cooperative characteristics of hoplite warfare fitted so neatly with the cooperative characteristics required of the good citizen (and also with the position of the head of the family), the hoplite continued as the mainstay of most Greek cities' armies even after it could have been superseded by other arms.

The traditional hoplite was a true citizen-soldier and (except at Sparta) not a full-time warrior. The hoplite provided his own equipment and therefore needed to possess a certain minimum level of wealth. Most worked (many as small- or medium-scale farmers) and were called up as required. The campaigning season was mainly during the spring and summer, as most soldiers needed to be back home in autumn to harvest their own crops. In winter the problems caused by weather and terrain made movement and fighting not only difficult but very unattractive.

The Spartan hoplite force was a professional one, though, which helps explain its well-deserved reputation as the preeminent hoplite force in Greece. The difference between the normal Greek hoplite and the Spartan one, freed to train for warfare by the existence of the helot serf-class, is well illustrated by an anecdote about Agesilaus campaigning at the head of a coalition force (Plutarch, *Moralia*, 213F-214B). Sparta's allies had complained that the number of soldiers the Spartans had sent on the expedition was too small in comparison to their own contributions. Agesilaus assembled the army and had it sit down, the Spartans in a group, the rest mixed in together. Asking the soldiers to stand when their occupation was named, he called out potters, smiths, carpenters, and a variety of other trades and jobs until the only men still sitting were his Spartiates. He then pointed out that Sparta was the only state which had provided soldiers—all the others had provided men whose primary trade was something else.

The classic warfare of mainland Greece from the seventh century to the early fifth century involved an army marching into enemy land and ravaging the crops and trees. This was usually enough to bring the inhabitants out to defend their land in a pitched battle between the two hoplite phalanxes. The result was decided in this single battle, with the losers retiring home. Either side could initiate hostilities again the next year, although a major loss of soldiers could keep a state quiet for sometime—as occurred with Argos after its crushing defeat by Sparta at the battle of Sepia in 494.

Warfare in the Greek west and east, though, was in some respects different from that on the mainland. Although the heavy infantryman was still the core of the army (and in fact the hoplite may have originated in Caria in the east), cavalry played a much more important role in these regions. The Greek cities of southern Italy and Sicily had a well-developed cavalry tradition and the greater expanse of territory in the east also led to modifications to the hoplite tradition in the Ionian cities. Almost all of these cities had been colonized from the sea and preserved a stronger naval tradition than most mainland Greek states (Corinth, Aegina, and Athens were notable exceptions to this). This meant that warfare in both the western and eastern Greek world tended to be less dominated by pure hoplite warfare and was more open to external influence.

The long Greek coastline and the numerous islands of the Aegean generally helped Greece develop a strong nautical tradition, but even so the prime military use of ships in early Greece was in the amphibious role—transporting soldiers to conduct landings. These would mostly take the form of raids, as large navies required extensive financial resources and access to the products required for shipbuilding. According to Greek tradition, the first true naval battle was between Corinth and its colony Corcyra circa 660. This apparently occurred during the struggle to establish control over trade and trade routes in the west, as well as over the question of a mother city's primacy over its colonies. This commercial aspect of naval warfare was to remain a constant for much of Greek history. Down to the fourth century at least, the naval powers of mainland Greece—Corinth, Corcyra, Aegina, and Athens—were also the main trading states. The same essentially went for the eastern and western Greek world, where Samos and Lesbos in the east and Syracuse in the west tended to dominate the seas. As late as the Roman period, Rhodes, the most important Greek naval nation-state, was a major commercial power. The Rhodian navy at this time was largely employed in protecting its status as a free port and keeping the seas free from piracy.

The early naval battles were fought between fleets of 50-oared ships called "pentecontors," but these were rendered obsolete with the development of the trireme in the second half of the sixth century. The trireme, in essence a floating projectile, was an exceptionally well-designed warship. Propelled by three banks of oars, with a skilled crew these trim vessels could reach speeds of nearly 10 knots in short bursts, imparting considerable velocity to the bronze ram fitted to the bow. The ram was used to either hole an enemy ship below the waterline or to help strip away the oars from one side. However, the trireme did require a skilled crew and this slowed its widespread adoption until the first half of the fifth century.

Greek Warfare in the Fifth Century B.C.

The early years of the fifth century saw the mainland Greeks challenged by a major foreign power, Persia—at the same time as Carthage was attempting to subjugate the Greek cities in Sicily. The first Persian attack in 490 was repelled by Athenian and Plataean hoplites on the plain of Marathon. The second invasion (480-479) was much larger and had an important influence on Greek warfare and history. The Persian invasion force comprised a large army and navy, which operated together. A simple traditional hoplite defense would not suffice against this combined naval and land threat, even though that is what the Persians expected from the Greeks. According to Herodotus, the Persians saw hoplite warfare as pretty rudimentary and unsophisticated. In the words of Mardonius:

> Whenever they declare war on each other they find the best-looking
> and most level ground and go there to fight. As a result the victors
> come off with great losses and I will say nothing about the losers for
> they are utterly destroyed. . . . if it is absolutely necessary to wage war
> against each other they each ought to find the most defensible spot
> and make their stand there. (Hdt. 7.9)

This may have been true when fighting each other—at least in the seventh and sixth centuries—but the Greeks were certainly not blinkered traditionalists when their survival was at stake and quickly rose to the challenge posed by the Persian invasion. Thirty-one Greek states united to resist the invaders with a combined military and naval force, with the Spartan hoplites and the new Athenian navy at its core. Contrary to Mardonius' reported opinion, the battle sites (including the naval ones) were carefully chosen so that the terrain would not allow the Persians to deploy their full strength against the much smaller Greek forces. The result was a decisive victory for the Greek coalition. At the same time the Carthaginian attack on Greek Sicily ended in disaster at Himera.

Although the postwar period saw a general return to traditional hoplite warfare in Greece, the middle and second half of the fifth century saw some changes, primarily an increasing use of light infantry and cavalry and the rapid expansion of Athenian naval power. By assuming the leadership of those Ionian Greeks who wished to be free of Persian rule, Athens laid the foundations of empire based firmly on naval not land power. In a relatively short time Athens created a naval alliance and then an empire which rivaled the land-based power bloc of Sparta and her Peloponnesian League.

The rivalry between these two powers led to several clashes, culminating in the Peloponnesian War of 431-404. By not marching out to defend their agricultural land during this war, the Athenians had considerable success in avoiding the expected cycle of hoplite battles against a vastly superior coalition. Instead, they replaced their crop losses by imports and retaliated with amphibious raids on the Peloponnese and annual land invasions of their small neighbor Megara after the Peloponnesian army had returned home. However, this tactic was really only available to a major naval power such as Athens, and most Greek states of the time either had to defend their agricultural land or run the risk of at least some of their citizens going hungry, or even starving, over the winter. In the event, Athens, weakened by a major loss in Sicily, by revolts of her subject states and her general unpopularity as an imperial power, and by the long-term after-effects of a virulent plague, succumbed to Sparta's strategy of developing her own navy—paid for by the Persians.

This war demonstrated the need for a strong financial base in order to wage anything other than the old-style hoplite clash. It also demonstrated that long, drawn out wars were possible—and that they demanded a much more sophisticated approach to strategy than before. Linked with this was the development of the ancient equivalent of "total war." As the Peloponnesian War dragged on, more and more of the participants' resources were required, along with new measures, to gain victory. Although it never became the most common treatment accorded to a defeated enemy in Greek warfare, the total destruction of an enemy city, with the execution of the males of military age and the selling of the rest of the inhabitants into slavery, became more commonplace during and after the Peloponnesian War. Warfare had moved on considerably from the seventh- and sixth-century model of resolving the issue in one short sharp engagement with a simple clash between two hoplite phalanxes on an agricultural plain.

Greek Warfare in the Fourth Century B.C.

During the fourth century the use of the non-hoplite arms continued to develop and the preeminent position of the hoplite was increasingly challenged. For much of this century the major mainland Greek states competed with each other for domination, or hegemony, over the rest of Greece. Sparta was the dominant power from 404 to 386, although increasingly challenged from then on by Thebes and a resurgent Athens, which created a new version of its fifth-century naval alliance. Boeotia, led by Thebes, emerged from the pack in 374 and effectively broke Spartan power at the battle of Leuctra in 371. The subsequent liberation of Messenia from Spartan domination ensured that Sparta would never regain her previous power and position. Thebes maintained her position as *hegemon*, or leader, until around 362.

During these struggles the pattern of traditional hoplite warfare underwent considerable modification, and it became increasingly difficult for the part-time citizen-soldier to meet the expanding needs and pace of war. War was waged for longer periods during the year and, partially as a result of this, much more use was made of mercenaries. Several states, notably Thebes, developed a small standing force to form the core of their citizen army. These bands of elite soldiers were known as *epilektoi* (the chosen ones) and were the first hoplites who could challenge the professional soldiery of the Spartans.

Another feature of this period was the increasing use of cavalry and of *psiloi*, or light armed troops, including archers and slingers. Generals such as Iphicrates of Athens experimented with new equipment, formations, and tactics for peltasts and other *psiloi*. By the mid-fourth century a Greek army on campaign was rather different from its average sixth- and fifth-century equivalent. The hoplite force was still the core of the army, but now was much more likely to include mercenaries and to be supplemented by an organized body of light infantry containing specialist slingers, archers, and javelineers—some of whom might well also be mercenaries. The army would also be supported by cavalry, who carried out a variety of tasks, from reconnaissance and raiding to providing flank security on the march and in battle—and pursuing a fleeing enemy or even protecting the withdrawal of their own infantry if the battle was lost. Rather rhetorically, the Athenian politician Demosthenes summed up these changes with his description of Philip II of Macedon "marching wherever he wishes, not leading a phalanx of hoplites but accompanied by an army of light-armed infantry, cavalry, archers, and mercenaries" (9 [*Third Philippic*] 49).

Naval power also played a greater role in the military activities of more Greek states in the fourth century. Toward the end of the fifth century

Sparta had been forced to develop as a naval power in order to defeat the Athenians in the Peloponnesian War. Using Persian money and capitalizing on Athenian weakness, the Spartans created a navy and used it in a two-pronged attack on Athens, detaching its island and coastal subjects in Ionia and preventing the vital shipments of grain from the Black Sea. From this point onward, any state serious about gaining hegemony over Greece needed a navy. Jason of Pherae developed one in the 370s and it was expanded by his successor Alexander. On the advice of Epaminondas, Thebes developed its own navy in the 360s—despite concerns about the expense. The Athenians also tried to revive their naval power in the same period with a new naval alliance, the Second Athenian Confederacy. Philip II of Macedon had a small naval force in the late 350s which carried out nuisance raids against Athens and her possessions. In 340 he possessed a rather larger force, which he used in conjunction with his control of the European shore of the Dardanelles to interfere with the vital Athenian grain imports from the Black Sea.

The trireme was still the main warship of choice during this period because of its effectiveness—despite its inability to operate far from land. However, the quinquereme, or five, was invented very early in the fourth century under Dionysius I of Sicily. Although initially used in limited numbers, often as the flagship, it made a brief appearance in larger numbers in the fleet of Sidon, a Phoenician city, in the mid-fourth century. Greater use was made of it in Alexander's fleet at the siege of Tyre and in Athenian service in the last quarter of the century. Along with other larger ship types, such as the four and the six, the five became increasingly popular at the end of the century, particularly in Antigonus I (Monophthalmus)' fleet, completed in 315.

Warfare was more complex and sophisticated in the fourth century, although in most places the trireme remained the mainstay of fleets and the hoplite the mainstay of armies. But, like the fleets which were starting to vary in some respects, the hoplite and hoplite tactics also underwent some changes. There is evidence that the hoplite equipment (or "panoply") grew lighter. The essential shield, spear, and sword were retained, but breastplates and greaves were often dispensed with or, in the case of the breastplate, replaced by a lighter equivalent—as was often the case with the helmet. This reduced the protection available to the hoplite but gave him greater mobility on the battlefield and allowed a quicker reaction against light troops harassing the hoplite phalanx.

Prior to the fourth century the only hoplite force sufficiently skilled to execute even moderate maneuvers on the battlefield was apparently the Spartan phalanx. This changed with the pool of experienced soldiers available after the Peloponnesian War and the creation of professional or semiprofessional bands of *epilektoi*. The Greek mercenaries of the Ten Thousand, serving in Asia under the Persian prince Cyrus in 401, were able to deal with obstacles and defiles by making fairly complex changes of formation on the march. Before the middle of the fourth century the army of at least one other state, Thebes, was also able to execute rather more than the normal basic maneuvers on the battlefield.

The Thebans, under the influence of their great generals Pelopidas and Epaminondas, also adopted a deeper hoplite formation. The traditional hoplite army was drawn up eight ranks deep, although this must have varied according to the number of troops available, the frontage that needed to be covered, and what the enemy forces were doing. The Thebans had always had a preference for more ranks than this, which gave increased depth and reduced the frontage. The increased depth added weight to a charge and the reduced frontage exposed fewer soldiers to the enemy at the start of the battle. This was occasionally a sore point in allied armies when the other states concluded that the Thebans were not shouldering their fair share of the frontage. In the fourth century the Thebans adopted a 40- or even 50-rank formation, giving them great weight and mass in the attack, especially as this was concentrated against a relatively small frontage of the enemy line.

The Macedonian Phalanx and Philip and Alexander of Macedon

The Theban reforms took the hoplite about as far as he could go and paved the way for the soldier who would eventually overcome and succeed him—the Macedonian phalangite. Credit for this reform goes largely to Philip II of Macedon, although he seems to have been influenced by the Thebans. Certainly the young Philip was a hostage at Thebes while the deep phalanx was evolving under Pelopidas and Epaminondas. Another influence may have been the peltasts, a type of light infantry from neighboring Thrace. One type of peltast (the other used javelins) was equipped with a small, crescent-shaped, wicker shield (the *pelte*—the origin of the name "peltast") and a longer spear than the hoplite. These peltasts probably fought in some sort of formation, probably similar to the hoplite phalanx; their equipment certainly would have enabled them to do so.

What apparently occurred under Philip II was a modification of the peltasts' equipment and formation. The spear was lengthened to become the 14- to 17-foot (4.5- to 5.5-meter)-long *sarissa*, the *pelte* was replaced by a small, round, convex shield, and the men were placed in a deep phalanx like the Theban one. The *sarissa* was long enough for the points of those wielded by the first five ranks to extend forward of the front rank of the formation. This presented to the enemy a hedge of spearpoints vastly more difficult to penetrate than the hoplite phalanx. The convex shield was strapped to the left arm and used to turn away enemy weapons. The *sarissai* of the rear ranks were carried upright until needed and this provided a forest of spears above the rear of the formation which helped deflect missiles.

However, the equipment is only part of the story, and its importance has been exaggerated by some modern authors. The Macedonian phalanx was not made invincible by it, as the two defeats of Philip II by Onomarchus' hoplites in 353 demonstrate. Like its hoplite counterpart, the Macedonian phalanx found it difficult to maintain formation on rough, broken, or steep terrain. On the flat, though, it usually proved superior to a hoplite formation. A major contribution to this was Philip's creation of a professional body of infantry. The high level of training and experience of the men was crucial to the success of the Macedonian phalanx. When such a phalanx was used intelligently in combination with the traditional Macedonian arm, the cavalry, it almost always proved irresistible against other types of Greek or Asian troops.

As employed by Philip II and Alexander, the phalanx fixed the enemy foot in position, inflicting damage and wearing it down, while the cavalry was maneuvered into a suitable position and used to strike the decisive blow. The end of the hoplite as the dominant arm in Greece came with the defeat of the allied Greek army by the Macedonian phalanx and cavalry at Chaeronea in 338. The hoplite did continue in use for some time after this, but the Macedonian-style phalanx and cavalry combination had become the dominant force on the Greek battlefield.

Alexander's conquest of the Persian empire exposed the Greeks to new influences and the armies of Alexander's successors in Asia underwent some changes. The Asian emphasis on cavalry accorded well with the traditions of the Macedonian nobility and royal family and was well suited to the greater campaigning distances in Asia Minor. The Hellenistic period initially saw an increasing emphasis on cavalry, although the Macedonian-style phalanx and massed infantry continued to play a major role and the cost of cavalry did eventually reduce its use. Siege warfare and the continued development of catapults and other mechanical devices (begun under Philip II) also contributed to the increasing complexity of war. The advancement of siege techniques also allowed Alexander the Great to dispense with his fleet. This fleet, supplied by the Greek naval powers, was not only of dubious loyalty but was also outnumbered by the opposing Phoenician navy. To overcome this problem, Alexander disbanded his fleet and neutralized the Phoenician navy by capturing its ports. This would have been out of the question with the comparatively primitive siege techniques of the fifth century.

Greek Warfare from the End of the Fourth Century B.C.

However, naval warfare was by no means rendered obsolete by Alexander's use of land power against naval bases in his conquest of the Persian empire. Navies were still required and naval warfare underwent a similar technological change. As noted above, the trireme was superseded by ever larger and faster ships such as the quinquereme, the four, the six, and even the nine and eleven. The initiator of this was Antigonus I (Monophthalmus), followed by his son Demetrius I (Poliorcetes). However, all the Hellenistic monarchies that followed Alexander made considerable use of naval power in their struggles with each other, and possession of the largest fleet composed of the biggest ships became a matter of prestige. Naval power was essential in controlling the maritime areas of southern Asia Minor, Cyprus, Syria, and Phoenicia. Nevertheless, the Hellenistic monarchs often saw naval warfare as subsidiary—necessary to protect trade and commerce and a matter of prestige, but with the decisive fighting normally taking place on land.

In general it is fair to say that Greek land warfare from the late fourth century on was no longer focused on the short, sharp, decisive hoplite clash between armies of citizen-soldiers. It was now more likely to be a year-round affair, waged by large armies directed by monarchs and backed up by the resources of extensive territories and rich possessions in Asia Minor. Armies now more typically consisted of the Macedonian-style phalanx, complemented by specialist light infantry, cavalry, a siege train, and logistic support. In the civil wars which followed the death of Alexander, rival Macedonian military leaders led large armies against each other with monotonous regularity. One difference, though, was that the average soldier was much more focused on booty and survival. The victors tended to enroll the opposing Macedonian infantry among their forces, as their legitimacy depended on the votes of the Macedonians in their armies. It was usually only the senior members of the opposition who suffered execution if defeated. The casualties in these wars were therefore remarkably light and were often concentrated among the companions and cavalry of the various rulers—the ones with the most to gain from victory and the most to lose from defeat.

In some respects, by the start of the third century, mainland Greece had become the battleground for more oriental-style dynasties, albeit of Macedonian or Greek origin, and the independent *polis* increasingly slipped into the background. Except in local clashes its forces were often too small to do anything much more than to serve, when ordered, as part of the larger dynastic armies. The tendency toward larger landed estates concentrated in fewer hands also reduced the number of free farmers, the basis of the hoplite citizen-soldier class. However, as a result, the period from the end of the fourth century also saw the rise of confederacies of Greek states like the Achaean and Aetolian Leagues—the latter from one of the areas regarded by Greeks of the Classical period as being rather backward, if not semibarbarian. These leagues allowed groups of cities to collectively compete on more even terms with the Hellenistic monarchies of Asia Minor and with Macedon itself, which somewhat declined after the Gallic invasions of the 270s.

The third century saw the western Greeks increasingly falling under Roman control. Independent Greek warfare in the west ended with the Roman occupation of the whole of Sicily at the end of the Second Punic War (218-202) between Rome and Hannibal. In Greece proper and the east, the second and subsequent generations of Hellenistic kings continued their ancestors' struggles for supremacy with each other and with the various Greek confederacies. In general, the Antigonids (based in Macedon) contested with the Greeks for control of mainland Greece, while the Seleucids of Syria and the Ptolemies of Egypt contested with each other for control of Greek Asia. However, they all readily interfered in each other's countries and affairs. Alliances shifted rapidly and the weakness of any monarch was rapidly seized upon by the other players—whether enemies or allies. Parthia was lost to Greek rule in the mid-third century, although at least nominal control was temporarily regained by Antiochus III (the Great) in the last decade of the century. At the start of the second century, Demetrius I of Bactria extended Greek rule over north India, restoring the limits of Greek control under Alexander the Great. In general, land warfare was regarded as the decisive medium but naval warfare continued to be important, at least down to the late third century.

It was these Hellenistic monarchies which waged the last truly independent Greek warfare against the encroaching Roman empire in the second and first centuries. The first real clash between Rome and the Hellenistic monarchies arose from the (ultimately ineffective) assistance given by Philip V of Macedon to Hannibal in his titanic struggle against Rome in the Second Punic War. This First Macedonian War (215-205) did not involve any real contact between Roman and Macedonian troops. The Romans fought it by proxy through their allies in Greece, the Aetolian League, and in the Adriatic a strong Roman naval presence (based largely on the five—as was Philip's navy) prevented Philip from transporting troops to

Italy. The war eventually simply petered out and died of inactivity. Philip and the Aetolians made peace in 206, with the Romans following suit a year later. However, the Romans remained incensed by Philip's offer to help Hannibal when they were at their lowest ebb after the battle of Cannae in 216. When Athens and Rhodes appealed for help against Philip in 201 the Romans responded quickly, leading to the Second Macedonian War (200-197).

Philip V was decisively defeated at the battle of Cynoscephalae in 197 and although the Romans withdrew their troops from Greece at the end of hostilities, the Macedonian defeat was the beginning of the end for independent Greek warfare. Antiochus III (the Great), ruler of Syria, landed in Greece in 192 and proclaimed its independence, but was unable to persuade the Greeks of his good intentions. Again they appealed to Rome, effectively making the Romans the arbiter of Greek affairs. A Roman army expelled Seleucus from Greece and then followed him to his home territory and defeated him there. The resultant Treaty of Apamea in 188 essentially ended the power of the Seleucid empire and foreshadowed the loss of the Greek east to Rome.

The demise of the Macedonian-style phalanx occurred at the battle of Pydna in 168, during the Third Macedonian War (171-167). In this hard-fought battle, the phalanx, little changed from Alexander's time, again proved ultimately ineffective against the more mobile and flexible Roman maniples. Unfortunately for the Greeks, Pydna proved to be the pattern for further Roman-Greek wars. Time and time again the armies of the Hellenistic monarchs proved incapable of defeating the Roman legions. Macedon was split into four republics, which the pretender Andriscus briefly reunited, sparking the Fourth Macedonian War (150-148). This ended in another defeat at Pydna (148) and the incorporation of Macedon into the Roman empire as a province. The final attempt to preserve some vestige of Greek independence ended when the Achaean League went to war with Rome in 147. Its defeat in 146 saw the whole of mainland Greece become part of the Roman empire and the end of independent Greek warfare there—except for a short-lived revolt in 88, crushed by Lucius Cornelius Sulla.

In the Hellenistic east, varying degrees of Greek independence continued for some time. So too did independent Greek warfare, although of the eastern-influenced kind that had emerged after Alexander's death. However, Rome increasingly dominated affairs even there. Success led to success and small kingdoms sought Rome's help against the larger ones, leading to several grateful Hellenistic monarchs bequeathing their kingdoms to Rome. Pressure from the Parthians had already detached the eastern satrapies from Syria. A series of wars between Rome and the weakened Seleucids and other dynasties, not inspired by any Roman master plan for conquest, destroyed the last Hellenistic monarchies.

By 62, Greek Asia Minor was absorbed into the Roman empire, and Egyptian affairs were frequently determined by Rome from about the same date. Egypt itself was made a Roman province after the Greek-influenced forces of Marcus Antonius and Cleopatra VII proved no match for Octavian's experienced army and navy at the battle of Actium in 31 and in the subsequent Roman invasion of Egypt. The irony of all this is that early Roman military formations, from which the flexible legions which conquered Greek civilization grew, were originally heavily influenced by the hoplite phalanx used by Greek colonists in Italy.

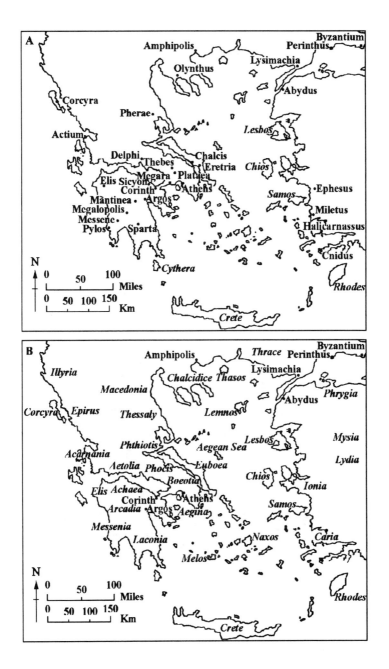

Map 1. Greece: a. Settlements. b. Regions

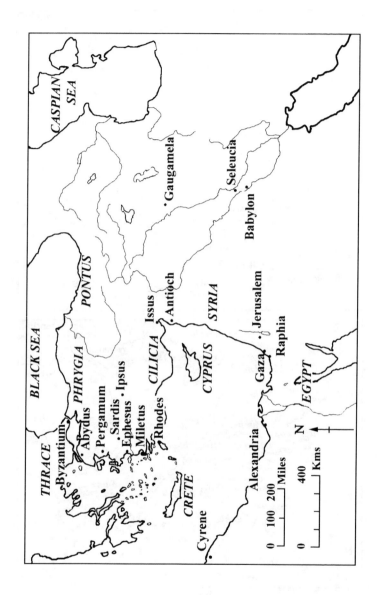

Map 2. Asia Minor

Map 3. Italy and Sicily

Figure 1. The Dendra Armor

THE DICTIONARY

- A -

ABYDUS (ABYDOS). A colony of **Miletus**, situated at the narrowest point of the Hellespont (the Dardanelles), on the Asian side. It was a strategic crossing point between Asia and Europe and therefore saw considerable military action. **Xerxes I** and his army crossed over to Europe from there in early 480, during the Second **Persian War**. After its revolt against **Athens** in 411 Abydus was also fought around during the **Peloponnesian War** of 431-404. In 200 its capture by **Philip V** of **Macedon** helped start the Second **Macedonian War** between **Rome** and Macedon. In 197 **Antiochus III (the Great)** of **Syria** used Abydus as a major base. It was the site of several battles:

(1) 411. A naval battle fought in 411 during the **Peloponnesian War** of 431-404; often known as the battle of Cynossema. An Athenian fleet of 76 **triremes** abandoned its siege of Eresus and sailed after a Peloponnesian fleet of 86 triremes that had slipped into the Hellespont from **Chios**. The Athenians were led by **Thrasyllus** and **Thrasybulus** and the combined Peloponnesian fleet, which included a Syracusan contingent, was led by the Spartan **Mindarus**. The Peloponnesians attempted to outflank the Athenian right wing and force the Athenian center onto the shore. Although this succeeded, partly because the intervening promontory of Cynossema blocked the view of part of the Athenian fleet, when the Peloponnesian fleet began to split up in pursuit, the Athenians rallied and routed the enemy. The Athenians lost 15 ships and captured 21. The victory at Abydus provided a great psychological boost to **Athens** which had had no real successes since the destruction of the **Sicilian expedition**.

(2) 411, winter. A naval battle, during the **Peloponnesian War** of 431-404. Fought between the Peloponnesian fleet under **Mindarus** and the **Athenian** fleet after (1) above. The battle lasted all day without a result until **Alcibiades** arrived with 18 ships and the Peloponnesians fled toward Abydus. There they drew up their ships on the beach and, supported by the **Persian** satrap, **Pharnabazus**, resisted from land. The Athenians captured 30 ships and withdrew to Sestus.

(3) 409. A land battle, during the **Peloponnesian War** of 431-404. Fought between an **Athenian** force of unknown size under **Alcibiades** and **Thrasyllus**, against "a large **cavalry** force" (Xen. *Hell.* 1.2.16) commanded by the pro-**Spartan Persian** satrap, **Pharnabazus**. The

Athenians won, and their cavalry and a 120-man **hoplite** contingent pursued Pharnabazus until darkness allowed his escape (this is one of only two recorded pursuits by Athenian cavalry—the other was at **Cerata** in the same year).

(4) 322, late spring. A naval battle during the **Lamian War**. The **Macedonian** fleet under **Cleitus** defeated a contingent of the Greek fleet. Although the details of this naval campaign are notoriously obscure, the Greek fleet was probably keeping open the grain-supply routes to **Athens** and interdicting Macedonian movement from Asia Minor to Europe across the Hellespont. Cleitus apparently concentrated the majority of the Macedonian fleet and began his piecemeal destruction of the coalition Greek naval force with this victory at Abydus. *See also* AMORGOS.

ACHAEA (ACHAIA/AKHAIA). The northernmost region of the Peloponnese, bordering the Gulf of Corinth (there is another, militarily less important region of the same name, Phthiotic Achaea, in southern **Thessaly**). It consisted of a narrow coastal strip and the northern part of the mountain range separating Achaea from **Arcadia**. In **Homer** the term "Achaeans" is often used to mean "Greeks," but in historical times the region was not particularly important until the refounding of the **Achaean League** in the early third century. The region produced both light and heavy infantry but was not noted for any particular military arm. Achaea remained neutral during the **Persian Wars**, was briefly dominated by **Athens** in the mid-fifth century (until given up under the terms of the **Thirty Years' Peace**), and, apparently reluctantly, joined **Sparta** in the **Peloponnesian War** (2) of 431-404. In the fourth century the region took an anti-**Macedonian** line, sending troops to **Chaeronea** in 338 and joining **Agis III**'s unsuccessful revolt against **Alexander the Great** in 331. The Achaean cities expelled their Macedonian garrisons over the period 280-275 and recreated the Achaean League. The military history of the region from then until its incorporation into the **Roman** empire is essentially the history of the Achaean League. *See also* ARATUS; PHILOPOEMEN.

ACHAEAN LEAGUE (ACHAIAN/AKHAIAN LEAGUE). Also known as the **Achaean** Confederacy. Possibly the earliest Greek federal league, although few details are known of its early period. It originally consisted of 12 cities from the northern Peloponnese, apparently survived the King's Peace (**Common Peace** [1]) of 386, but collapsed sometime shortly after 323. From the beginning it allowed non-Achaean cities as members. Refounded around 284, it had a common organization, including a federal army, led by a *strategos* elected from the whole league (there were originally two *strategoi*, reduced to one around 255). The *strategos* could not serve two terms in a row and it was common for prominent leaders such as **Aratus**, **Lydiadas**, and **Philopoemen** to serve in alternate years. **Polybius** states (29.24) that the league could put 30,000-40,000 men in the field in 169/8. In the third century the league was strengthened by the admission of **Sicyon** and then Megalopolis and other **Arcadian** cities. From this point on, it was a powerful force in Greek affairs, often opposing **Macedon** and in a semiconstant state of hostilities with **Sparta**. At various times in the third and second centuries it was also involved in wars against **Athens**, **Argos**, and the **Aetolian League**. In 146, after an unsuccessful military challenge to **Rome**'s dominance in Greece, the league was dissolved. However, it was probably reconstituted in a reduced form shortly afterward.

ACHAEUS (AKHAIOS; d. 213). An uncle and general of **Antiochus III (the Great)**. He was a successful general, campaigning against **Attalus I (Soter)** of Pergamum in 223-220 and recovering the territory in Asia Minor that Attalus had gained from **Antiochus Hierax**. However, in 220 Achaeus revolted against Antiochus and proclaimed himself king of the area regained from Attalus (Polyb. 4.48). Antiochus besieged Achaeus in Sardis, capturing the city but not the citadel in 215 (Polyb. 7.15-18). **Polybius** (8.17-23) preserves a detailed and dramatic account of Achaeus' betrayal, capture, and brutal execution, which was followed by the surrender of the citadel in 213.

ACHILLES (ACHILLEUS/AKHILLEUS). The main character in **Homer**'s *Iliad*. Son of Peleus and the sea nymph Thetis, Achilles was made invulnerable, except for his heel, when his mother dipped him into the River Styx. Achilles led the Myrmidons during the **Trojan War**. His performance as a warrior was so good that the tide turned against the Greeks when he refused to participate in the fighting after a quarrel with **Agamemnon**. Achilles reemerged when his best friend Patroclus was killed, fighting in Achilles' armor, while trying to stiffen the Greek resistance. In revenge, Achilles killed the Trojan hero **Hector**, mutilated his body, and dragged it behind his **chariot** around the walls of Troy. Achilles was later killed when an arrow fired by the Trojan prince Paris struck him in the heel. He was regarded by later Greeks as an archetypal hero, partly because he went to Troy even though it was foretold that he would be killed there.

ACTIUM (AKTION). The naval battle, fought on 2 September 31, which
ended the war between Octavian (Augustus) and his opponents Antony
(Marcus Antonius) and **Cleopatra VII (Philopator)** of **Egypt**. The war
had broken out in summer 32, when Octavian capitalized on Roman
hostility toward Antony's relationship with Cleopatra to secure sufficient support to go to war. Antony in the meantime had been gathering
forces in **Ephesus** and from there crossed to Greece with Cleopatra and
ended up at Actium at the mouth of the Ambraciot Gulf, south of Epirus. Not especially clear descriptions of the campaign and battle are
given in **Plutarch**, *Antony*, 61-67 and Cassius Dio, 50. Both armies
were approximately the same size, but Antony had 500 ships while Octavian had 600. Although Antony had some big **warships**, including
fours, eights, and tens, Octavian's fleet was better equipped and more efficient, and his admiral, Agrippa, was one of the best **Rome** ever produced.

The fairly hostile accounts portray Antony as totally besotted with
Cleopatra, who, having persuaded him to stake everything on a sea battle instead of his experienced legions, then deserted him during the battle, causing Octavian's victory. While Cleopatra's presence certainly did
not help the morale of Antony's officers and men, her role in the loss is
probably much less than the sources suggest. Antony's intent may well
have been to row out with around 200 of his best ships (the rest were
burned) with their sails on board (these were normally stored on land
prior to an engagement), and then take advantage of the local winds to
make a sudden run for it back to Egypt. However, while the first part of
this worked, when it came time to make the break, Cleopatra's 60 Egyptian ships, followed by Antony, succeeded, but the remainder of the fleet
failed to achieve a clean break and were sunk, captured, or got back into
Actium. If this was the plan, though, it seems very strange that Antony
was prepared to abandon his troops. Plutarch's account stresses the in-
credulity of his army at being left behind—they actually held out for
several days, believing he was going to return. However, when their
commander slipped away during the nigh, they realized it was all over
and surrendered.

This battle marked the end for Egypt, the last semi-independent
Hellenistic monarchy. When Octavian reached Alexandria the following
year, Antony's troops deserted to him and Antony and Cleopatra committed suicide. Egypt was looted and incorporated into the Roman empire.

AEGITIUM (AIGITION). A town in **Aetolia**, site of a battle in 426 during the **Peloponnesian War** (2) between a joint force from **Athens**,

Locris, and **Messenians** from Naupactus, commanded by **Demosthenes** (1) and the Aetolians. Demosthenes was trying to neutralize Aetolia as a threat to Naupactus and open a way for a land attack on **Boeotia** via **Phocis**. Contemptuous of the Aetolians, who were equipped as *psiloi* (light troops), Demosthenes had pressed into the area without awaiting Locrian troops which were also equipped as *psiloi*. The Aetolians knew of the invasion plans and had mustered troops from throughout the region.

After Demosthenes had stormed Aegitium, the combined Aetolian army attacked from all sides out of the hills. Demosthenes' troops beat them off for some time, but when their supporting **archers** had run out of arrows and the **hoplites** had become exhausted from continually but unsuccessfully charging out at the enemy, the Athenians and their allies broke and ran. In the pursuit they suffered heavy casualties from the Aetolians. Some were trapped in a dried watercourse, others in a forest which the Aetolians fired. **Thucydides** comments, "every type of disaster in a rout happened to the Athenian force" (3.98.2). Around 120 of the 300 Athenian *epibatai* (marines) died, as did Demosthenes' fellow *strategos* (general), Procles, and a great number of their allies. After the engagement Demosthenes remained in the area for some time, afraid of a hostile reception back at Athens. This was one of the first triumphs of *psiloi* in the Peloponnesian War and Demosthenes' experiences here led him to become an innovative user of light troops. *See also* LECHAEUM; PYLOS.

AEGOSPOTAMI (AIGOSPOTAMOI). An open beach on the European side of the Hellespont (the Dardanelles), off which the Athenians and Peloponnesians fought a naval battle in 405, during the Peloponnesian **War** (2). The Athenian fleet of 180 **triremes**, under the command of **Conon** and other generals, took station at Aegospotami, watching the Peloponnesian fleet under **Lysander**, which was based at Lampsacus and interdicting the movement of grain ships from the Black Sea to **Athens**. Lampsacus was a much better location than Aegospotami, where the Athenians were moored on the beach and had to travel some distance to Sestus to procure food. According to **Xenophon**, the Athenian generals ignored **Alcibiades**, who was in exile in the area, when he warned them of their dangerous position (*Hell.* 2.1.25-26).

For four successive days the Athenian fleet sailed out to do battle. On each occasion Lysander stayed in the harbor—but with his boats fully manned and ready, crews concealed by side screens—and sent a few fast ships to follow the Athenians when they returned to Aegospotami at the end of the day. On the fifth day, when the fast ships had signaled that the Athenian crews had disembarked and were scattered in the area looking for food, Lysander ordered the attack. Caught completely unprepared, only eight ships under Conon, and the state galley, *Paralus*, escaped. The remaining 171 triremes were captured and the camp overrun by Lysander's land forces. In revenge for recent Athenian atrocities against enemy sailors, Lysander executed all the Athenians, except one (the general Adimantus, who had spoken out against these actions), and freed all the other **prisoners**. This defeat effectively destroyed Athenian naval power, led to the immediate loss of **Byzantium** and **Chalcedon**, and opened the way for the blockade and ultimate capture of Athens and the end of the war.

AELIAN (AELIANUS). Author of a treatise (the *Tactics*) on the **Macedonian phalanx** and its tactics. Probably written under Trajan (A.D. 98-117), the *Tactics* borrows heavily from **Asclepiodotus** and has little independent value. *See also* AENEAS TACTICUS; ONASANDER.

AENEAS TACTICUS (AINEIAS TAKTIKOS). Writer of military manuals; possibly to be identified with Aeneas of Stymphalus, **Arcadian League** general in 367. Aeneas wrote several manuals but only the one dealing with the defense of cities and fortifications survives. This was written circa 356 and is interesting for some of the examples given and the fact that Aeneas considered the main threat to a besieged city to be internal. Much of the treatise is therefore devoted to internal security precautions. *See also* AELIAN; ASCLEPIODOTUS; ONASANDER.

AETOLIA (AITOLIA). A generally mountainous region north of the Corinthian Gulf and southwest of **Thessaly**, noted for its *psiloi* (light troops). Regarded by many other Greeks as a backward region, it was still organized on tribal rather than *polis* (city-state) lines in the fifth century. In 426, during the **Peloponnesian War** (2), it successfully repelled an invasion attempt by the Athenian general **Demosthenes** (1) at **Aegitium**. He was apparently trying to protect Naupactus, bring the region over to **Athens**, and secure a route into **Boeotia** from the west (Thuc. 3.94-99). In 413 Aetolian **mercenaries** are recorded as part of Athens' **Sicilian expedition** (Thuc. 7.57.9). The region played a peripheral role in Greek history until the formation of the **Aetolian League** in the first half of the fourth century greatly increased its power. From then until its surrender to **Rome** in 189 the military history of Aetolia is really the history of the Aetolian League.

AETOLIAN LEAGUE (AITOLIAN LEAGUE). Also known as the
Aetolian Confederacy. A federal league established in **Aetolia** in the first
half of the fourth century, it was apparently a true *sympoliteia*, with each
community enjoying equal rights and contributing to a common treasury. Over time, non-Aetolian states were added as associate members,
without full rights. The members also contributed troops to a league
army, which was led by an annually-elected general. One of the peculi-
arities of the league was that it regarded the taking of **booty** from anyone
involved in any hostilities as quite legitimate—even if a formal state of
war did not exist between the Aetolians and the other party. This helped
gain them a reputation for piracy and unscrupulous profiting from wars.

In the third century the Aetolian League dominated **Delphi** and the
Amphictionic League and played a major part in saving the sanctuary
(and central Greece) from the **Gauls** in 279. Initially allies of **Rome**, the
league did most of the fighting during the First **Macedonian War** (215-
205) but made peace with **Philip V** of **Macedon** in 206 without Roman
approval. The league did assist Rome in the Second Macedonian War
(200-197) but was rewarded with only a small part of **Thessaly** instead
of the whole country. Because of this, the Aetolians invited **Antiochus
III (the Great)** to intervene in Greece. In the ensuing war with Rome,
the Aetolian League was forced to make terms in 189, which limited the
league to Aetolia and made its remaining members subject allies of
Rome. From this point on, the league effectively ceased to exercise any
real power in Greece.

AGAMEMNON. King of Mycenae; leader of the Greeks in the **Trojan
War.** **Homer's** *Iliad* represents him as a proud and at times arrogant
leader and a brave warrior. At the very start of the expedition he sacrificed his own daughter Iphigenia when the seer Calchas prophesied that
only this would persuade the gods to lift the contrary winds preventing
the Greeks from sailing to Troy. Agamemnon led the Greeks to initial
victory, forcing a landing at Troy and establishing the siege. However,
he later used his position as commander to take **Achilles'** female slave,
Briseis, a prize of war, to replace one of his own. This led to Achilles'
withdrawal from the fight and placed the whole expedition in jeopardy.
On his return home after the war, Agamemnon was murdered by his wife
Clytemnestra, who had never forgiven him for sacrificing Iphigenia.

AGATHOCLES (AGATHOKLES; 361-289). A native of Himera who
became tyrant in 317 and then king (c. 307) of **Syracuse. Diodorus
Siculus** describes him as a man "of the humblest origins who involved

not only Syracuse but all of **Sicily** and Libya in the greatest misfortunes" (19.1.6).

His family emigrated to Syracuse around 340 and in the mid-320s Agathocles distinguished himself as a commander against Acragas and the Bruttians in **Italy**. However, he fell out of favor with the ruling oligarchy and served as a **mercenary** for **Tarentum** (Diod. Sic. 19.3.1-3; 19.4.1-2; cf. Justin, 22.1.7-16). He seized power in Syracuse in 317 (after a recall and another exile), staging a massacre of the aristocrats and wealthy using a force of 3,000 non-Syracusan exiles and the poorer classes in the city (Diod. Sic. 19.6-9).

In 315 he staged an unsuccessful attack on Messana and the following year faced a coalition of enemies, including Acragas, Gela, and Messana, who brought out the **Spartan** prince Acrotatus to command them. He proved a failure and the coalition made peace with Agathocles, brokered by the **Carthaginian** commander on the island, under the terms that Heraclea, Himera, and Selinus remained under Carthaginian control; the rest were to be autonomous under Syracusan leadership (Diod. Sic. 19.65; 19.70-71). Agathocles, however, used the peace to extend his domination by capturing some cities and allying with others. His success and the associated increase in wealth and power enabled him to employ a mercenary force of 10,000 infantry and 3,500 **cavalry** (Diod. Sic. 19.72).

However, in 312 this success provoked the Carthaginians (along with Syracusan exiles and other local enemies) to attack him. They inflicted a suffered a major defeat on Agathocles at the Himeras River in 311, overcoming a reverse in the initial stages of the battle by the effectiveness of 1,000 **slingers**. This loss caused the defection of many of Agathocles' subject and allied cities and led to a Carthaginian siege of Syracuse (Diod. Sic. 19.102-110).

In an interesting example of the indirect approach, Agathocles responded to the siege of Syracuse by attacking Carthage. In 310, he broke out of Syracuse by sea and managed to land in Africa with a force of 12,000 infantry (6,000 of whom were mercenaries—3,000 Greeks, the remainder Samnites, Etruscans, and Celts) and an unspecified number of cavalrymen (without their horses). He burned his ships and advanced inland, taking two Carthaginian towns and then defeating a hastily raised Carthaginian army outside Carthage itself. The speed and audacity of his advance took his unprepared enemy by surprise (Diod. Sic. 20.3-13). The interesting military situation was now that a Carthaginian army was besieging Syracuse while Agathocles was besieging Carthage. He concentrated on the surrounding areas, capturing several cities and causing some of Carthage's subjects to come over to him. He then defeated the

Carthaginians (now reinforced by troops recalled from Sicily) in a dawn attack outside Tunis (Diod. Sic. 20.17-18).

Hampered by a lack of troops, in 309/8 Agathocles allied with Ophellas of Cyrene. He brought a force consisting of 10,000 infantry, 600 cavalry, and 100 **chariots** (one of the features of Agathocles' African campaign was the use of chariots by both sides) and which included quite a number of **Athenians** and other mainland Greeks. When Ophellas arrived, Agathocles killed him and took over his army (Diod. Sic. 20.40-43). The following year he assaulted Utica, fixing Utican **prisoners** on his siege tower when the city refused to surrender. Despite a determined resistance, the city fell and Agathocles slaughtered many of the inhabitants. He followed this by storming Hippu Acra, making himself master of the coastline and much of the interior.

However, taking advantage of a Syracusan victory over the Carthaginians, Acragas now made a bid for hegemony of Sicily. It achieved considerable initial success under the banner of restoring freedom to those cities under Carthaginian or Syracusan domination (Diod. Sic. 20.30.1-32.2). Feeling that his position in Africa was secure, Agathocles returned to Syracuse with a small force, arriving just after the Syracusans had defeated Acragas in a major battle (Diod. Sic. 20.54-56).

During Agathocles' absence, the situation in Africa deteriorated rapidly, with the loss of two large independent contingents in separate actions, leading to a large-scale defection of local allies. The campaign in Sicily, however, went well. Agathocles personally defeated the Carthaginians in a naval engagement and one of his generals inflicted a major land defeat on Acragas. The situation restored, Agathocles returned to Africa in 307 but was defeated in battle, losing around 3,000 men, with more killed in a major panic the night after the battle—coincidentally the Carthaginians lost many men in a similar occurrence that same night (Diod. Sic. 20.59-67). Agathocles then abandoned his army (and sons) and fled back to Syracuse. His sons were killed by the enraged soldiers, who then either surrendered under terms or were killed, ending any threat to Carthage (Diod. Sic. 20.68-69).

On his return to Sicily, Agathocles sacked Segesta, an allied city, to raise money. The inhabitants were killed or enslaved. In revenge for the death of his sons, he then massacred the relatives of those still in Africa (Diod. Sic. 20.71-72). In 305 he defeated a coalition of his Sicilian enemies in battle, despite being outnumbered, executed all those who had surrendered under a promise of safety, and pacified most of Sicily (Diod. Sic. 20.89-90).

The remainder of Agathocles' reign (he had proclaimed himself king around 307) was spent extending his power in Italy and the Adriatic.

Around 300 he defeated **Cassander's Macedonian** forces outside **Corcyra** and secured the island for himself. Soon afterward he was defeated in Italy, losing 4,000 men, and was forced to withdraw to Syracuse. Around 295 he married his daughter Lanassa to **Pyrrhus I** of Epirus, giving him Corcyra as a dowry (Diod. Sic. 21.2-4; Plut., *Pyrrhus*, 9). He used the pretence of escorting his daughter to Epirus for the marriage to surprise and capture Croton—he destroyed part of the wall with siege **artillery** and tunneling; the male inhabitants were killed and Agathocles garrisoned the city (Diod. Sic. 21.1.4). The following year he had mixed success against the Bruttians in Italy (Diod. Sic. 21.8).

Agathocles died in 289, supposedly poisoned by a Segestan slave revenging the sack of his city. His chosen successor had just been murdered by another son and Agathocles handed power back to the people on his deathbed. After his death the citizens tore down his statues and confiscated his property (Diod. Sic. 21.1.16); Syracuse and Agathocles' Sicilian empire soon lapsed into anarchy.

Agathocles was clearly an able ruler and skilled general, whose performance was marked by clever use of ruse and stratagem, as well as by daring attacks by both night and day. He was also quite ruthless toward his enemies (and friends), although this aspect of his military and political activity may be exaggerated. The main extant account of his reign, Diodorus Siculus, is based on the very hostile history of one of Agathocles' opponents, Timaeus. Despite Diodorus' recognition of this hostility (21.17), it does appear to have affected his treatment of Agathocles.

AGESILAUS II (AGESILAOS; 444-360). Eurypontid king of **Sparta**, 399-360; son of **Archidamus II**. Agesilaus succeeded his half-brother, **Agis II**, as king through **Lysander's** influence but soon proved his independence. Agesilaus led a Spartan army (some 18,000 non-Spartiate **hoplites** and a small force of **cavalry**, supported by around 100 **triremes**) against the **Persians** in Asia Minor in 396. Raising troops locally, especially cavalry and *psiloi* (light troops), Agesilaus trained them to a high level of efficiency in the winter of 396/5. In 395 he won a series of successes against **Tissaphernes** and then **Pharnabazus**, but allowed himself to be distracted from the main strategic goal, **Caria**, by the rich **booty** available in Pharnabazus' satrapy. The same year, partly helped by Persian gold, Greek opposition to Sparta resulted in the outbreak of the **Corinthian War**. Agesilaus was recalled after the Spartan naval loss at **Cnidus** and, successfully traversing a hostile **Thessaly** (no mean feat), defeated the **Boeotians** and allies at **Coronea**. Agesilaus proved the most competent Spartan general during the remainder of the war, winning successes near **Corinth** and in 389/8 forcing Acarnania to come to terms.

Following the end of the Corinthian War in 387, Agesilaus followed a pragmatic policy of furthering Sparta's interests by weakening potential enemies—even though this involved several actions in breach of the autonomy clause of the King's Peace (**Common Peace** [1]). He condoned **Phoebidas'** seizure of the Cadmea at **Thebes** in 382 and Sphodrias' failed raid on Piraeus in 378. His two invasions of Boeotia in 378 and 377 were well planned (he seized passes well in advance to guarantee his route) but achieved little. Ill health reduced his role in military activities for a while but in 371 his refusal to let the Thebans sign the Restatement of the King's Peace (Common Peace [2]) on behalf of Boeotia led to the battle of **Leuctra** and the defeat of his fellow king, Cleombrotus. Agesilaus played a prominent role in the defense of Sparta against Boeotian attacks after Leuctra and also campaigned in foreign service in Asia Minor and **Egypt** (364 and 361) in order to raise money for the Spartan treasury. He died in 360 on his way back from Egypt.

AGESIPOLIS I (d. 380). Agiad king of **Sparta**, 395-380. Son of **Pausanias** (2), he came to the throne as a minor. In 388 he led an invasion of **Argos**, despite Argive pronouncements of a sacred truce. He did this on the basis of a reply from the oracles at **Olympia** and **Delphi** that a sacred truce was invalid when a sacred period was not calculated at the normal time and only identified when hostilities were imminent. During this invasion Agesipolis did considerable damage and declined to withdraw when there was an earthquake. He argued that, as it had not occurred before or as he was invading but only once the army was well inside Argive territory, it was therefore a sign of encouragement. However, Agesipolis did abandon plans to construct a permanent fort at a pass into Argos because of unfavorable omens. In 385 Agesipolis invaded **Mantinea** and when facing a long siege (the city was well stocked with grain), dammed a river which flowed through it, causing it to rise and damage the upper courses of the walls, which were made of mud brick. The Mantineans surrendered and the city was broken up into four separate villages in 384. Agesipolis died of fever in 380 while campaigning in the Chalcidice—after **ravaging** the territory of **Olynthus** and storming Torone. His body was placed in honey and returned to Sparta for a full royal burial. *See also* SACRIFICE.

AGIS. The name of several Eurypontid kings of **Sparta**. The most important militarily are:

(1) Agis II (d. c. 399). Son of **Archidamus II**; king, 427-400/399. During the **Peloponnesian War** of 431-404 Agis led invasions of Attica in 427 and 426. He was one of the signatories of the Peace of **Nicias** in 421. In 418 Agis led the Spartan army against **Argos** but concluded a four-month armistice which was heavily criticized at home because it led to the loss of **Orchomenus** (**Arcadia**). Sent back again into Argos, he defeated the Argives and their allies at the battle of **Mantinea** (1). On the resumption of the Peloponnesian War in 413, Agis was tasked to occupy **Decelea** in Attica and bring the war to **Athens**. In 404, in combination with **Pausanias** (2) and **Lysander**, he was involved in the total blockade of Athens that led to its surrender. Agis played a part in the negotiations, although **Xenophon**'s account (*Hell.* 2.2.10-24) has Lysander as the dominant participant. Following this, Agis led the Spartans in war against **Elis**. Although he withdrew his first invasion force because an earthquake was interpreted as an ill omen, his subsequent campaigns forced Elis to make to terms. Agis died of natural causes around 399.

(2) Agis III (d. 331). King, 338-331; focal point of Greek resistance to **Macedon** while **Alexander the Great** was campaigning against **Persia**. Agis received 30 talents of silver and 10 **triremes** from the Persians, which he used to raise a large force, including 8,000 **mercenaries**. Issuing a general call for revolt, he managed to gain the support of parts of the Peloponnese. Megalopolis, however, remained loyal to Macedon and while besieging it in summer 331, Agis was forced to fight against **Antipater**'s much larger army. Agis was defeated and killed, ending Greek resistance.

(3) Agis IV (c. 262-241). King, c. 244-241. He attempted to reverse the decline of Sparta by returning to the Lycurgan constitution of earlier times. He had some success with radical reforms in 243 and 242 and apparently had made some progress toward restoring the professionalism of the Spartan army. However, Agis' reforms were reversed in 241 while he was away with **Aratus** and the **Achaeans** preparing to meet an **Aetolian** invasion of the Peloponnese. Agis and his men were dismissed without action (the Achaeans fearing his revolutionary ideas) and on his return home he was arrested and executed.

AGOGE. See SPARTA

ALCIBIADES (ALKIBIADES; c. 450-404). An aristocratic Athenian general and politician, prominent during the **Peloponnesian War** of 431-404; son of Cleinias. Brought up by his guardian, **Pericles**, and taught by Socrates, Alcibiades served in the **cavalry** at **Delium** in 424.

Following the Peace of **Nicias**, he was largely responsible for persuading **Athens** to join **Argos** in an anti-Spartan coalition in the Peloponnese. This policy ended in failure with the Spartan victory at **Mantinea** in 418 and contributed to the Spartan animosity toward Athens that helped cause the renewal of the Peloponnesian War in 413.

Alcibiades successfully proposed an Athenian expeditionary force to **Sicily** in 415, another factor which led to the renewal of war with **Sparta** in 413. Prior to the expedition's departure, Alcibiades was implicated in the mutilation of the Hermae (sacred statues located all over Athens). His plea to have the matter heard before he left was rejected, leaving him vulnerable to his political enemies after his departure. In Sicily, Alcibiades proposed a sensible plan that could well have led to success, but he was recalled to Athens before it could be implemented, on charges of having profaned the Eleusinian Mysteries. En route for Athens, Alcibiades deserted to Sparta. His recall left effective command in the hands of **Nicias**, who was very cautious and sick and did not believe in the expedition. This played a major part in the destruction of the expeditionary force.

Alcibiades rapidly adapted to Spartan ways and significantly damaged Athens' prospects by suggesting that Sparta send a general to assist **Syracuse**, renew the war against Athens, and fortify **Decelea** in Attica. In 412 he led a mission to **Ionia** that persuaded a considerable number of Athens' allies in the area to revolt. Shortly afterward he fell under suspicion at Sparta and deserted to **Tissaphernes**, the local **Persian** satrap. Again quickly adapting to Persian ways, Alcibiades persuaded Tissaphernes to adopt a policy of keeping both Athens and Sparta weak. He then attempted to rehabilitate himself at Athens by persuading Tissaphernes to support the Athenians. Although he failed, his maneuverings contributed to the overthrow of the democracy at Athens and the temporary establishment of oligarchic rule. Ironically, the army and fleet at Samos, which remained true to the democracy, then elected him *strategos* (general). He dissuaded them from sailing against Athens, something which almost certainly would have given victory to Sparta, and played an active part in the subsequent campaigns against the Peloponnesians in Ionia and the Hellespont. However, he was never able to deliver on his promises of Persian support and in 410 was arrested by Tissaphernes. He escaped a month later and was largely responsible for the decisive Athenian naval victory off **Cyzicus** in 410.

That same year the democracy was restored in Athens and Alcibiades was continually reelected general to conduct operations in the Hellespont and Ionia. In 407 he judged it safe to return to Athens and was again elected *strategos* for the following year. The error of a subordinate at **Notium** in 406, while Alcibiades was elsewhere, led to his final exile from Athens. In 404 he was murdered on his estate in Phrygia. **Thucydides** viewed him as a brilliant, if erratic, man whose wild and expensive lifestyle alienated his fellow citizens, commenting: "Although in terms of public matters he (Alcibiades) conducted the affairs of the war best, everyone was upset at his private conduct and, handing affairs over to others, the city was shortly afterward ruined" (6.15.4). *See also* SICILIAN EXPEDITION; THIRTY TYRANTS.

ALEXANDER (OF MACEDON). The name of several kings of **Macedon**, the most famous being **Alexander the Great** (see separate entry). The others of most military interest are:

(1) Alexander I (d. c. 452). King, c. 495-452. Alexander I played an ambiguous role during the Second **Persian War**. He warned the Greek force sent to Tempe in 480 that its position was untenable given the size of the Persian army and advised it to withdraw (Hdt. 7.173). He subsequently led **Macedon** in **medizing** and provided troops to garrison the **Boeotian** cities to ensure their loyalty to **Persia**. In winter 480/79 he was sent on an unsuccessful mission to persuade **Athens** to defect to Persia (Hdt. 8.136-144). The night before the battle of **Plataea** he rode over to the Greek lines and warned them of the Persian intention to attack at dawn (Hdt. 9. 44-45). The story of his warning (and the earlier assassination of Persian envoys [Hdt. 5.17-21]) may well be part of his successful propaganda effort to be accepted as a Greek (cf. Hdt. 5.22; this also involved hellenizing the Macedonian court).

After the Second Persian War, Alexander extended his kingdom, largely at the expense of the **Thracian** tribes. He acquired territory up to the upper part of the Strymon River (the later site of **Amphipolis**) between 478 and 476, securing important mines, which led to a marked increase in Macedonian prosperity (Hdt. 5.17). This brought him into conflict with Athens, which had captured Eion at the mouth of the Strymon.

Although he had previously been on very friendly terms with the Athenians (Hdt. 8.136), Alexander clearly now regarded them as a threat and worked actively against them in the region. Alexander gave asylum to **Themistocles** when he fled southern Greece (Thuc. 1.137.1), and in 462 after Athens' failure at Drabescus against the Edones (who had recovered the area from Alexander), the commander, **Cimon**, was charged with taking bribes from Alexander (Plut., *Cimon*, 14.2). The region saw continual struggles between Macedon, the Thracian tribes, and Athens (at least until 461 when Athens' main effort shifted to **Egypt**). Numismatic evidence suggests that Alexander lost control of the strategic mines for several years after 460 (see Hammond, *A History of Macedonia*, 2, 107-108, 114). Having considerably strengthened Macedon, Alexander died circa 452, either assassinated or possibly killed in battle (cf. Curt. Ruf. 6.1.26). Although the Macedonian **cavalry** was of generally high quality during this period, Alexander's military resources were hampered by a fairly primitive society and a lack of heavy infantry.

(2) Alexander II (d. 368). King, 370-368; elder brother of **Philip II**. Although Alexander II's reign was very short, it has some important military aspects. He was forced to buy off the Illyrians (Justin, 7.5.1) but in 369 he was called in by the Aleuadae to intervene in **Thessaly** against **Alexander of Pherae**. He took the opportunity to seize and garrison both **Crannon** and Larissa for himself; however, it is doubtful whether he had the resources to hold down even parts of Thessaly for an extended period. In the event, a rebellion back home (led by Ptolemy of Alorus, probably a younger son of Amyntas II) forced him to withdraw and a **Boeotian League** expedition under **Pelopidas** ejected these garrisons and then advanced into **Macedon**. Pelopidas settled affairs there, getting Ptolemy to recognize Alexander as king and Alexander to restore some exiles. Pelopidas also set up an alliance with Alexander, who sent the future Philip II as a hostage to **Thebes** to guarantee his good faith (Diod. Sic. 15.61.2-5, 15.67.3-4; Plut., *Pelopidas*, 26). Alexander was shortly afterward assassinated, probably by Ptolemy. Alexander overreached himself in Thessaly but was almost certainly responsible for setting up the **Foot Companions** (*pezhetairoi*) alongside the much older Companion Cavalry (*FGrH* 72 fr. 4). This, and the **hoplite** training under Alexander and Perdiccas III (Justin, 11.6.4-5), began the important development of the Macedonian infantry brought to fruition by Philip II.

ALEXANDER (OF MOLOSSIA/EPIRUS). The name of two kings of Molossia, which later became the major state within the Epirote alliance. From the time of **Pyrrhus I** (319-272), the kings of Molossia also ruled Epirus.

(1) Alexander I (d. 331). King, 342-331. Alexander was the brother-in-law of **Philip II** of **Macedon** and uncle of **Alexander the Great**. Alexander attained the throne of Molossia in 343/2 with the help of Philip II and married Philip's daughter Cleopatra in 336, further linking Molossia/Epirus and Macedon (Diod. Sic. 16.72.1, 16.91.4-6). In 334 he took an army to **Italy** to fight the Lucanians on behalf of **Tarentum**. Alexander was initially successful, defeating the Bruttians and Lucanians in several battles and capturing Heraclea, as well as several Lucanian and Bruttian towns. He was apparently acting independently from the Taren-

tines when killed near Pandosia (probably late in 331) by a Lucanian exile serving with him (Livy, 8.24).

(2) Alexander II (d. c. 240). Son of **Pyrrhus I** of Epirus; king of Molossia and Epirus, 272-c. 240 (although in exile c. 262-260). Details of his reign, particularly the chronology, are obscure and mainly derived from Justin's epitome of Pompeius Trogus. At some point Alexander was successful in defending his kingdom from a major Illyrian attack but subsequently failed in an invasion of **Macedon**. This was in the latter part of the **Chremonidean War**, perhaps 262, and occurred while **Antigonus II (Gonatas)** was distracted by his campaign against **Athens** and **Sparta**. The attack was possibly at the instigation of **Ptolemy II (Philadelphus)**, but there is no real evidence for this and it was certainly to Alexander's personal advantage. However, Alexander was decisively beaten and forced into exile in Acarnania for around two years. Around 260 he was restored to his throne with **Aetolian League** military assistance. Probably around 243 (a rather more likely date than 260) he was involved in the partition of Acarnania with the Aetolian League (Polyb. 2.45; 9.34).

ALEXANDER OF PHERAE (d. 358). Nephew of **Jason of Pherae**; tyrant of Pherae in **Thessaly**, 369-358. Following his accession Alexander was opposed by Larissa and other Thessalian cities, which refused to recognize him as *tagus*. Opposition from **Thebes** further weakened his control and **Pelopidas** invaded Thessaly several times, initially checking Alexander in 369 and assisting in the establishment of the Thessalian League as a counterweight to Pherae. In 368, newly allied with **Athens**, Alexander seized Pelopidas on a visit to Thessaly and fought off a poorly led Theban army sent to rescue him. The following year an army under **Epaminondas** was more successful and secured Pelopidas' release. In 364 Alexander's Thessalian opponents again invited Thebes to intervene and Pelopidas defeated him in the decisive battle of **Cynoscephalae** (1). After a second success later the same year or early the next, Thebes drastically reduced Alexander's territory and forced him to join the **Boeotian League**.

The literary tradition is hostile to Alexander, portraying him as a cruel leader. Although this may be exaggerated, he was certainly less diplomatic and skillful in dealing with the other Thessalian cities than his uncle, Jason. **Xenophon** describes him as "a lawless plunderer by both land and sea" (*Hell.* 6.4.35). He was assassinated by his wife's brothers, at her instigation, in 358. The eldest of these, Tisophonus, succeeded him as tyrant.

ALEXANDER THE GREAT (356-323). King of **Macedon** from 336 to 323. Son of **Philip II** and Olympias (a member of the royal family of Epirus), he succeeded to the throne as Alexander III without any difficulties when his father was assassinated in 336. Although Alexander has often been suspected of complicity in his father's murder, no link has ever been proven and many others had as much or even more motive. Aristotle was his tutor and Alexander received an excellent military education serving with his father. In 340 his first independent command led to the conquest of a rebellious **Thracian** tribe, the Maedi. The following year he participated in Philip's campaign against the **Scythian** king Atheas. In 338 Alexander commanded the **cavalry** on the left wing against the Greeks at **Chaeronea**.

On his accession, Alexander cowed those Greek states which were inclined to challenge Macedonian control of Greece (**Thessaly**, Ambracia, **Thebes**, and **Athens**) by a rapid show of force and then reasserted control in the north. In a quick campaign he defeated the Triballi and Getae, tribes who lived either side of the lower Danube, and the Dardanians and Talauntians in Illyria. Pausing only to put down a revolt by Thebes (which was destroyed on the decision of the **Hellenic League** [2] council), Alexander then took over his father's well-advanced preparations for the invasion of **Persia**.

In 334 Alexander crossed the Hellespont to join with **Parmenio** and the advance forces dispatched by Philip II. Alexander commanded 160 **triremes** (largely provided by the Greek states) and an army of around 40,000 men. This included 32,000 infantry and 5,100 cavalry, plus an unknown number of troops already in Asia (numbering 10,000 in 335, but not necessarily at that strength in 334). Of these, members of the Hellenic League provided 2,100 cavalry and 7,000 **hoplites**; a further 5,000 **mercenary** hoplites also served. Alexander advanced inland with one month's provisions and soon defeated the Persian army of the Hellespont at **Granicus**. He then marched down the coastline of **Ionia**, liberating the Greek cities from Persian rule and enrolling them in the Hellenic League. He occupied Sardis and **Ephesus** and captured **Miletus**. Lacking money to pay for the fleet, which in any case was considerably less experienced than the much larger Persian fleet of 400 ships, Alexander sent the bulk of it home. He retained the small Macedonian naval contingent to secure his communications across the Hellespont and began to neutralize the Persian navy by capturing all its ports.

Alexander conquered Caria and, over the winter of 334/3, southern Asia Minor and then moved north through Pisidia into Phrygia. At this time the Persians had been having some success at sea, occupying Samos and **Chios**, but suffered a setback with the death of their admiral

Memnon and the recall of the Greek mercenaries with the fleet to join **Darius III**'s army in Persia. Alexander, after a bout of sickness, advanced through Cilicia to meet the Persian army. In November 333 Alexander destroyed this Persian force at the hard-fought battle of **Issus**. This opened up the coast of **Syria** and Phoenicia and Alexander occupied most of it without resistance. However, the major port of **Tyre** refused to submit and it took a seven-month siege before the city was captured in July 332. This effectively spelled the end of Persian naval resistance. At this point he rejected a generous peace offer from Darius to cede all territory west of the Euphrates and pay a large indemnity. By November 332 Alexander had traversed **Palestine**, successfully besieged **Gaza**, and occupied **Egypt** without a fight.

In spring 331, back in Tyre, Alexander made the military, financial, and administrative arrangements for his new conquests in the eastern Mediterranean. He then led his army (now of 40,000 infantry and 7,000 cavalry) into Mesopotamia to find and defeat Darius III. In October 331, weakened by his loss of access to Greek mercenaries, Darius was decisively defeated at **Gaugamela** but managed to escape to Media. Alexander accepted the surrender of Babylon, then occupied Susa (and the Persian treasury) and captured the Persian heartland, setting fire to the royal palace at Persepolis. In the same year, Alexander's commander in Greece, **Antipater**, suppressed an Odrysian rising and then a Persian-financed rebellion, led by **Agis III** of Sparta.

Alexander's pursuit of Darius III into the northeast of the Persian empire ended in July 330 with the great king's assassination by one of his own satraps, Bessus. Bessus was executed in 328 but Alexander was forced to deal with a serious revolt led by Spitamenes in Bactria and Sogdiana. It was not until spring 327 that Alexander was in full control of the entire Persian empire. In the summer of the same year he invaded India via modern-day Afghanistan.

After some hard fighting against the hill tribes in the north, Alexander, supported by the local ruler Taxiles, entered India via the Khyber Pass and defeated **Porus** at the **Hydaspes River** in 326. Moving east he reached the Hyphasis River (modern Beas). There his troops refused to go any further and Alexander reluctantly turned back. Moving south down the Indus River, Alexander fought a bitter campaign against the local tribes, treating them very harshly when they resisted. In mid-325 he turned northwest along the coast, losing many men and camp-followers in the crossing of the Gedrosian desert. He did not reach Susa until spring 324. Alexander spent some time reorganizing his empire but died in June 323 before the process was complete.

Although Alexander inherited an excellent army and some good senior officers from his father, there is no doubting his military genius. This is evident on the tactical level from his major victories over Greeks, Persians, and Indians. His success here was due to charismatic leadership, careful **reconnaissance**, sound planning, masterly use of infantry and cavalry in combination, and a willingness to take calculated risks—as well as to the quality of his troops. Alexander also proved tenacious and skilled at besieging cities and strongholds, often (as at Tyre and the Aornian Rock) thinking laterally about ways to solve the problem. Alexander's decision to neutralize the Persian navy from the land and his relentless concentration on Darius as his main focus illustrate his abilities at the strategic level. Although often generous toward enemies (especially in the early part of his career), he did make use of massacre and enslavement to secure his rule.

His policies of founding new cities had the effect of aiding the spread of Hellenism in the east. Ironically, his policy of adopting local customs and institutions to secure his rule not only caused problems among his Macedonian and Greek troops but also helped increase eastern influence upon the Greeks in Asia Minor which was such a feature of the Hellenistic monarchies. Although his achievements as a conqueror were great, his early death meant that his empire had not been consolidated and it rapidly fragmented as his successors, the *diadochoi*, struggled for power and position. *See also* BUCEPHALUS; HALICARNASSUS; SIEGE WARFARE.

AMBUSH. There was a strong, if rather ambivalent, tradition of ambush in ancient Greek warfare. Ambush could encompass a range of activities on very different scales. These include hiding troops to attack small enemy parties and pre-positioning a hidden force to attack an enemy army from the rear or flank during a battle. Slightly less clear-cut are larger-scale actions such as unexpectedly attacking an isolated enemy force (as at **Lechaeum** in 390) or even enticing an enemy army on and waiting with a whole army deployed to meet them (as at the **Elleporus River** in 388). Although the latter two types border on battle, the element of planning to surprise an enemy force on the move is an essential element of ambush.

In the small-scale **early Greek warfare**, ambush must have been fairly common—used against raiders, pirates, or personal rivals. For example, it was the solution adopted by Penelope's suitors to solve the problem of Telemachus' increasing opposition to their presence in the house (*Odyssey* 4.660-674). The courage required in an ambush was seen as being of a special kind and was greatly valued in Homeric warfare (for a list of examples, see Pritchett, *Greek State at War,* vol. 2, 178). As the scale of warfare and the emphasis on the pitched battle increased, the ambush almost certainly became less common. With the introduction of **hoplite** tactics, the emphasis was on putting as many men into the **phalanx** as possible, not detaching parties for other duties. Probably at around the same time, a distaste for ambush developed. Hoplite warfare focused on two armies battling it out face to face—with a clear winner—and ambush or other ruse came to be seen as dishonorable. A good (although literary) expression of this occurs in a fifth-century Athenian play, the *Rhesus,* ascribed to Euripides, and set during the **Trojan War.** Here, the title character says of **Odysseus**:

> the man who goes against his enemy face to face is worth
> the epithet brave, not he who kills his enemy by stealth. I
> shall take alive this man, whom you say goes like a thief
> and weaves ambushes, and shall place him at the exits of
> your gates, impaled through the back as a feast for the
> winged vultures (lines 510ff.).

Polybius also links ambushes with deceit and trickery on several occasions (3.81.9; 4.8.11; 13.3.2-7; 18.3.1-4), while, of course, demonstrating that they still occurred—because they offered considerable opportunities for success. There are quite a few examples of ambush during and after the Classical period, particularly in conditions of more open warfare and with *psiloi* (light troops) and **cavalry.** **Xenophon** argues that hiding *hamippoi* among the cavalry provided a good opportunity to ambush the enemy (*Hipparch.* 8.19) and **Onasander** (6.7-8) describes cavalry scouting ahead to detect ambushes. The fifth-century Athenian *strategos* (general) **Demosthenes** (1) made use of ambush on several occasions during the **Peloponnesian War** of 431-404, and the destruction of the *mora* of **Spartan** hoplites at Lechaeum in 390 is a good example of a fairly large-scale ambush. In 389 **Dionysius I** of **Syracuse** destroyed an Italiote League army in a major ambush at the Elleporus River and **Agathocles** of Syracuse initiated the battle of Himeras River in 311 with an ambush (Diod. Sic. 19.108.3-5). **Pyrrhus I** of Epirus lost his son to a Spartan ambush in 272. *See also* CHABRIAS; ELLOMENOS; IPHICRATES; OLPAE; PHILOPOEMEN.

AMORGOS. An island in the Cyclades, site of a major naval battle during the **Lamian War** in May/June 322. Although the details of the naval campaign of 322 are obscure, the battle of Amorgos probably followed the Greek naval loss at **Abydus** in early spring. The Macedonian admiral, **Cleitus**, had apparently concentrated his fleet and used it in succession against the separate contingents of the Greek fleet. Amorgos was a major defeat for the Greeks (the Athenians lost 170 ships) and, along with the loss at the Malian Gulf later in 322, removed the Greek naval threat to **Macedon** and led to the demise of **Athens** as a major naval power.

AMPHICTIONIC LEAGUE. Also known as the Delphic Amphictiony, it was the most powerful amphictiony (a league concerned with the maintenance of a cult and its temple). The Amphictionic League was originally based around the cult of Demeter at Anthela (near **Thermopylae**) but was strongly associated with the cult of Apollo at **Delphi** from the First **Sacred War** (c. 595-590). It included several tribes/states (originally 12) of central and northern Greece, and because of this, and the importance of the Delphic oracle, the league became the closest thing to a panhellenic organization prior to the **Persian Wars**. However, it was subject to domination by major powers, including **Thessaly** in the sixth century, **Thebes** and then **Macedon** in the fourth, and **Aetolia** in the third. These frequently used the league as a means of exerting political, moral, and/or military pressure on opponents. The Amphictionic League did claim to pass measures binding on all the Greeks (although these could be ignored by more-powerful states) and exerted a direct influence on the military history of Greece through its role in the declaration of Sacred Wars.

AMPHIPOLIS. An Athenian colony in the territory of the Edonians in **Thrace**, situated on the Strymon River, northeast of the Chalcidice, that replaced an existing Thracian town, Ennea Hodoi. Amphipolis was strategically located astride the best route inland from the coast and the main coast road from Thrace into **Macedon**. It was also near the gold and silver mines at Mount Pangaeus and abundant forests with timber suitable for shipbuilding. There were two attempts to colonize Amphipolis prior to its successful foundation in 437/6. In 497 Aristagoras of **Miletus** and his party were destroyed by the Edonians, and the same fate befell a 10,000-strong Athenian-led expedition, annihilated north of the site at Drabescus in 465. The successful expedition of 437/6 was led by Hagnon, son of **Nicias**, but included non-Athenians, who were in a majority by 424 (Thuc. 1.100, 4.102, 4.108; Hdt. 5.126, 9.75). The city was the jewel of the Athenian empire but was lost to **Athens** during the **Peloponnesian War** of 431-404. In 424 **Brasidas** arrived unexpectedly with an army and persuaded the citizens to surrender by offering very moderate terms and by threatening their property outside the walls, which they had not had time to bring inside. The Athenian commander, **Thucydides**, happened to be absent at Eion, the port at the mouth of the Strymon, and was unable to get back before Brasidas had secured the city (Thuc. 4.102-8).

The Athenians failed in an attempt to recover the city in 422 and although it was supposed to be returned under the terms of the Peace of **Nicias** in 421, its inhabitants refused to be handed over to Athens and the city effectively maintained its independence (Thuc. 5.1-11, 5.18.5, 5.21). It subsequently allied with **Olynthus** and resisted all attempts to return it to Athenian control, including a major attack by **Timotheus** in 360. In many respects the Athenians' obsession with the return of Amphipolis dominated their foreign policy in the middle of the fourth century. In 357 Amphipolis was captured by **Philip II** of Macedon and remained part of Macedon until 167 when **Rome** made it a free city at the end of the Third **Macedonian War** (Diod. Sic. 16.8.2-3; Livy 44.29).

Amphipolis was the site of a major battle in 422, described in detail by Thucydides (5.6-11) and in a stock literary battle description by **Diodorus Siculus** (12.74). The Athenian politician **Cleon** commanded an army of 1,200 Athenian **hoplites** and 300 **cavalry**, as well as an unspecified, but larger, number of allies. After a moderately successful approach in which he recaptured several cities which had revolted, Cleon conducted a **reconnaissance** to Amphipolis with his whole army. Brasidas had about 2,000 hoplites, 300 Greek cavalry, 1,500 Thracian **mercenaries**, large numbers of Edonian **peltasts** and cavalry, and an additional 1,000 Myrcinian and Chalcidian peltasts.

Cleon viewed the city and apparently decided to return to Eion to await reinforcements. However, he appears to have mishandled this move by underestimating the distance between his men and the city and wheeling the right wing of his **phalanx** so that its unshielded side was exposed. Judging his moment nicely, Brasidas launched a charge from the city gates and caught the Athenians in a confused state in midmaneuver. The Athenian left wing broke and ran immediately. The right wing made a stand but was routed, largely by the enemy cavalry and light troops, with the loss of 600 men, including Cleon. According to Thucydides, Brasidas and six of his men died. This figure seems very low, but Thucydides explains that it was more of a rout than a battle. In addition, most of the serious fighting was on the right wing and involved missile fire against the hoplites, and this would also have reduced Brasidas' casualties. Thucydides' account contains some bias against Cleon, but is coherent and generally plausible. This battle led directly to the Peace of Nicias and to Athens' permanent loss of Amphipolis.

AMPHISSAEAN WAR. *See* SACRED WARS (4)

ANDRISCUS (ANDRISKOS; d. 148). A pretender to the Macedonian throne who claimed to be the son of the last king, **Perseus**. In 150 he briefly united the four republics and sparked off the Fourth **Macedonian War** with **Rome**. He was initially successful, overrunning **Thessaly** in 149, but was defeated at **Pydna** (2) by Quintus Caecilius Metellus in 148, captured in **Thrace**, and subsequently executed in Rome after Metellus' triumph. The loss of this war led to the incorporation of **Macedon**, along with Epirus and **Thessaly**, into the Roman empire in 148.

ANTALCIDAS (ANTALKIDAS also [more correctly] ANTIALKIDAS; fl. 400-367). A **Spartan** officer who became Sparta's **Persian** expert. In 392, during the **Corinthian War**, he was sent to negotiate with Tiribazus, the Persian commander in Asia Minor, persuading him that Sparta had no designs on Persian territory and recognizing Persian sovereignty over the Greek states in Asia Minor. Tiribazus arrested the **Athenian** general **Conon**, secretly provided money to Antalcidas, and requested guidance from the Persian king. However, the king sent **Struthas** to take command and he followed a pro-Athenian line.

In 387, Antalcidas was sent out as **navarch** and this time brought the Persian king over to support Sparta. With Persian financial support and skilled naval action, he attained local naval superiority and prevented ships from traveling from the Black Sea to Athens. This pressure on its crucial supply lines forced Athens to come to terms and in 386 the **Common Peace** (1), the King's Peace—also known as the Peace of Antalcidas—was signed. This ended the Corinthian War and left Sparta in a very favorable position, allowing it a virtual free hand in Greece. Antalcidas was involved in at least two further missions to Persia. The first, in 372, was successful and led to the Restatement of the King's Peace (Common Peace [2]). The second, in 367, was a failure as the Thebans, post-**Leuctra**, proved more attractive allies for Persia. Antalcidas is supposed to have committed suicide after this diplomatic rebuff.

ANTALCIDAS, PEACE OF. *See* COMMON PEACE (1)

ANTIGONUS (ANTIGONOS). Name of three members of the Antigonid dynasty of **Macedon**, one of the major powers in the Hellenistic period. They are:

(1) Antigonus I (Monophthalmus) (c. 382-301). Antigonus Monopthalmus ("One-eye") was the founder of the Antigonid dynasty and reigned as king from 306 to 301. A **Macedonian** noble, he served under **Alexander the Great** in Asia and was made satrap of Phrygia in 323. However, in 321 he went into exile when threatened by **Perdiccas** (2) and persuaded **Antipater** and **Cassander** that Perdiccas was aiming at eliminating them. In 320 Antipater gave Antigonus military command in Asia against **Eumenes of Cardia**. Antigonus secured rapid initial success against Eumenes, in 319 capturing his satrapy, laying siege to him in Nora, and defeating his allies Alcetas and Attalus.

Following Antipater's death in 319, Antigonus became a major player in his own right among the *diadochoi*, joining Cassander, **Ptolemy I (Soter)**, and **Lysimachus** against Antipater's chosen successor, Polyperchon. Antigonus' part in the coalition was to defeat Eumenes, now given a new lease on life by his appointment as Polyperchon's general in Asia. The difficult campaign included the indecisive battle of **Paraetacene** in 317, but at **Gabiene** in 316, Eumenes was betrayed by his *argyraspides* and subsequently executed. After a brief period of consolidation, during which **Seleucus** fled to **Egypt**, Antigonus was left in control of the bulk of Alexander's Asian empire—including the treasure.

Antigonus' success led to the formation of a coalition against him, comprising Cassander, Lysimachus, Seleucus, and Ptolemy. To overcome his naval weakness Antigonus set about building a fleet and sponsored a maritime league, the League of Islanders. He also sent financial aid to Polyperchon against Cassander, successfully campaigned in **Syria** and Phoenicia, and ensured local support by proclaiming all Greek cities free and autonomous. His run of successes was ended by the defeat of his son, Demetrius—later **Demetrius I (Poliorcetes)**—at **Gaza** in 312. This allowed Seleucus to recover his satrapy and led to peace with Cassander, Lysimachus, and Ptolemy in 311 on the basis of the status quo. The war with Seleucus ended in 309 with a decisive defeat of Antigonus' army (he was not leading it) somewhere in the eastern part of the empire.

From this point on, Antigonus was involved in a continual struggle with Ptolemy for naval supremacy in the Aegean. He sent Demetrius to liberate **Athens** in 307 and the following year Demetrius gained control of Cyprus after defeating Ptolemy's navy there. Antigonus then had himself crowned king (the first of the *diadochoi* to do so), along with his son Demetrius. Antigonus was less successful against **Rhodes**, where Demetrius failed to capture the city and had to be recalled by his father to meet Cassander's threat to Athens in 304. On Antigonus' behalf Demetrius restored the situation, securing **Corinth** and its isthmus.

Antigonus and Demetrius revived **Philip II's Hellenic League** in 302 and very nearly destroyed Cassander. However, Cassander joined Ptolemy, Seleucus, and Lysimachus and in 301, aged over 80, Antigonus was killed at the battle of **Ipsus** when Demetrius pursued the enemy **cavalry** from the field, leaving the Antigonid **phalanx** open to a flank attack.

(2) Antigonus II (Gonatas) (c. 320-240/39). Son of **Demetrius I (Poliorcetes)**; king, 283-240/39. From 287 Antigonus Gonatas (the epithet is of unknown meaning) acted as his father's commander in Greece, where the Antigonids controlled **Corinth** and the isthmus, parts of Attica—but not **Athens**—and had garrisons in the Peloponnese, at Piraeus, and at Chalcis and Demetrias. He became king on his father's death in 283, his main asset a strong fleet. In a series of fortunate events for him, two of his main competitors, **Lysimachus** and **Seleucus I (Nicator)**, were killed and **Macedon** was ravaged by **Gauls**. In 277 Antigonus marched north and destroyed a strong army of Gauls near Lysimachea in **Thrace**. Newly popular as the savior of Greece, Antigonus was recognized as the rightful king of Macedon, which he secured by the end of 276.

Antigonus suffered a major setback in 274 when **Pyrrhus I** of Epirus inflicted several defeats on him, gained the desertion of a large number of Antigonus' Macedonian troops, and captured **Thessaly** and Upper Macedon. However, when Pyrrhus was campaigning in the Peloponnese in 272 Antigonus recaptured most of his lost possessions and followed Pyrrhus south. His presence at **Argos** seems to have forced Pyrrhus to attempt to seize the city by a coup de main aided by traitors. Pyrrhus' death in this attack left Antigonus as undisputed master of most of Greece. He wisely concentrated on consolidating his position in Macedon and Thessaly, exerting varying measures of control in Greece through garrisons at Corinth, Piraeus, Chalcis, Demetrias, and other key locations.

Antigonus was constantly engaged in a struggle for naval supremacy against **Ptolemy II (Philadelphus)**. Despite Ptolemy's financial assistance to them, Antigonus defeated Athens and **Sparta** in the **Chremonidean War** (268-261). He did so by a clever campaign, using his garrison in Corinth to prevent his enemies from linking up. **Areus I** of Sparta was defeated and killed in battle near Corinth in 265 and Athens capitulated in 262/1. Antigonus built on this success with a major naval victory over Ptolemy off Cos, perhaps in 261 or around 258. He subsequently allied with Antiochus II (Theos) against Ptolemy in the Second **Syrian War**, which led to a strengthening of Antigonus' naval position in the Aegean.

Ptolemy, however, kept up his intrigues in Greece and in 250/49 helped spark the revolt of Alexander, Antigonus' governor in Corinth. The loss of Corinth and Chalcis, which Alexander also held, was a serious strategic blow to Antigonus—effectively cutting him off from his garrisons in the Peloponnese. It was not until 246-245, with another major naval victory over Ptolemy (off Andros) and his recovery of Acrocorinth (Corinth's citadel), that Antigonus restored the situation—albeit temporarily. Acrocorinth was seized by **Aratus** in 243 and Antigonus was unable to recover it, despite making peace with the **Achaean League** in 241/0. Although he had a troubled reign, particularly at the start, Antigonus was very successful in restoring the Antigonid dynasty's power and securing its position in Greece.

(3) Antigonus III (Doson) (c. 263-221). Antigonus Doson (the meaning of the epithet is uncertain, but perhaps means "he who will give") reigned from 227 to 221. He ruled as guardian for the future **Philip V** from 229 but assumed the throne himself, probably around 227. Prior to this, Antigonus had successfully overcome the first military challenges of his reign—repelling an invasion by Dardanians, recovering **Thessaly**, which had revolted, and ending the threat from the **Aetolian League** by some territorial concessions.

In 224 (after negotiations probably initiated in 227/6) Antigonus allied with the **Achaean League**, up until then a staunch enemy, against **Sparta**—newly resurgent under **Cleomenes III**. In return for his help, the Achaean League handed over Acrocorinth (the citadel of **Corinth**) and control of the city to Antigonus. He quickly proved his worth as an ally—**Ptolemy III (Euergetes)** ended his crucial financial support to Cleomenes when Antigonus handed over territory to him in Asia. At the same time he led an army of 20,000 men south, rapidly recaptured **Arcadia**, shored up the Achaean League, and again restored the **Hellenic League** of **Philip II**, with himself as *hegemon*.

In 223 Antigonus, leading a joint **Macedonian**-Achaean League force, captured Tegea after a siege and seized both **Orchomenus** (Arcadia) and **Mantinea**. The following summer he led the allied army into Laconia, decisively defeated Cleomenes at **Sellasia**, and occupied Sparta itself—the first foreigner to do so. According to **Polybius** Antigonus treated the Spartans "with generosity and humanity" and restored the previous constitution (2.70). He died the following year from the effects of a burst blood vessel (probably connected with tuberculosis) suffered while defeating Illyrian invaders of Macedon in battle. An able king, Antigonus' short reign left his nephew and successor Philip V well positioned on his accession. *See also* ARATUS; CLEOMENIC WAR; HELENIC LEAGUE (3).

Figure 2. The Argos Armor

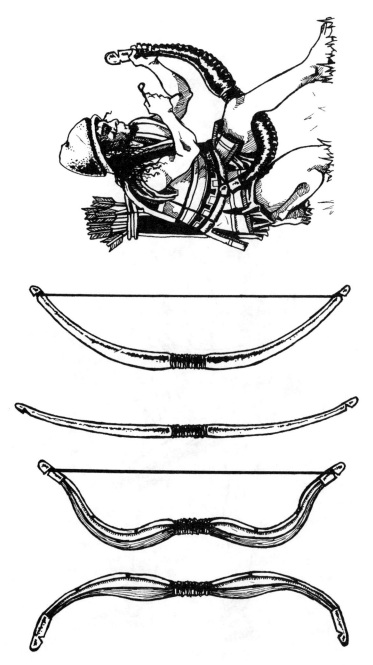

Figure 3. Greek Bows

ANTIOCHUS (ANTIOCHOS/ANTIOKHOS). The name of several members of the Hellenistic-period Seleucid dynasty, rulers of an empire centered on **Syria**. The most important militarily are described below. *See also* SELEUCID ARMY.

(1) Antiochus I (Soter) (324-261). Son of **Seleucus I (Nicator)**, Antiochus reigned from 281-261, although he ruled the eastern half of the empire from around 293/2. **Syria** was in revolt when he succeeded his father and it took several years of hard fighting to secure his rule. This may explain his renunciation of Seleucid claims in the west, although pressure from **Egypt** was probably also a factor. His treaty with **Antigonus II (Gonatas)** set the boundary between the two at the **Thrace-Macedon** border and established good relations between the Seleucids and Macedon for some time. Antiochus was also under pressure from his eastern provinces throughout his reign—his attention seems to have been focused more on the west, where he was hard pressed by the **Gauls**, Egypt, and several local separatist dynasts. His relative neglect of the east probably hastened its loss under his successors.

Antiochus received the epithet Soter ("Savior") for his defeat of the Gauls around 275 in the "**Elephant** Victory," but probably lost territory as a result of the First **Syrian War**. In 262, near Sardis, **Eumenes I of Pergamum** defeated Antiochus' expeditionary force sent to recover Pergamum. The circumstances of Antiochus' death in 261 are unknown.

(2) Antiochus III (the Great) (c. 242-187). Son of **Seleucus II (Callinicus)**, Antiochus reigned from 223-187. He succeeded to a kingdom in serious trouble—**Attalus I (Soter)** of Pergamum still retained the territory in Asia Minor he had gained from **Antiochus Hierax** (see [5] below), Bactria and Parthia were independent, and the stability of Media, Persis, and Babylonia was suspect. Antiochus addressed all these problems vigorously and generally with considerable skill. Asia Minor was recovered by his uncle, **Achaeus**, whose subsequent revolt was crushed after the dogged siege of Sardis (220-213). Antiochus campaigned in the east from 212 to 202, conquering Armenia and reducing Bactria and Parthia to vassal states (cf. Polyb. 10.27-31, 48-49). His expedition reached into India, securing a new supply of precious war **elephants**, and earned him the epithet "the Great."

However, his expansion south met with mixed success. Antiochus lost the Fourth **Syrian War** (219-217) at **Raphia**, although retaining some of the territory he had conquered earlier. Part of the reason for his loss was that he was fighting on several fronts—he was simultaneously dealing with Achaeus' revolt in the north—but he also made mistakes in his conduct of the war. Following his triumph in the east, he captured

Coele-**Syria** and **Palestine** from **Ptolemy V (Epiphanes)** during the Fifth Syrian War (202-195).

Antiochus' major error was to misjudge the power of **Rome**, whose attention he first attracted in 203/2 by a secret agreement to split the overseas possessions of Ptolemaic **Egypt** with **Philip V** of **Macedon**. Suspicion was further increased (perhaps without real cause) by Antiochus' conquest of Coele-Syria and Palestine (202-198) and by his invasion of **Thrace** in 196. After a series of flawed and failed negotiations, Antiochus invaded Greece in 193, initiating war with Rome.

Antiochus was fairly quickly defeated, first at **Thermopylae** (3), and then at **Magnesia** on his home territory—again confirming the superiority of Roman legions over Hellenistic armies. The war was ended by the Treaty of **Apamea**, which effectively ended Seleucid Syria's role as a Mediterranean power. Antiochus attempted to rebuild his fortunes by campaigns in the east but in July 187 he was killed by an outraged mob of his own subjects while plundering an Elamite temple.

(3) Antiochus IV (Epiphanes) (c. 215-164). Son of **Antiochus III (the Great)**; reigned from 175-164. According to **Polybius** (26.1), Antiochus was rather an eccentric personality, fond of parties, associating with artisans and workers, drinking with foreigners, and earning from his subjects the unofficial epithet of Epimanes or "Crazy" (a pun on his official epithet Epiphanes or "Renowned"). Under the terms of the Treaty of **Apamea** he spent from 188-176 as a hostage at **Rome** and made some good contacts there, although this did not help him once the Romans decided he had become too powerful. The historical tradition is fairly hostile to him, probably because of his harsh (or firm, depending on one's perspective) treatment of **Palestine**. However, he did well to maintain the status quo of the Seleucid empire against the policies of Rome and some able local opponents.

In 169/8, during the Sixth **Syrian War**, Antiochus captured **Cyprus** and most of **Egypt**. In 168, though, freed from the distraction of the Third **Macedonian War**, the Romans ordered him to withdraw. Sensibly, he did so. Blocked in the west by Roman suspicion of his power, Antiochus turned toward the east, where Bactria and Parthia were dangerously destabilizing influences. In 167, during his preparations for his eastern expedition, the Jews revolted. Antiochus acted swiftly and ruthlessly and by 165 was able to set out for the east, the rebels largely pacified. Unfortunately for the stability of his kingdom, he died in Media in 164 en route for Parthia, leaving a nine-year-old successor.

(4) Antiochus VII (Sidetes) (c. 159-129). Antiochus reigned from 139-129 and was known as "Sidetes" ("the Sidonian") because he had been brought up in the city of Side. He was an energetic and able ruler

who restored order in **Syria** and recovered both **Palestine** and (temporarily) Babylonia. He came to the throne when his brother, Demetrius II (Nicator), was captured by the Parthians. In 138 he defeated and killed Tryphon, the pretender to the throne in Antioch. After a short period of consolidation, he recovered Palestine in 135-134, capturing Jerusalem after a siege but treating the rebels leniently. In 130 he recaptured Babylonia from the Parthians and placed troops in winter quarters in Media, apparently planning a campaign in 129 to rescue his brother. However, in early spring 129 he had to rush to the aid of these troops when the cities of Media revolted and the Parthians launched their own attack. With a scratch army Antiochus was forced to face the main Parthian army and was defeated and killed. The Parthians returned his body to Syria in a silver casket. Antiochus' defeat permanently detached the eastern provinces from the Seleucid empire.

(5) Antiochus Hierax (c. 263-226). Antiochus Hierax ("Hawk") ruled Seleucid Asia Minor from around 243 to 228. He was the second son of Antiochus II (Theos). In return for the assistance of his mother, Laodice, during the Third **Syrian War** (246-241), Antiochus' brother, **Seleucus II (Callinicus)**, was forced to recognize his independent rule of Asia Minor. By a series of judicious alliances with Pontus, **Bithynia**, and the **Gauls** (Galatians), he became strong enough to repulse Seleucus' attempt to restore control over him (the "War of the Brothers," around 239-236). Seleucus' initial success was negated by a decisive defeat near Ancyra and he was forced to make peace with Antiochus.

However, Antiochus' alliance with the Galatians drew him into conflict with **Attalus I (Soter)** of Pergamum. Attalus' defeat of the Gauls outside Pergamum left Antiochus vulnerable and over the period 229-228 Attalus systematically defeated him and drove him from his possessions. Antiochus then allied with his aunt, Stratonice, in an attempt to wrest **Syria** from Seleucus. He invaded Mesopotamia, while she raised a revolt in Syria; both actions were unsuccessful. Antiochus went into exile and was subsequently killed in **Thrace**. His main contribution to the Seleucid empire was to destabilize it internally and lose it significant territory in the north.

ANTIPATER (ANTIPATROS; 397-319). Macedonian general, active through the reigns of **Philip II** and **Alexander the Great**; one of the *diadochoi*. Antipater was left in charge of the home front during Alexander's conquest of Asia, supplying Alexander with reinforcements and maintaining order in Greece. During this period Antipater was faced with two serious revolts. The first was in **Thrace**, led by **Memnon**; the second and more dangerous was from a Greek coalition led by **Agis III** of **Sparta**. Antipater decisively defeated Agis at Megalopolis in 331.

After Alexander's death in 323, Antipater was initially hard-pressed by the anti-Macedonian Greek coalition in the **Lamian War**. Short of troops because of the reinforcements he had been sending to Asia, Antipater's small army was defeated by the Greeks near **Lamia** when his **Thessalian cavalry** deserted during the battle. His decision to retire to Lamia proved extremely sound. Macedon itself was in a very vulnerable position, but the Greek forces could not afford to bypass his army in Lamia and had to commence siege operations. Although a relief force under Leonnatus was defeated, Antipater managed to break out and in 322 ended the war by defeating the Greeks at **Crannon**. The Greeks were hampered by reduced troop numbers, but Antipater also handled the battle well. He had his infantry engage the Greek foot before the Thessalian cavalry had the chance to defeat his inferior cavalry force.

Antipater's settlement of Greece reversed Alexander's generally lenient policy. He abolished the democracy at **Athens**, placed a garrison in **Munychia**, and insisted on dealing with the rebel states individually. This effectively ended the role of the **Hellenic League** and ensured that Greece remained divided. He did, though, make easy terms with the **Aetolian League** in order to free up troops to meet an expected attack from **Perdiccas** (2).

In concert with **Craterus** and **Ptolemy I (Soter)**, Antipater invaded Asia in 321. He was elected regent with full powers by the Macedonian army in assembly following Perdiccas' death. Later that year Antipater left **Antigonus I (Monophthalmus)** as "general of Asia" and returned to Macedon. He died of natural causes in 319.

ANTIPHILUS (ANTIPHILOS). The **Athenian** commander of the Greek coalition forces against **Macedon** in the **Lamian War** in 322. Antiphilus succeeded to the command on the death of **Leosthenes** at the siege of **Lamia**. He won a victory over Leonnatus at **Melitaea** but failed to prevent **Antipater** from evacuating Lamia, linking with Leonnatus' undefeated **phalanx**, and withdrawing to Macedon. Hampered by a steady loss of his troops, who returned home in considerable numbers after the success at Melitaea, Antiphilus was defeated at **Crannon** in summer 322.

APAMEA, TREATY OF. Signed in 188, this treaty ended the war between **Rome** and **Antiochus III (the Great)** of **Syria** and greatly benefited Rome's allies in the war, **Rhodes** and Pergamum. The main terms (given in **Polybius**, 21.45) required Antiochus to pay a huge war indemnity of 15,000 talents and repay a 400 talent debt to **Eumenes II (Soter)**; to relinquish his claims to Thrace and evacuate Asia Minor west of the Taurus Mountains and the River Halys; to surrender all his war **elephants** and not replace them; to surrender all but 10 of his **warships**; and to return all **prisoners**. In return Rome ended the war and agreed not to attack Antiochus again or assist anyone else who did so—as long as he fulfilled all the provisions of the treaty. Any disputes were to be settled by arbitration. Although the power of Rhodes and Pergamum were considerably increased (they gained the territory Antiochus had to relinquish), their long-term position was in fact weakened by Rome's involvement in Asia Minor.

ARATUS (ARATOS; 271-213). General and political leader of the **Achaean League**. Born in **Sicyon** but educated in **Argos** as an exile after his father's assassination in 264, he returned at age 20, ended the tyranny at Sicyon, and enrolled the city in the Achaean League. From 245 he was normally elected *strategos* of the league every second year (the maximum tenure possible under its constitution). **Polybius** (2.40) credits Aratus with a policy of uniting the Peloponnese, which involved expelling the **Macedonians** and ending tyranny. In executing this policy, Aratus took Acrocorinth in 243 and persuaded **Corinth** to join the Achaean League. He formed an anti-Macedonian alliance with the **Aetolian League** in 239. Megalopolis joined the Achaean League in 235 and Argos followed suit in 229. This increase in Achaean power led to opposition in the Peloponnese and in 228 **Cleomenes III** of **Sparta** and the Aetolian League combined against the Achaean League, triggering the **Cleomenic War**. Under considerable pressure following defections from the league and defeats at Lycaeum, Ladocaea, and Hecatombaeum, in 224 the Achaean League reluctantly struck an alliance with **Antigonus III (Doson)**, ending 20-odd years of opposition to Macedon. On Antigonus' death Aratus and the league smoothly transferred their allegiance to his successor, **Philip V**.

Aratus died in 213, perhaps from tuberculosis, although he and many others believed he had been poisoned by Philip V. Although an able and generally successful statesman, Aratus was not a particularly good general. In antiquity he was renowned for a lack of physical bravery (although **Plutarch** states that this was put about by "flatterers of tyrants" [*Aratus*, 29.5-6]) and Polybius drily remarked "that the Peloponnese was full of **trophies** marking his defeats" (4.8). His actions at **Caphyae** in 220 demonstrate some problems with his military leadership and almost led to his dismissal as league general (Polyb. 4.9-14).

Aratus wrote a history of the Achaean League that was praised by Polybius but unfortunately does not survive.

ARBELA. *See* GAUGAMELA

ARCADIA (ARKADIA). The mountainous central region of the Peloponnese, north of Laconia and west of **Argos**. Its most important cities were **Mantinea, Orchomenus**, and Tegea. The area produced hardy soldiers, including **hoplites**, and was a major source of **mercenaries** from the fifth century down to the third century. **Homer** (*Iliad* 2.603-604, 609) records Arcadian participation in the **Trojan War**, describing it as a region where "men are hand-to-hand fighters." Tradition also relates that they supported **Messenia** against **Sparta** in the **Messenian Wars**, although not always wholeheartedly—king Aristocrates of Orchomenus was executed by his fellow citizens for betraying the Messenians at the battle of the Great Foss.

The Arcadian cities and towns did not always act in concert, and this proved a weakness in their long struggle with Sparta. By about 560 Sparta had defeated Tegea, bringing it and several other Arcadian cities into the **Peloponnesian League**. Arcadia fought under Sparta in the Second **Persian War**. However, Spartan domination was never complete and in the second quarter of the fifth century a resurgent Arcadian nationalism forced Sparta to fight two major battles: against Tegea (supported by Argos) at Tegea, and against a combined Arcadian force at **Dipaea**. Spartan victories on both occasions kept Arcadia generally quiet for some time—although Mantinea did attempt to assert its independence during the **Peloponnesian War** of 431-404. Its defeat at the battle of Mantinea (1) forced its return to the Peloponnesian League in 417. In 385/4 Sparta dismantled the city of Mantinea, returning its population to the original villages which had merged to form the city. From 370 the military history of Arcadia initially revolved around the **Arcadian League**, formed with **Theban** assistance after the battle of **Leuctra**. However, after Tegean-Mantinean rivalry caused the collapse of the league, Arcadia made its most important military contribution through the **Achaean League**, in which it played a major role.

ARCADIAN LEAGUE. Formed in 370, after the battle of **Leuctra**, by 369 the league included most of **Arcadia**. The new city of Megalopolis was founded to serve as the federal capital. League members contributed troops to a federal army, possibly of 10,000 **hoplites**, which was led by an annually elected *strategos* (general). Its early years involved a sustained struggle, assisted by **Thebes**, for survival against **Sparta**, but the league fairly soon adopted a more independent line toward Thebes. In 366/5 the league allied with **Athens**. Prone to disunity, it suffered a collapse in 364 with a major split between Tegea and **Mantinea**. The Tegean faction requested, and secured, Theban assistance. The **Boeotians** invaded the Peloponnese in 362, supported by **Argos**, Messene, Tegea, Megalopolis, and contingents from central Greece, and won the battle of Mantinea (2) against Sparta, **Elis**, Athens, and Mantinea. From this point on, united action was difficult, although "the Arcadians" allied with **Philip II** of **Macedon**, generally supporting his interests—despite remaining neutral in 338 for the battle of **Chaeronea**. In 245, Megalopolis, the former federal capital, joined the **Achaean League**, in which it soon came to play a leading part. *See also* TEARLESS BATTLE.

ARCHELAUS. A Greek who commanded the forces of **Mithridates VI (Eupator)** of Pontus in the First **Mithridatic War**. He rapidly conquered **Bithynia**, massacred the Italians on Delos, and landing in the Piraeus, secured **Athens**. From there Archelaus won over the Peloponnese and most of central Greece. However, he was starved out of Athens by Lucius Cornelius Sulla during the siege of 87-86. Later in 86 Archelaus linked up with the main Pontic army, which had frittered away the opportunity of taking Sulla in the rear while Archelaus was besieged in Athens. He was defeated by Sulla at the hard-fought battles of **Chaeronea** (2) and **Orchomenus (Boeotia)**. In 83, when he fell under suspicion of treacherous dealings with Sulla, Archelaus fled to **Rome**. He assisted Lucius Licinius Lucullus against Mithridates at the start of the Third **Mithridatic War**. Archelaus was an able general who had the misfortune to meet Sulla with a poorer quality army.

ARCHERS, FOOT. Despite their apparent potential, archers never played a major role in Greek warfare. While this was in part due to the relatively poor penetrative power of the arrow against **armor**, it was also influenced by the stigma often attaching to archers. Few heroes in the **Homeric** literature use the **bow** as their main **weapon** (Philoctetes and Paris are exceptions) and there is a hint of disapproval in the portrayal of bowmen that increased in later periods. When the **hoplite** dominated Greek warfare, archers were generally looked down upon as men who lacked the bravery to engage in hand-to-hand combat. A fairly typical utterance is "the measure of a man's bravery is not archery; rather he who stands fast in his rank and gazes unflinchingly at the swift gash of the spear [is a brave man]" (Euripides, *Heracles*, 190-192). This attitude may have existed as early as the **Lelantine War** if the evidence for the ban on missile weapons during this conflict has any value. The attitude probably developed even further with Greek exposure to foreigners such as the **Persians** and **Scythians**, who used the bow and not the spear as their principal weapon.

Although there was prejudice against the archer in classical Greece, many armies included them when they could, normally deploying them on the flanks of the hoplite **phalanx**, although where the **terrain** permitted they could be deployed behind the phalanx and shoot over it. Few states had sufficient archers of their own to develop useful levels of combat power in the field, and **mercenary** archers, particularly from Crete, were frequently employed. However, **Athens** maintained a body of Scythian **slave** archers as police and in the fifth century also maintained 1,600 citizen archers, from the proceeds of empire. Athenian **triremes** normally had four archers serving on board. **Sparta** raised its first body of archers in 424 in response to Athenian seaborne raids. Under **Philip II** and **Alexander the Great**, and their successors, archers became more prominent (although again generally used on the flanks of the heavy infantry). This was due in part to the more sophisticated all-arms approach to warfare, but in the Hellenistic kingdoms also reflected the greater availability of archers in Asia Minor. *See also GASTRAPHETES; HIPPOTOXOTAI.*

ARCHIDAMIAN WAR. *See* PELOPONNESIAN WAR (2)

ARCHIDAMUS (ARCHIDAMOS/ARKHIDAMOS). The name of several Eurypontid kings of **Sparta**. The most important militarily are:

(1) Archidamus II (d. 427). King, 469-427. Archidamus performed well during the helot revolt (the Third **Messenian War**) after the earthquake of 464. Nothing is known of his career from then until just before the outbreak of the **Peloponnesian War** of 431-404. According to **Thucydides**, Archidamus unsuccessfully tried to persuade the Spartans not to declare war on **Athens** in 431 because they were not ready (1.80-85). He led the **Peloponnesian League** invasions of Attica in 431, 430, and 428. In 429 he led the league forces against **Plataea** instead because of the plague in Athens. The first stage of the Peloponnesian War (2) is named after him.

(2) Archidamus III (d. 338). King, 361/0-338; son of **Agesilaus II**. Because of the often poor health of his father, Archidamus played a major role at **Sparta** for 10 years before becoming king. In 371 he led the force that brought back the Spartan survivors of **Leuctra**. He successfully attacked Caryae and **ravaged** Parrhasia in the campaign of 368, which ended with his victory over the **Arcadians** in the **Tearless Bat-**

tle. In 364 his garrisoning of the fort at Cromnus, which threatened the road from **Messenia** to Megalopolis, achieved its aim of forcing the Arcadians to withdraw from **Elis**. However, the garrison was such a threat that the Argives and allies besieged it with a large force. Archidamus was wounded leading one unsuccessful relief effort and a second was only partially successful. Some of the garrison escaped, but the fort was captured, along with more than 100 Spartans. In 362 Archidamus distinguished himself during the **Theban** invasion of the Peloponnese by driving the Theban army out of the outskirts of Sparta.

During the Third **Sacred War** Archidamus commanded the Spartan contingent sent in 346 to **Thermopylae** to block **Philip II** of **Macedon**'s route south. This plan failed when the **Phocians** refused to hand over the fortifications there as agreed. In 343 Archidamus led a force of **mercenaries** to **Italy** to help **Tarentum** against the Italians. After several years of inconclusive campaigning, he was killed in battle at Mandionion fighting the Lucanians.

ARCHILOCHUS (ARCHILOCHOS/ARKHILOKHOS; fl. c. 680-640). Greek poet and **mercenary** soldier. Born on Paros, he was probably part of the second wave of settlers on Thasos and fought in **Thrace**. Archilochus died in battle against **Naxos**. Although throwing away one's **shield** in battle (in order to run away faster) was firmly frowned upon, Archilochus made a joke of it. He wrote: "Some Saian rejoices in the shield, an unblemished item, which I unwillingly abandoned beside a bush—but I saved myself. What do I care for that shield? Sod it. I'll buy another one, just as good." He also penned the lines: "I do not like a big general with long striding legs or who is proud of his hair . . . let there be a small general . . . going firmly on his feet, full of heart."

ARES. Greek god of war (also called Enyalios—originally a separate deity who seems to have become subsumed by Ares). Ares, one of the lesser Olympian deities, was regarded as the manifestation of war and violence and was never a particularly popular god. He was regarded as favoring no particular city and was a deity to be feared rather than respected. As Enyalios, dogs were sacrificed to him at **Sparta**—a common feature of purification and propitiation rites with gods associated with magic and an irrational or baleful effect on human life. Probably originating in **Thrace**, his worship is generally associated with northwest Greece, although he was also worshipped in **Thebes** and perhaps **Athens**.

AREUS I (d. 265). Agiad king of **Sparta**, 309-265. His ambition was apparently to restore Sparta as a major power. In 281 he recreated the **Pe-**

loponnesian League and allied with Antiochus I (Soter). In 280 he marched north, causing Argos and Megalopolis to expel their Antigonid garrisons. Areus then invaded Aetolia but was heavily defeated and the Achaean members of his new alliance defected. Although he failed in his aims, Areus' actions caused several states to revolt against Antigonus, weakening his position in Greece. In 272 Areus had to hurry back from campaigning in Crete to prevent the fall of Sparta (for the first time now protected by fortifications) to Pyrrhus I of Epirus. Areus was assisted in this by Antigonus, who regarded Pyrrhus as the more serious threat. When Pyrrhus withdrew, Areus followed him and killed his son in an ambush. Areus aligned Sparta with Athens and Egypt against Antigonus, leading Sparta against him in the Chremonidean War (268/7-262/1). Areus probably conducted three unsuccessful campaigns to break through Antigonus' forces in the Isthmus of Corinth and link up with Athens. In 265, during his third attempt, Areus was killed near Corinth. Apart from his ambition and the fortification of Sparta, Areus is noted for his adoption of a court and coinage in the style of the Hellenistic monarchies.

ARGAEUS (ARGAIOS; d. 359). A claimant to the Macedonian throne when Perdiccas III was killed by Bardylis I in 359. That year, backed by Athens, Argaeus landed with 3,000 mercenary troops at Methone, supported by an Athenian naval force. He marched inland, but the expected local uprising failed to materialize at Aegeae and he attempted to withdraw. His rival, Philip II, overtook and fairly easily defeated Argaeus before he could reach the coast. Argaeus was either killed in the battle or executed afterward. Philip helped to secure a treaty with Athens by releasing the prisoners from this action without ransom.

ARGINUSAE (ARGINOUSAI). Located just southeast of Lesbos, the Arginusae Islands were the site of a major Athenian naval victory over Sparta in August 406, during the Peloponnesian War (2) of 431-404. In June 406, the Spartan navarch Callicratidas had attacked an Athenian fleet of 70 ships under Conon, capturing 30 of them and blockading the rest in Mytilene. A scratch Athenian relief force of more than 150 triremes, whose crews included metics (foreign residents), cavalrymen (normally exempt from naval service), and freed slaves, arrived in the area and anchored at the Arginusae Islands. A thunderstorm prevented Callicratidas' original plan to surprise them by a night approach, and the next day he faced the Athenians with 90 ships. Accepting that the enemy crews were better than their own, the Athenians drew up in a double line to prevent the Spartans from using the *diekplous* maneuver.

Despite being outnumbered, Callicratidas, ships drawn up in a single line, refused to retire and was killed early in the battle. The Peloponnesians were heavily defeated, losing 75 ships, and the blockade of Mytilene was raised. The Athenians had 13 ships sunk and 12 disabled. According to **Xenophon**, although the Athenian *strategoi* made provision for the rescue of the disabled ships and men in the water, a storm arose and none of the survivors could be recovered. All six of the eight *strategoi* involved who returned to Athens were tried and executed for this failure (*Hell.* 1.6.35, 1.7). *See also* NAVAL WARFARE; THRASYBULUS; THRASYLLUS.

ARGOS. A large and populous city in the northeast Peloponnese, located in the Argolid (along with other settlements such as **Mycenae** and Nauplia). Argos was inhabited from the early Bronze Age and was of some importance in the Mycenaean period. Although the evidence is scanty, during the Dark Age Argos seems to have been the dominant power in the Peloponnese. Argos' power reached its height under **Pheidon** but shortly afterward (mid-seventh century?) **Sparta** displaced it as the premier state in the region. From this point on, Argos often remained neutral in Greek affairs, but every generation or so unsuccessfully challenged Sparta's position. A continual bone of contention was the disputed border area of Thyreae (or Cynuria). Sparta acquired this area around 547, after the **Battle of the Champions**.

During the Second **Persian War**, Argos was officially neutral but was suspected of favoring **Persia**. It claimed the **Delphic** Oracle advised this stance because the city was still recovering from the very heavy losses suffered against Sparta in the battle of **Sepia** in 494 (Hdt. 7.148). After the Second Persian War, Argos became a democracy and around 462/1 allied with **Athens** and **Thessaly** against Sparta. Argos fought against Sparta in the First **Peloponnesian War**, providing 1,000 troops at the battle of **Tanagra**. However, in 451, Argos abandoned its alliance with Athens and concluded a thirty years' peace with Sparta—a treaty which (rather unusually) was honored by both sides for the duration.

On the outbreak of the Peloponnesian War (2) in 431, Argos remained neutral under the terms of her treaty with Sparta. However, on its expiry in 421, Argos began to display a more openly hostile attitude toward Sparta. Argos joined with Athens, **Elis**, and **Mantinea** to form an anti-Spartan bloc. The coalition force was defeated at the battle of Mantinea (1) in 418—despite the good performance of Argos' newly formed body of *epilektoi*, 1,000 picked **hoplites** trained at public expense. Argos again joined an anti-Spartan coalition during the **Corinthian War** and in 394 provided 7,000 hoplites at **Nemea** and an unknown number at **Coronea** (where they performed poorly, according to Xen. *Hell.* 4.3.17). In 392, apparently as the dominant partner, Argos united with **Corinth** to give their citizens equal and joint citizenship. This posed a considerable threat to Spartan interests, as the new state controlled the Isthmus of Corinth, the only land route out of the Peloponnese. Argos absorbed Corinth sometime between 390 and 388 but in 386 the union was dissolved by Sparta, backed by the threat of Persian intervention, after the signing of the King's Peace (**Common Peace** [1]).

Argos later joined with **Thebes** against Sparta, fighting at Mantinea (2) in 362, and subsequently supported **Philip II** of **Macedon**. Following **Chaeronea**, Argos was rewarded with some Spartan territory. Despite supporting the anti-Macedonian coalition during the **Lamian War**, Argos generally took a pro-Macedonian line during this period. In 272, **Pyrrhus I** of Epirus was killed there during an attempt to detach the city from **Antigonus II (Gonatas)**. Argos joined the **Achaean League** in 229 and despite temporary occupations by Sparta under **Cleomenes III** and **Nabis**, remained a member until the **Roman** annexation of Greece in 146.

ARGYRASPIDES. Literally "Silvershields"; the name given to **Alexander the Great**'s 3,000-strong force of **hypaspists** just before he entered India. They retained the name after his death to distinguish them from other hypaspist forces in the successor kingdoms. The *argyraspides* ended up in the army of **Eumenes of Cardia**, but cost him his life when they deserted to **Antigonus (I) Monophthalmus** at the battle of **Gabiene** in 316. Shortly afterward, Antigonus ended its existence as a unit by splitting its members up. *See also* SELEUCID ARMY.

ARISTIDES (ARISTEIDES; d. post 467). An Athenian statesman and soldier, prominent during the **Persian Wars** and in the foundation of the **Delian League**. Aristides was a *strategos* (general) at **Marathon** in 490 and supported **Miltiades'** plan to attack the Persians. Ostracized (exiled) from **Athens** in 482 during a political struggle with **Themistocles**, he was recalled to resist the Persians. As *strategos* from 480-479 he was at **Salamis** (informing the Greeks that the Persians had blocked off their retreat and leading the **hoplites** who cleared the Persian troops from Psyttalea) and led the Athenian contingent at **Plataea**. He also took part in the postwar operations in the Hellespont area. Along with **Cimon**, Aristides apparently played a major part in encouraging the Greeks of the islands and Asia Minor to transfer their allegiance from **Sparta** to Athens, capitalizing on the arrogant behavior of **Pausanias** (1). Because of his renowned honesty, Aristides was nicknamed "the Just" and was given the task of assessing the tribute level for the members of the Delian League.

ARMOR. Armor was worn by both **cavalry** and footsoldiers, including *epibatai* (marines), in a variety of configurations throughout ancient Greek warfare. The full heavy infantry protective panoply consisted of a **helmet**, breastplate or corselet, arm protectors, greaves (leg protectors), and **shield**, although not all heavy infantrymen were necessarily equipped with all items. There was also a tendency toward the use of lighter protection for the infantry as warfare became more fluid.

The Dendra armor (figure 1) represents the level of protection available to a wealthy **Mycenaean** around 1400. It could have been worn without a shield, especially if combined with some sort of lower arm protection, but must have been very expensive. It is also very heavy. This would have restricted the wearer's mobility and may mean it was only used by **chariot**-borne infantry. Only about 200 years later the Mycenaean "Warrior Vase" (Athens, National Museum 1426) shows infantry equipped fairly uniformly with what later became the standard Greek heavy infantry panoply—helmet, shield, greaves, and a cuirass (probably of the bell-type) with *pteryges* (lit. "wings"—strips of leather or other material hanging from the base of body armor to protect the groin area or from the shoulder to protect the upper arm).

With the introduction of the **hoplite**, this defensive panoply became the standard in Greece throughout the Archaic and most of the Classical periods and was worn even after that. The **Argos** panoply (figure 2) shows an early version, including the bell cuirass, dated to around 750. There were different patterns of helmets and body armor, and the wearing of greaves and/or arm protectors was probably very much a matter of individual preference. However, the general tendency during the Classical period was to reduce the protection available to the heavy infantryman in order to increase his mobility. The first step toward this was a change from the bell cuirass and its successor, the muscle cuirass (which dispensed with the outward curve at the base of the cuirass and increased the realism of the anatomical details on the surface of the armor). Corselets made of leather or stiffened linen (plate 1) are depicted in art from the mid-sixth century and became the most common type of body armor from the start of the fifth century (from which time they could incorporate metal scales sewn on to increase protection). From the middle of the fifth century, perhaps under the influence of the Third **Messenian War** and certainly of the **Peloponnesian War** (2) of 431–404, a lighter style hoplite appears—first in the **Spartan** army but then elsewhere, including **Athens** and **Boeotia**. This panoply consisted of the shield and a felt (or leather) *pilos*-type helmet (figure 4) but no body armor. However, not all hoplites adopted this and the fuller panoply coexisted with the lighter type and even appears to have made something of a comeback in the fourth century.

The introduction and spread of the **Macedonian phalanx** perpetuated the lighter panoply, with helmet and small shield being the commonplace protection (although some sort of body armor is not uncommon). This was cheaper (useful if the state had to provide it) and the close formation and the hedge of *sarissai* in front and overhead provided some collective protection against missile weapons. In the Hellenistic period some changes occurred, caused by the increased exposure to Eastern practices and equipment with **Alexander the Great**'s conquest of the **Persian** empire. On one level, the increased wealth available led to fancier armor with elaborate decoration—especially for the **officers**—but the lighter-style panoply continued in use, especially for state-equipped troops (although elite groups and bodyguards were usually more fully equipped). Body armor now ranged from the muscle cuirass through the linen or leather corselet (although by this time available in a metal version), the use of metal scales or chain over all or part of a corselet, and padded or quilted jackets.

In general, Greek cavalry maintained a higher level of protection for a longer period than was the case for the infantry. Protection did not need to be sacrificed for mobility, as this derived from the horse and volunteer cavalry were drawn from the wealthy classes who could afford to equip themselves with good quality armor. The fairly standard cavalry panoply was a helmet, usually of a more open-faced design than the original infantry helmets (figure 4), a cuirass (often bell-shaped to allow free movement of the hips and legs while mounted) or corselet, and heavy boots to protect the lower legs and feet (greaves are occasionally attested in sixth-century art but soon disappear). However, exactly what a cavalryman wore was generally determined by his personal preference (and wealth), so there was some variation.

Horse armor is also attested, ranging from the face protector (*champfrein*) used in the Archaic and Classical periods to the fuller protection available in the late Classical and the Hellenistic periods. The tendency with horse armor was to increase rather than reduce the level of protection. Padded and, later, scale armor protection was available for the horse's chest and sides (sometimes also combining leg protection for the rider). Not all cavalry wore this by any means, but there was a significant development of heavy cavalry of this type in the Hellenistic period.

ARRIAN (c. A.D. 90-175). Full name Flavius Arrianus Xenophon. He wrote the best surviving ancient work on **Alexander the Great**'s campaigns the *Anabasis*, (called the *Campaigns of Alexander* in the Penguin translation). He was born in Nicomedia in **Bithynia**, studied philosophy, and had a distinguished military career in the **Roman** imperial service. After holding the consulship, he was appointed governor of Cappadocia by Hadrian and defeated the great Alan invasion of A.D. 134. He was recalled sometime before A.D. 138 and retired to **Athens**, where he wrote several books, including a manual of **cavalry** tactics and an account of the invasion of the Alans. Arrian's main work is the history of Alexander, which is regarded as the most accurate of the surviving ancient accounts. Although a late author, he used the earlier writings of **Ptolemy I (Soter)** and Aristoboulos, contemporaries of Alexander. His practical experience of warfare and administration in the East undoubtedly helped him analyze the numerous contradictory accounts of Alexander's campaigns. Arrian was prepared to criticize Alexander, but in general preserves a favorable account of him. He has been accused of glossing over some of Alexander's more brutal actions, probably influenced by his main sources, who were companions of Alexander.

ARTAPHERNES (also ARTAPHRENES). A **Persian** noble, half-brother of **Darius I** and satrap of Sardis. On Aristagoras of **Miletus**' advice, he organized the expedition against **Naxos** circa 500. The failure of this enterprise was a major factor in Aristagoras' decision to foment the **Ionian Revolt**, and Artaphernes' call on **Athens** to restore **Hippias** helped influence its decision to participate in this. Artaphernes successfully defended the central stronghold of Sardis against the **Ionian** rebels circa 498, although other parts of the city were destroyed by fire. He played a major role in the subsequent suppression of the revolt and in the transition to peace afterward. His son (also called Artaphernes) was joint commander with **Datis** of the Persian expedition that ended in defeat at **Marathon** in 490.

ARTEMISIA (fl. 480). Ruler of **Halicarnassus**, Cos, Nisyra, and Calydna under **Xerxes I**. She accompanied him on his invasion of Greece in 480. **Herodotus** preserves a fairly detailed and favorable account of her actions (presumably because of his Halicarnassan origins). Like **Demaratus**, she is portrayed in Herodotus as someone who gave Xerxes excellent advice, which he ignored. She distinguished herself in the battle of **Salamis**. *See also* PERSIAN WARS.

ARTEMISIUM. A promontory on the northwestern end of **Euboea**. In 480 it was the site of the first naval engagement between the Greeks and **Persians** during the Second **Persian War**. The Greek fleet at Artemisium was linked with the army at **Thermopylae** (1) as part of the joint land/naval defense of Greece. A Greek fleet of 324 **triremes** (including 180 **Athenian** ships) and seven **pentecontors** under **Eurybiades** faced a Persian fleet of just over 800 warships stationed at Aphetae on the mainland opposite. On arrival, 15 Persian ships were captured after mistakenly heading for the Greek fleet instead of their own. A further 30 Persian ships were captured in a preliminary engagement the next day. That same night, in a storm off "the Hollows," the Persians lost the squadron of 200 ships they had sent around the east coast of Euboea to attack the Greek rear. The next day the Persians attacked but were unable to break the Greek fleet. Although the result was a stalemate, **Herodotus** states that "many Greek ships and men were lost and many more Persian ships and men" (8.16). Shortly afterward, the Greek fleet received the news of their army's defeat at Thermopylae and withdrew. *See also* THEMISTOCLES.

ARTILLERY. Ancient Greek artillery is divided into two categories, nontorsion and torsion.

Nontorsion artillery developed from the composite **bow** (figure 3a, c) via the *gastraphetes* (invented at **Syracuse** in 399). **Athens** apparently possessed bolt-firing engines by 371/0 and the fortifications at Messene (**Messenia**), begun in 369, seem to have been designed to accommodate such machines. Perhaps soon after this the addition of a stand and a winch to the *gastraphetes* allowed the bow section to be substantially increased in size. This allowed a larger projectile to be fired with improved accuracy and over a greater range. The stonethrowers (*lithoboloi*) which Onomarchus used against **Philip II** in 353/2 may have been the improved *gastraphetes*. The design was later further improved by connecting the **weapon** section to the stand using a universal joint, which allowed for rapid traversing and speedier target acquisition. Early nontorsion artillery fired either a stone shot of about 5 lb. (2.5 kg) or a bolt, but slightly later versions may have fired a stone shot as large as 40 lb. (20 kg). However, without the use of torsion, this type of artillery was limited both in terms of size of projectile and range. Despite this, nontorsion weapons continued in use because they were considerably cheaper and less complicated to manufacture than torsion machines and provided effective projectile fire (whether bolts or stone shot) out to a respectable maximum range of around 300 yards (275 meters).

Torsion artillery was probably developed in the 340s and 330s when it was realized that there was an upper limit on the weight of projectile and range nontorsion artillery could achieve. It was powered by sinew or hair (animal or human) rope, tightly wound in several layers inside two connected wooden frames to produce a compact mass. Each of these bundles could be tightened by twisting it with levers (hence the name "torsion artillery"). A wooden arm was forced through each bundle of sinew rope and a bowstring secured at either end. When the bowstring was drawn back and released, the power generated from the torsion of the sinew bundles was considerably more powerful than could be achieved by nontorsion artillery.

Over some time a series of developments resulted in a high level of technical achievement. These included the use of washers to improve the operation of the tightening levers, amalgamating the frames of each arm into a single frame with compartments, and increasing the arc through which the arms could move. Smaller, more easily transported machines were used on the battlefield, while larger ones were used against city walls during sieges. All of the early machines fired bolts of varying sizes. The ability to fire stone shot came with a return to the separate frames for each arm and a change to the angling of the torsion and arm assembly (for these developments, see the diagrams in Marsden's *Greek and Roman Artillery: Historical Development*, 19ff.).

By a combination of trial and error and theoretical calculation, probably in Ptolemaic **Egypt** in the first half of the third century, Greek engineers developed a formula for the ideal ratios between the different parts of a catapult. This formula, preserved in extant technical treatises, allowed machines of varying sizes to be constructed while retaining optimum performance. The best torsion artillery could probably attain ranges of 400-500 yards (370-460 meters) and the larger stone-throwing machines could be effective against walls. Torsion artillery could also be extremely accurate. In 304, **Rhodian** artillerymen dislodged some of the iron plates from **Demetrius I (Poliorcetes)**'s siege tower and then successfully targeted the wooded frame with flaming bolts. *See also* GAZA; MANTINEA (3); SIEGE WARFARE; WOUNDS.

ASCLEPIODOTUS (fl. 1st century). An academic military theoretician who wrote on Greek and **Macedonian** tactics. His book *Tactics* is the earliest extant military manual. Although very theoretical—it has no historical examples—and discussing tactics and formations which were no longer used, the *Tactics* does preserve some useful information, in particular about the **phalanx** and the **cavalry**. *See also* AELIAN; AENEAS TACTICUS; ONASANDER.

ASCULUM. A battle fought in the spring of 279 between **Pyrrhus I** of
Epirus and the Romans during the war between **Tarentum** and **Rome**.
After Pyrrhus' victory at the **Siris River**, the Romans rejected his offer
of peace terms and dispatched both consuls south with their armies and
large allied contingents, probably totaling around 40,000 men. The Roman army met Pyrrhus' force at Asculum at the base of the Apennines in
southeast **Italy**, where Pyrrhus was apparently aiming to free Samnium
from Roman control. The size of Pyrrhus' army is unknown, but in addition to the survivors of his 28,000-odd Greeks from Siris River, he
had received substantial local reinforcements and his numbers were
probably roughly equal to the Romans.

The battle was fought on rough and wooded **terrain**, which reduced
Pyrrhus' ability to maneuver his **elephants** and **cavalry**. According to
Hieronymus, the better tradition preserved in **Plutarch** (*Pyrrhus*, 21),
the battle was fought over two days. The first was indecisive because of
the limitations imposed by terrain and vegetation, but on the second
Pyrrhus seized the rough terrain with a detachment of troops and was
then able to force an engagement with the Romans on more open
ground. As at the Siris River, he fixed the Roman infantry with his
phalanx but this time used his elephants, interspersed with **archers** and
slingers, to break their formation. One of the Roman consuls and 6,000
soldiers died; Pyrrhus lost 3,505 men. Plutarch states that Hieronymus
records Pyrrhus' losses from the king's own commentaries and, if so,
they may record only Greek deaths. The other tradition of the battle, that
of Dionysius, has only one engagement, over a single day, with the total
casualties for both sides as 15,000. Whichever figure is correct (and the
lower tally is again preferable here), Pyrrhus was unable to replace the
loss of his experienced Greek troops. He is reputed to have remarked
about Asculum that "if we defeat the Romans in one more battle we
shall be utterly lost" (Plut., *Pyrrhus*, 21.9). Following the victory
Pyrrhus withdrew to Tarentum, garrisoned the city, and crossed to **Sicily**. The Romans followed up his departure by successful attacks on Tarentum's allies. *See also* BENEVENTUM.

ASSINARUS RIVER (ASSINAROS). The location of the final defeat of
the **Athenian Sicilian Expedition** in 413. The Athenian army was continually harassed by the **Syracusans** and allies during its withdrawal
and was finally brought to bay at the Assinarus River and forced to surrender. Although quite a few managed to break out and escape to Catana,
Thucydides estimates that the total of those captured in Sicily was "not
fewer than 7,000" (7.87.5).

ATHENIAN ARMED FORCES. Information concerning **Athens'** military organization is generally better than for most states, especially during the Classical period. For example, **Thucydides** (2.13.6-9) gives figures for all available troop types and **triremes** in 431 and this can be supplemented by other sources, including inscriptions. In common with most states, Athens had no standing army—all service was regarded as a civic duty and soldiers reported when called up for the duration of the campaign. The navy was rather different, though, especially in the heyday of empire (479-404). Given the nature of naval service, and the need to patrol sea lanes fairly continuously, parts of the fleet must have spent long periods of the year at sea.

Overall command was exercised by 10 *strategoi* (generals), who were elected to an annual term of office, commanded both land and sea forces, and were accountable to the people. In the fifth century *strategoi* were often both politicians and soldiers but in the fourth there was an increasing specialization in military matters. Nevertheless, there was often a political element to both election and service as a general even then. Command below the *strategoi* was exercised by a variety of subordinate **officers** within the various arm of service (navy, **hoplite** force, **cavalry**).

The navy was arguably the most professional of the Athenian services and in the fifth century was the largest and best of the Greek navies. Paid for by the profits of empire, the fleet was generally on continuous operations, whether combat, blockading, or routine patrolling. This led to a high level of efficiency and the Athenian navy was only defeated at the end of the fifth century when weakened by losses from the **Sicilian Expedition** and faced by a coalition of Peloponnesian and rebel states, financed by **Persia**. The state provided the hulls of the ships (300 triremes in 431), but the wealthiest Athenians served as trierarchs (captains), providing the crew and equipping the ships as a civic duty. The navy was very small in the years following the loss of the **Peloponnesian War** of 431-404, and even with its resurgence as a naval power, Athens was never able to afford the navy it desired and which it had maintained in the fifth century. Although for most of the fourth century the Athenians still possessed impressive numbers of ships in the state dockyards, they generally had trouble equipping and manning all of them. Despite this, they were usually able to maintain a fleet of around 200, and they remained the major naval power in Greece until 322, when their naval power was broken with the loss of 170 ships at the decisive defeat at **Amorgos**.

The Athenian hoplites were not noted for their military prowess, but performed creditably on most occasions. Following Cleisthenes' reforms

(c. 506), the hoplite force was organized into 10 tribal *taxeis* (regiments), each 1,000 strong, each commanded by a **taxiarch**, and each further subdivided into *lochoi* (companies), which were commanded by a *lochagos*. These 10,000 hoplites were presumably the numbers on the rolls each year, but others were available for an emergency or a full mobilization. Cleisthenes' geographical distribution of the political/administrative units of the tribes may in part have been influenced by the location of the roads in Attica and the desire to mobilize groups from the same tribe along the same main routes into Athens. At the outbreak of the Peloponnesian War in 431, the number of hoplites available was about 13,000 in the field army and a further 16,000 on garrison duty and available for home defense (the latter drawn from the youngest and oldest hoplites). Some of these would also serve as *epibatai* (marines) in the navy (normally 10 per ship). *See also* ZEUGITAI.

Athens had no organized *psiloi* (light infantry) as late as 424 (Thuc. 4.94.1), as the navy had a higher priority for the poorer citizens from whom this arm was recruited. Any need for light troops was apparently met by the ad hoc mobilization of those too poor to afford hoplite equipment and who were not serving with the fleet. However, Athens, unusually for a Greek *polis* (city-state), did maintain a corps of citizen **archers** in the fifth century. In 431 there were 1,600 of them—a very respectable number, although 1,200 would have been required to provide the four-man contingent of bowmen for each trireme.

The Athenian cavalry was of generally good quality, despite its reputation for poor discipline. In the sixth century the Athenian mounted troops were predominantly **mounted infantry**, with some true cavalry, whose numbers were supplemented when required by allied cavalry from **Thessaly**. From the time of Solon's reforms (around 594) each naucrary (an administrative unit that provided ships and crews for the fleet) also supplied two cavalrymen, providing a total force of 96.

This system may have survived down to around 483 or sometime shortly afterward, when the numbers were probably raised to 300. Sometime between 445 and 438 the numbers were further increased to 1,000, with financial support from the state for the maintenance of the mounts. This was the canonical strength of the Athenian cavalry corps in the Classical period. The corps was commanded by two **hipparchs** and subdivided into 10 tribal squadrons or *phylai*, each commanded by a **phylarch**. Down to sometime between 395-365 it was supplemented by 200 *hippotoxotai* (mounted archers), probably consisting of a core of Scythian **mercenaries**, supplemented by citizens. The *hippotoxotai* were apparently then replaced by *hamippoi* (light infantry trained to operate with the cavalry). The corps was reduced to 300 again at the end of the fourth century under **Macedonian** influence and its last recorded engagement was in 279 against the **Gauls**.

ATHENS. Athens, the dominant settlement in Attica, first came to prominence in the seventh and sixth centuries, mainly for its internal troubles caused by rural poverty and the struggles of aristocratic clans for dominance. In the mid-sixth century Athens was subjected to several attempts at tyranny, culminating in Pisistratus' successful coup, backed by **mercenary** and allied soldiers, in 561. A period of stability followed (including an alliance with **Plataea** and successful war against **Thebes** in 519) until in 511/10 Pisistratus' son, Hippias, was expelled from Athens by a **Spartan** army led by **Cleomenes I**. However, Cleomenes' subsequent attempt to use troops to support his favored candidate for power at Athens failed. His force was besieged on the Acropolis and had to evacuate under terms. The subsequent reforms (c. 506) of the winning Athenian, Cleisthenes, established a democracy at Athens and apparently created the elected office of *strategos* (general) as part of a military reorganization. Around 506 the new democracy was put to the test when the Spartans organized a concerted attack. A Peloponnesian force **ravaged** southwest Attica, while the **Boeotian League**, led by Thebes, seized parts of northwest Attica and Chalcis attacked in the northeast. On the same day, in two separate battles in **Boeotia** and **Euboea**, Athens defeated the Boeotian League and Chalcis.

At the start of the fifth century, Athens was not regarded as a major power in Greece. However, the Athenians contributed 20 **triremes** to assist their fellow Greeks during the unsuccessful **Ionian Revolt** and then defeated the resultant **Persian** punitive expedition at **Marathon** in 490. This instantly elevated Athens' status and **Themistocles'** naval reforms led to a real increase in power in the 480s. These included fortifying Athens' port, the Piraeus, and the building of a fleet of 200 triremes. With these ships, Athens formed the backbone of the Greek naval defense against Persia during the Second **Persian War**. The Athenians made a tremendous contribution to Greek victory in this war—surrendering naval command to Sparta in order to preserve unity, abandoning their own city rather than submit to the Persians, providing more than half of the ships in the Greek navy, and contributing troops at the battle of Plataea. This contribution, particularly to the victory at **Salamis**, placed Athens at the forefront of Greek states—especially as Thebes had **medized** and **Argos** had remained neutral.

After the expulsion of the Persians from Greece in 479, Athens used her naval power and newly acquired prestige to form the **Delian League**, an anti-Persian alliance of island and **Ionian** states, and to attack Persian territory in Asia Minor. This alliance was transformed over the next 30 or 40 years into an Athenian empire, with states converted into subject rather than voluntary allies. The tribute from the empire, and the increased military power it brought, increasingly placed Athens on an equal footing with Sparta—up until then largely unchallenged as the dominant military power in Greece.

The growth in Athenian power and her expansionist attitude, combined with Spartan disquiet and pride, led to two major clashes. Athens came very close to establishing a land empire (incorporating the Megarid and Boeotia) during the First **Peloponnesian War** (459-445). However, the Athenians overreached themselves with simultaneous ventures against the Persians and suffered a major loss in **Egypt** in 454. When the Athenians lost the battle of **Coronea**, they made peace, largely on the basis of the prewar status quo. The Peloponnesian War (2) of 431-404, saw the whole resources of the Delian League pitted against Sparta and the **Peloponnesian League**. Despite some innovative strategies, particularly **Pericles'** policy of avoiding land battle in the early stages of the war, Athens was ultimately defeated. By the end, this war had approached something close to total war and Athens was lucky to be allowed to survive—several of Sparta's allies, notably Thebes and **Corinth**, argued for the city's destruction but Sparta refused.

At the end of the fifth century Athens had suffered major casualties, been stripped of its empire, and undergone considerable internal strife. The democracy was overthrown and only restored with some difficulty after an oligarchic revolution which resulted in the fairly brutal rule of the **Thirty Tyrants**—who had come to power with the assistance of elements of the Athenian **cavalry**. It took some time for Athens to recover, and it was never again as powerful as it had been from 479-431. Nonetheless, the Athenians continually tried to restore their state to its former glory. During the first quarter of the fourth century Athens staged a fairly remarkable recovery, fighting against Sparta in the **Corinthian War** and helping to free Thebes from Spartan control after its seizure by **Phoebidas**. The Long Walls, which connected the city to the Piraeus and had been demolished by the Peloponnesian League in 404, were rebuilt in the late 390s, signifying Athens' intention to rebuild itself as a naval power. In 378/7 Athens took a further step in this direction by forming the Second Athenian Confederacy. Although deliberately avoiding the most unpopular aspects of the Delian League, it was a military alliance involving some of the members of the old league. It was specifically aimed at ending Spartan domination of Greece.

Despite these efforts, Athens had been so weakened by the loss of the Peloponnesian War that, on its own, it could not seriously challenge either Sparta or a newly resurgent Thebes. Although still important, in many respects Athens was now a second-ranking player in the competition for Greek supremacy. When Thebes emerged as dominant after the battle of **Leuctra** in 371, Athens allied with Sparta in an attempt to maintain some balance of power in Greece. Hopes of a new naval empire were dashed in 357 with the outbreak of the **Social War**, when Athens' allies, angered and disturbed by increasingly high-handed Athenian activity in the Aegean, went to war to preserve their freedom. This war ended in 355 with Athens recognizing the independence of the main members of the confederacy. For most of the rest of the century Athens was obsessed with recovering **Amphipolis** and her possessions in the Chalcidice and the northern Aegean—and with the growing Macedonian threat. Ultimately outmaneuvered both diplomatically and militarily by **Philip II** of **Macedon**, Athens, along with Thebes and other allies, was defeated at **Chaeronea** in 338. Following **Alexander the Great**'s death, Athens played a leading role in the **Lamian War**, which attempted to overthrow Macedonian domination. This failed in 322 and **Antipater** installed a garrison in Athens. The city changed hands several times between then and the end of the century during the struggles among the *diadochoi*.

In the third century, Athens played a decreasingly important role in Greek military affairs. It was besieged and captured by **Demetrius I (Poliorcetes)** in 296-295 and its last real leadership role against Macedon was during the **Chremonidean War** (268/7-262/1), an unsuccessful revolt against **Antigonus II (Gonatas)**. Athens was garrisoned by Macedon from then until 229 and so exercised no independent action.

In the second century, Athens was from time to time embroiled in various Greek quarrels, but the climate of the time allowed no single city-state to exert much influence. Athens helped **Rhodes** and Pergamum persuade **Rome** to intervene against **Philip V** of Macedon, leading to the Second **Macedonian War**. Despite adopting a fairly pro-Roman policy from then on, Athens supported **Mithridates VI (Eupator)** in the First **Mithridatic War**. As a result, the city was besieged and, in 86, sacked by the Romans under Lucius Cornelius Sulla. From this point onward, Athens rested on its intellectual laurels, not its military ones. *See also* ATHENIAN ARMED FORCES.

ATTALUS I (SOTER) (ATTALOS; 269-197). King of Pergamum, 241-197. Attalus' main claim to fame was his refusal to pay tribute to the **Gauls** (Galatians) and their subsequent decisive defeat by him prior to 230. This established his reputation, gained him the sobriquet Soter ("Savior"), and probably led to his assumption of the title of king. He followed this success with a counterattack on **Antiochus Hierax**, the Gauls' ally, and over the period 229-228 conquered a large part of Seleucid Asia Minor. This represented the furthest extent of his territory, as he lost most of these conquests to the Seleucid general **Achaeus** between 223 and 220 (Achaeus was acting in the name of the future **Antiochus III [the Great]** but then seized the recovered territory for himself). However, Attalus recovered several districts from Achaeus in 218, and in the winter of 217/6 allied with Antiochus, who was planning to end Achaeus' independent reign. Although the details are uncertain, Pergamum may have profited territorially from Achaeus' capture and execution in 213.

Hostility to **Philip V** of **Macedon** seems to have been a major feature of Attalus' policy after 220. **Polybius** (10.41-42) records Attalus as providing support to the **Aetolian League** against Philip during the First **Macedonian War** (including naval activity in Greece in 208). In 202, Philip's activities in the Chersonese and the Propontis led to an alliance of Pergamum, **Byzantium**, and **Rhodes** against him. Philip **ravaged** Pergamum in 201, but was unable to take the city and had to raise the siege when he lost a major naval battle at **Chios** against the allies. Despite this victory, Attalus appealed to **Rome** for help and, along with Rhodes and **Athens**, helped persuade Rome to initiate the Second **Macedonian War** against Philip. During this war, Attalus provided naval assistance to supplement Roman land power. He died from a stroke shortly before the end of the war, a capable and energetic diplomat and soldier who had considerably increased the power of Pergamum. In the long term, though, his appeal to Rome helped lead to the Roman acquisition of Asia Minor.

- B -

BARDYLIS I (c. 448-358). King of the Dardanians circa 400-358. Bardylis created a powerful kingdom from several Illyrian tribes northwest of **Macedon** in the early fourth century. In 393/2 he invaded Macedon, defeating King Amyntas III and temporarily occupying Lower Macedon. Amyntas was restored to power in 391 with the assistance of the Aleuadae of **Thessaly**. In 385/4 Bardylis successfully invaded Molossia with support from **Dionysius I** of **Syracuse** but was subsequently expelled by the **Spartans**. In 383/2 his forces again invaded Macedon and defeated Amyntas, but this time Amyntas recovered his territory within three months. By the 370s Macedon was apparently paying tribute to Bardylis. Circa 360, Bardylis was raiding into Molossia again, and early in 359 he defeated and killed Perdiccas III of Macedon, along with 4,000 of his men, and occupied large sections of Macedon.

In 358 Bardylis was defeated by **Philip II** of Macedon at an unknown location, probably in Lyncestis. Both armies had 10,000 infantry, and the Macedonians had 600 **cavalry**, the Illyrians 500. Although recently successful against the Macedonians, Bardylis formed his army into a square. Philip broke this with his infantry, probably advancing with his right wing forward, Theban style, and used his cavalry to strike the decisive blow, killing 7,000 of the enemy, probably including Bardylis. This effectively ended the power of the Dardanians.

BATTLE OF THE CHAMPIONS. Fought circa 547 between 300 selected **Spartans** and 300 selected soldiers from **Argos**, this battle was designed to settle the ownership of the long-disputed border region of Cynuria. Instead of a full-scale battle, the two sides agreed that 300 **hoplites** from each side would fight and the victor would keep the territory. Unfortunately a lack of clarity in the rules led to a disputed result. According to **Herodotus** (1.82), darkness ended the battle when only three men were still alive—two Argives and one Spartan. The two Argives returned to their city when night fell, but the Spartan, Othyrades, remained on the battlefield, stripping the enemy corpses of their **armor**. On the next day the Argives claimed victory because they had the greater number of survivors, the Spartans because Othyrades had remained in possession of the battlefield. The argument degenerated into combat and the Spartans won the ensuing battle between the two armies. Rather than return to Sparta, Othyrades is supposed to have killed himself through shame at being the only survivor.

BENEVENTUM. A battle fought in 275 between **Pyrrhus I** of Epirus and the Romans during the war between **Tarentum** and **Rome**. This was the final battle in Pyrrhus' Italian campaign and was separated from the previous one, **Asculum**, by more than two years' campaigning in **Sicily** against **Carthage**. On his return, Pyrrhus was in a much worse position than when he left **Italy** in 278. In addition to the losses he had incurred on Sicily and on his way back to Tarentum, successful Roman campaigns had detached or conquered the majority of Tarentum's allies.

Pyrrhus chose to engage the consular army of Manius Curius at Beneventum (in Samnium, between Capua and Asculum) before it could be reinforced by Cornelius Lentulus' army in Lucania. The numbers on both sides are uncertain, but Curius' army may have been the standard two-legion army plus allies, totaling around 24,000 men, while Pyrrhus' army was probably smaller. **Plutarch** (*Pyrrhus*, 25) preserves a brief account of the battle, in which Pyrrhus planned to attack the Romans from two directions, sending a handpicked force of troops and **elephants** on a night approach march. Unfortunately for Pyrrhus, this force became lost and arrived in daylight. After routing the flanking force and capturing some of its elephants, Curius engaged the main army on the plain. It was a hard fight, but the Romans neutralized Pyrrhus' elephants with **javelins**, causing some of them to stampede through the Greek ranks, and forced him to retire. Pyrrhus withdrew to Tarentum, left a small garrison there (withdrawn in 272), and retired to Greece. *See also* SIRIS RIVER.

BITHYNIA. A region of northwest Asia Minor, east of Chalcedon and on the southern shore of the Euxine (Black Sea). In the Classical period its inhabitants were of **Thracian** origin and often waged war against the Greek colonies in the area. It was part of the **Persian** empire after Cyrus the Great's conquest of Lydia in the mid-sixth century. However, the region became independent under a local dynasty around 440 and from then on vigorously maintained its independence against all comers, including **Macedon**, the *diadochoi* (especially **Lysimachus**), and the Seleucids. In 278/7, King Nicomedes I (whose name indicates Greek influence in the royal family) invited the **Gauls** in to help him in an internal struggle with his brother and considerably strengthened his kingdom. However, his son and successor was killed by these same Gauls, who were finally defeated by Nicomedes' grandson, Prusias I, who came to the throne in 228. The following century, a series of wars with Pergamum resulted in the loss of some Bithynian territory and the diplomatic intervention of **Rome**. In 88 Nicomedes IV (Philopator) was defeated and exiled by **Mithridates VI (Eupator)** of Pontus. Restored to his throne by the Romans in 85/4, Nicomedes bequeathed Bithynia to Rome on his death 10 years later.

BOEOTIA (BOIOTIA). A rich agricultural region to the north of Attica and stretching from the Euboean Gulf to the Corinthian Gulf. It was a strategic location—anyone traveling by land between northern Greece and the Peloponnese had to pass through Boeotia. Its location on a plain, and its relative proximity to **Thessaly**, led to the development of good quality **cavalry** in the region. For much of Boeotia's history the most important *polis* (city-state) in the region was **Thebes**, which tended to dominate the federal organization, the **Boeotian League**. Although the existence of the Boeotian League gave a considerable amount of unity to the region, its cities were fairly frequently opposed to each other and did not always act in unison.

Boeotia's development as a major Greek power was also hampered by hostile relations with neighbors such as **Athens** and **Phocis**. Its reputation (particularly that of Thebes) suffered a major blow during the Second **Persian War** when all of Boeotia, apart from **Plataea** and Thespiae, **medized** after the battle of **Thermopylae** (1). During the fifth century the region was frequently at war with Athens and in the early fourth century with **Sparta**. Under Thebes' leadership it attained dominance in Greece from 371 but gradually lost control after the deaths of **Pelopidas** in 364 and **Epaminondas** in 362. The region came under **Macedonian** control after the battle of **Chaeronea** in 338 and under **Roman** control after the defeat of the **Achaean League** in 146.

BOEOTIAN LEAGUE (BOIOTIAN LEAGUE). A federal league of Boeotians, generally, but not always, dominated by **Thebes**. It was issuing federal coinage before 550, but suffered a decline with the **medizing** of Thebes in the Second **Persian War** and its subsequent dissolution. The league was restored by **Sparta** around 457 and allied with it in the **Peloponnesian War** of 431-404 but refused to join the Peace of **Nicias** in 421. It was increasingly out of step with Sparta from the end of hostilities in 404 and by 395 was at war with Sparta. At about this time its federal army consisted of 11,000 **hoplites** and 1,100 **cavalry** led by 11 boeotarchs, below which were *strategoi* or generals. The numbers of troops per district (1,000 hoplites and 100 cavalry) is approximate and the whole of **Boeotia** could probably raise considerably more troops than that if required. At this time (and probably at the outbreak of the Peloponnesian War [2] in 431), Thebes controlled four of the 11 districts. The league was dissolved under the King's Peace (**Common Peace** [1]) of 386 and Thebes' citadel, the Cadmea, was subsequently occupied by Sparta. In 379 the Thebans regained control of their city and soon afterward reconstituted the Boeotian League on a democratic basis (although Thebes controlled three of the seven districts). The league played an important part in Greek history and exercised hegemony over much of Greece in the period from its defeat of Sparta at **Leuctra** in 371 down to its own defeat at **Chaeronea** in 338, following which the league was dissolved by **Philip II** of **Macedon**.

BOOTY. Collecting booty was an important aspect of Greek warfare. It damaged the enemy, enriched the state, was sometimes the main source of income for a force, and could be of considerable importance to the individual soldier or sailor—particularly in times of irregular **pay**. When the enemy secured his baggage train at the battle of **Gabiene**, **Eumenes of Cardia**'s *argyraspides* ("Silvershields") deserted him in order to preserve their possessions, accumulated over several years of hard campaigning.

Booty could encompass any movable item, including livestock and people (who could either be ransomed or, more commonly, sold as **slaves**). During the Decelean War phase of the **Peloponnesian War** of 431-404 the Boeotians even looted the roof tiles from Attic country houses. The sums acquired could be quite considerable: the Athenians collected 80 talents' worth of booty from the **Persians** after **Marathon**, **Alexander the Great** gained 440 talents when he captured **Thebes** in 335, and **Antigonus III (Doson)** secured 300 talents when he took **Mantinea** in 223. King **Agesilaus II** of **Sparta** is recorded as taking 1,000 talents of booty during his Asia Minor campaign in 394. Although individual acquisition took place, the normal practice was for the general to collect together as much of the booty as possible. This belonged to the state (or was distributed among the states in allied forces) and the general was held accountable for it. However, he might use part of it to reward acts of bravery and might be authorized to sell it on the spot to raise pay for his soldiers. Traditionally one-tenth of the total was dedicated to the gods as a thanks offering, often in the form of elaborate statues but also of captured weaponry, gold weapons, or even a new temple. According to **Pausanias** (3) the temple of Zeus at Olympia was built by the Eleans from the proceeds of booty (5.10.2). *See also* AETOLIAN LEAGUE.

BOW. The ancient Greeks used two types of handheld bow: the simple wooden bow and the composite bow (see figure 3). The composite bow was the more powerful of the two—and the more difficult to string and use. It was built from three layers of material. The central core was wood, backed by a layer of horn and faced by a layer of sinew, both glued in place. The construction was designed to maximize the elasticity of the **weapon** immediately above and below the hand grip, while minimizing it at the hand grip itself and the tips. The tension in the bow meant that, when unstrung, the tips pointed forward from the handgrip. It took considerable strength to string such a bow, a fact recognized as early as **Homer**—in the *Odyssey* none of Penelope's suitors were able to string **Odysseus'** bow, only he could do so (21.144ff.). The effective range of the bow was around 150-250 yards (140-185 meters). The arrows had reasonable penetrative power at closer ranges but were not so good over longer ranges or against good **armor**. *See also* ARCHERS; *GASTRAPHETES*.

BRASIDAS (d. 422). A **Spartan** general during the **Peloponnesian War** (2) of 431-404. He was an innovative commander, "quite an able speaker—for a Spartan," and with "a reputation in Sparta for vigor in everything" (Thuc. 4.81.1, 4.84.2). In the summer of 431, with only 100 **hoplites**, he charged right through an Athenian army outside the weakly fortified Methone (in Laconia), reinforcing the defenders and preventing its loss. As a result, he was the first Spartan to be officially congratulated during the war. In summer 429 he was one of three commissioners advising the fleet facing **Phormio** in the Gulf of Corinth. An excellent plan almost brought decisive victory to the Peloponnesian fleet, but poor execution saw the Athenians turn the tables during the pursuit phase. Immediately after this he was one of the commanders behind the daring Peloponnesian naval raid on **Salamis** which caused great panic in **Athens**.

In 427 Brasidas was adviser to the Peloponnesian fleet, which won a victory off **Corcyra**, and he suggested a vigorous follow-up against Corcyra itself, which was in a state of panic and confusion (Thuc. 3.79.3). His advice was rejected and the opportunity lost. During the Spartan attempt to expel the Athenians from **Pylos** in 425 Brasidas played the main role in pressing the naval landing home, but was **wounded** and lost his **shield**. He also saved **Megara** from loss to the Athenians in 424 by swift action when its port, Nisaea, fell.

Later in 424 he was sent out to **Macedon** and **Thrace**, leading 700 helot hoplites and a force of **mercenaries**. His first success was to traverse a hostile **Thessaly** by speed and bluff. In the Chalcidice he used a mixture of force, threat, and diplomacy to win over a large part of the Athenian empire there, posing a major threat to the integrity of the empire as a whole. He secured **Amphipolis**, Athens' prize possession in the area, and did well against the large Athenian expeditionary force sent to oust him from the region. He was killed in 322, during his victory over the Athenian general **Cleon** at the battle of Amphipolis. After his death the people of Amphipolis **sacrificed** to him as a hero and honored him as the founder of their city, removing all trace of the real founder, the Athenian Hagnon.

BUCEPHALUS (BOUKEPHALOS; also BUCEPHALAS). Alexander the Great's favorite horse, named after the ox-head marking on his head. Alexander won him in a bet with his father after successfully riding him when others could not (the young Alexander had noticed that the horse was shying at his own shadow). When Bucephalus died after the battle of the **Hydaspes River** in 326, Alexander founded the city of Bucephala there in his honor.

BYZANTIUM (BYZANTION). Byzantium was founded by **Megara** (joined by other cities) in the mid-seventh century on a strategic site on the European side of the Bosporus, opposite **Chalcedon**. The city rapidly grew rich from fishing, its excellent harbor, and the levying of tolls on ships passing between the Aegean and the Euxine (Black Sea).

Byzantium's strategic location made it an attractive possession for the great powers. It was taken by the **Persians** about 512 and remained under their control until freed in 478 after the Second **Persian War**. It then joined the **Delian League** but unsuccessfully revolted in 440 and was made a subject state of the Athenian empire in 439. During the **Peloponnesian War** of 431-404, Byzantium revolted again and was garrisoned by **Sparta** in 411. In 408, **Athens** briefly recovered the city after a siege, but it fell into Spartan hands again after **Aegospotami** in 405. In 400/399 (still under Spartan control) Byzantium was almost sacked by the **Ten Thousand**.

During the fourth century the city reverted to its alliance with Athens, although temporarily joining **Thebes** in 364 and fighting Athens in the First **Social War** (357-355). At the end of this war Athens recognized Byzantium's independence. Supported by Athens, it survived a memorable siege by **Philip II** of **Macedon** in 340-339, but was in alliance with Macedon by 335, when it assisted **Alexander the Great** in his expedition against the Triballi. Byzantium managed to remain neutral during the early struggles among the *diadochoi* but suffered severely at the hands of the **Gauls** from around 277, having to pay considerable sums to them for most of the rest of the century to preserve their territory. To pay for this, they levied increasingly heavy tolls on shipping in the Bosporus until Rhodian military activity in 220/19 forced them to stop. In 220 Byzantium allied with **Attalus I (Soter)** of Pergamum and **Achaeus**. In the late third century, in conjunction with Attalus and **Rhodes**, Byzantium opposed **Philip V** of Macedon's naval ventures in the northern Aegean. In the second century Byzantium received favored status from **Rome** for supporting it during the **Macedonian Wars** (despite initially backing **Andriscus** in 150 at the start of the Fourth Macedonian War). *See also* CLEARCHUS.

- C -

CALLIAS, PEACE OF. A peace treaty between the **Delian League** and the **Persians** made around 449/8 that formally ended the hostilities begun with the Second **Persian War** (480-479). The existence of this peace has been much debated—from as early as the fourth century, when Theopompus declared it a forgery. Most scholars now accept that if a formal peace of this name was not concluded circa 449/8, then a de facto cessation of hostilities occurred around then and was formalized in 424/3. The peace is named after the Athenian Callias. Its main clauses (reconstructed from a variety of scattered references) were probably that the Greek cities in **Ionia** were to be autonomous, but to pay tribute to Persia; there was to be a demilitarized zone on land, with Persian armies restricted to east of the Halys; the Ionian cities were to demolish their walls; and there would be a demilitarized zone at sea (no Persian naval movements past Cyanaea [at the mouth of the Euxine?] and Phaselis and the Chelidonean Islands, off the southern coast of Asia Minor).

One of the objections to the existence of the treaty is the differences in the naval limits recorded in the various sources. However, although the Chelidonean Islands are about 25 miles (40 kilometers) from Phaselis, the crucial distance is that they are only about 8 miles (13 kilometers) west of Phaselis. Based on **Thucydides** 6.13.1, it seems quite possible that two limits were given in the treaty—one for ships hugging the coast and the other for those moving further out to sea—and that each of our sources record a different one of these (which are relatively close to each other).

Whether or not a formal peace was concluded around 449, hostilities between **Athens** and Persia essentially ceased from this period. On balance, the evidence tends to support the existence of the peace, although this still remains a controversial question nearly 2,500 years after it was first raised by Theopompus.

CALLICRATIDAS (KALLIKRATIDAS; d. 406). A **Spartan navarch** who vigorously prosecuted the naval campaign against **Athens** in the Aegean in the latter stages of the **Peloponnesian War** of 431-404. Taking a much more independent line from **Cyrus the Younger**, Callicratidas raised support from the Greek cities in Asia Minor. In June 406 he caught the Athenian admiral **Conon** at sea, capturing 30 of his ships and penning the other 40 up in Mytilene. He then sank 10 of a 12-**trireme** relief force. In August 406, leaving 50 ships to maintain the blockade, he faced a further Athenian relief force of 150 ships at **Arginusae** with a fleet of only 90 ships. His initial plan, to surprise the Athenians by advancing on them at night, failed due to bad weather and Callicratidas' fleet was destroyed in open battle. Early in the engagement he was lost overboard while his ship was ramming another.

CALLIMACHUS (KALLIMACHOS/KALLIMAKHOS; d. 490). Athenian **polemarch** at **Marathon**, 490. Despite the institution of the office of *strategos*, Callimachus appears to have been commander in chief of the Athenian forces at Marathon—it was his deciding vote which committed the Athenians to battle. Callimachus took the traditional polemarch's place on the right flank of the front rank of the **hoplite phalanx** and was killed during the battle. He was commemorated in the paintings of the battle in the Stoa Poikile ("Painted Stoa") and in several epigrams, two of which are extant.

CAPHYAE (KAPHYAI). A battle fought in 220 between an **Achaean League** army led by **Aratus** and an **Aetolian League** force led by Scopas and Dorimachus, which sparked off the Second **Social War** of 220-217. The Aetolians had invaded **Messenia** and Aratus had assumed command as Achaean *strategos* five days early in order to lead the league forces to help. Aratus ordered the Aetolians to withdraw from Messenia, and when they began to do so he disbanded his army, except for around 3,000 infantry and 300 **cavalry** and another contingent of unknown size, with which he tailed the Aetolians. When Dorimachus realized this, he marched toward Megalopolis to test Achaean resolve and at the same time allow his **booty** to be embarked without hindrance.

Near Caphyae Dorimachus was challenged by Aratus, who apparently mismanaged the action. Aratus allowed the enemy to traverse a plain, which would have been suitable **terrain** for his heavy infantry, and sent his *psiloi* (light troops) and cavalry to attack the Aetolian rear guard of *psiloi* and cavalry as they were moving into the foothills. The Aetolian rear guard closed up to their main body in good order and Aratus, mistaking this for a rout, ordered his heavy infantry to pursue. At this point the Aetolian cavalry and *psiloi* charged their pursuers. The Achaean cavalry and *psiloi* were driven back in confusion onto their own heavy infantry, who were advancing across the foothills in individual companies, causing them to retire. Under pressure from the more mobile enemy cavalry and *psiloi*, the Achaeans broke and the only thing that prevented the force from annihilation was the existence of several friendly towns nearby where they could seek refuge. This unexpected victory allowed the Aetolians to renew their march across the Peloponnese with impunity, attacking Pellene and the territory of **Sicyon**, and eventually leaving via the Isthmus of Corinth. **Polybius** 4.9-13 provides a detailed account of the action.

CARTHAGE. Carthage was a Phoenician colony in North Africa (modern Tunisia), probably established around the second half of the eighth century—although the traditional founding date is late ninth century. The Carthaginians became a major maritime and trading power in the western Mediterranean, in the process coming into conflict with the Greek colonies in Spain, southern France, **Italy**, and **Sicily**. Given its reliance on trade, Carthage's navy was manned by citizens and of a high quality. The Carthaginian army was often good but generally of not quite such a high quality as the navy. Its core was citizen, but a large part of it was usually formed from **mercenaries** (including Greeks from time to time) and subject peoples from Numidia and later Spain, providing good light **cavalry** and light to heavy infantry, respectively. Generals were elected and often enjoyed a good degree of autonomy in the field, although they were accountable to the oligarchic government at home.

Around 535 the Carthaginians joined with the Etruscans to prevent Greek expansion west, defeating the fleets of Phocaea and Massilia at the battle of Alalia off Corsica. This allowed Carthage to consolidate its position in southern Spain, Sardinia, and the west of Sicily. Sicily was the main area of contact between Carthaginians and Greeks and was fought over by them almost continuously until 241, when **Rome** ended the Carthaginian presence there.

Carthage made several major concerted efforts to gain control of the whole island of Sicily, almost succeeding on several occasions. A large Carthaginian invasion force under Hamilcar was defeated by **Gelon** at Himera in 480, while from 409 to 405 the Carthaginians were able to extend their control over most of the island. However, **Dionysius I** of **Syracuse** recaptured most of this in the early fourth century, restricting Carthage to the western end of the island. This position was confirmed by the war with **Agathocles** at the end of the century, which almost led to the capture of Carthage by the Greeks.

In 280 Carthage provided naval support to Rome, a new regional power, against **Pyrrhus I** of Epirus but came into conflict with Rome in 264 over Carthaginian attempts to establish control of Messana on the eastern coast of Sicily, opposite Italy. Carthage subsequently lost all three Punic Wars against Rome (despite gaining the support of some Greek cities in Sicily and Italy in the Second Punic War) and was destroyed in 146 after the Third Punic War.

CASSANDER (KASSANDROS; c. 358-297). Son of **Antipater**, he joined **Alexander the Great** in Asia in 324. On his father's death in 319, the regency of **Macedon** passed not to him but to Polyperchon. Allying with **Lysimachus** and the future **Ptolemy I (Soter)** and **Antigonus I (Monophthalmus)**, Cassander invaded Greece and Macedon. Appointed regent in 317 by Eurydice, wife of the retarded Philip III (Arrhidaeus), by 316 he was in control of Macedon and most of Greece as regent to Alexander IV (Eurydice and Philip III having been killed by Olympias). He was confirmed in this position in 311 after the battle of

Gaza in 312 brought an end to a war fought by Cassander, Ptolemy, and Lysimachus against Antigonus. Shortly afterward Alexander IV and his mother died or were killed by Cassander, leaving him in complete control of Macedon.

From around 309/8 Cassander was mainly concerned with protecting his possessions in Greece from loss. Both Antigonus and Ptolemy actively encouraged trouble there and at various times sent over troops. While campaigning unsuccessfully against Epirus in 307, Cassander lost **Athens** to Antigonus' son, **Demetrius I (Poliorcetes)**. At about the same time, though, he was more successful in the Peloponnese, where he now restricted Ptolemy's garrisons, set up in 309/8, to **Corinth** and **Sicyon**. In 305 Cassander proclaimed himself king of Macedon. Still having problems with Demetrius, who had gradually extended control over much of the Isthmus of Corinth, in 302 Cassander joined Lysimachus, **Seleucus I (Nicator)**, and Ptolemy in coalition against Antigonus. Perhaps hoping to secure the withdrawal of Demetrius, Cassander dispatched troops to Asia Minor under the command of Lysimachus. These troops, as well as later reinforcements, participated in the victory at **Ipsus** in 301. Cassander spent the remaining years of his reign consolidating his kingdom, but when he died in 297 his surviving sons were too young to succeed and Macedon was soon split by dynastic rivalries.

CASSANDREIA. *See* POTIDAEA

CASUALTIES. Assessing casualty rates in ancient Greek warfare is often a difficult task because of problems with the data. Troop numbers recorded by the ancient sources are sometimes patently exaggerated, given in suspiciously round figures, or simply unavailable, and the same is true of casualty figures. For example, the only extant account of the battle of **Chaeronea (2)** in 86 gives the **Roman** casualties as 14 (two of whom turned up safe the following day), while claiming that only around 10,000 of the Pontic army of more than 90,000 men survived (Plut., *Sulla*, 15, 19).

Disparate casualties could occur, for example, at **Marathon** and **Plataea** during the **Persian Wars**, when relatively heavily armored **hoplite** forces operating on their home ground defeated much more lightly protected **Persian** forces (Hdt. 6.117; 9.70—although he undoubtedly exaggerates the Persian losses at Plataea). It also seems likely that **Alexander the Great**'s experienced and generally well-protected army inflicted disproportionate casualties on the Persians and Indians in its Asian campaigns, although probably not as disproportionate as the sources state. In warfare between the *diadochoi* and the Hellenistic monarchs, there was a tendency to try to preserve the **Macedonian** members of a defeated enemy army in order to enroll them in the victor's forces.

However, the situation was rather different in straight hoplite-versus-hoplite fights. P. Krentz ("Casualties in Hoplite Battles," *Greek, Roman and Byzantine Studies* 26 (1985): 13-20) analyzed the evidence from 472 to 271, demonstrating that the average ratio of casualties of winner to loser was 1:2.9 (the range being from 1:1.7 to 1:5.2). The winner generally lost between 5 percent and 10 percent of his force, the loser generally between 10 percent and 20 percent. These were fairly serious percentages, given the "small town" nature of most Greek *poleis* (city-states).

In general terms, hoplite casualties were usually higher when the victor had **cavalry** and/or *psiloi* (light troops) available for the pursuit, but there were some ways the loser could avoid casualties. Individuals who could afford to do so could reduce their chances of injury or death by wearing **armor**. They could also enhance their chances of survival during a battlefield withdrawal by retiring calmly and resolutely, thereby discouraging enemy interference. Socrates achieved this at **Delium** in 424, although he was also assisted by **Alcibiades**, who provided some mounted cover (Plato, *Symposium*, 221a-b; Plut., *Alcibiades*, 7.4).

The use of cavalry to inhibit pursuit was the main collective method for reducing casualty rates in a defeated army. For example, in the first battle outside **Syracuse** in 415, during the **Sicilian Expedition**, the **Athenian** victory led to few enemy casualties because the Syracusan cavalry prevented a proper pursuit (Thuc. 6.70). Athenian hoplite casualties were also reduced at the battles of **Mantinea** in 418 and 362 by the protective action of the friendly cavalry (Thuc. 5.73.1; Diod. Sic. 15.85.7). Fighting an action close to friendly cities also allowed a defeated force to seek refuge, as happened with the **Achaean League** forces under **Aratus** at **Caphyae** in 220 (Polyb. 4.12). *See also* WOUNDS.

CATAPULT. *See* ARTILLERY

CAVALRY. Cavalry was generally an undervalued and secondary (though not negligible) arm in ancient Greek warfare. From **Philip II of Macedon** onward, though, it had a more important place and was at times used to strike the decisive blow on the battlefield. The stirrup was unknown in ancient Greece, so a cavalryman did not have a particularly secure seat and this precluded the use of cavalry on hilly or rough **terrain**. This was not a big limitation on the cavalry's use in combat, though, as both the **hoplite** and the Macedonian **phalanx** also had problems with rough or steep terrain. Nevertheless, without stirrups or a saddle with a high cantle, the use of shock action in the traditional sense against infantry was curtailed.

Still, cavalry could be very effective against infantry who were out of **formation** or scattered. It could also be employed to good effect against the flank or rear of a hoplite phalanx, or against a phalanx disrupted by crossing a river or broken terrain. With sufficient numbers, cavalry could also disrupt a phalanx with **javelins** and then pursue the individual hoplites as they fled. Other uses of cavalry were the traditional ones of **reconnaissance**, **raids**, and protection of infantry on the march.

In the traditional Greek battle down to the mid-fourth century, the cavalry was almost always deployed on the flanks of the hoplite phalanx, providing protection against opposing cavalry and light infantry. However, cavalry charges were used by **Pelopidas** against **Alexander of Pherae** at **Cynoscephalae** in 364 and by **Epaminondas** against the Spartans at **Mantinea** in 362. Philip II and **Alexander the Great** developed this further, often using their phalanx to fix the enemy and delivering the decisive blow with their cavalry. This tactic continued into the Hellenistic period, although often not as well executed as it was under Philip and Alexander.

Most Greek cavalrymen of the sixth and fifth centuries were equipped with two javelins and a sword. Although most artistic representations prior to the fourth century show them without much protection, there is good evidence that cavalrymen could and did wear breastplates and **helmets**. Cavalry could also be equipped with a thrusting spear (*kamax*) or with both a spear and two javelins. The Macedonian cavalry used by Philip and Alexander in a decisive battlefield role were heavier cavalry, equipped with helmet, breastplate or cuirass, and often carrying the *sarissa* or *xyston*, a lance around 14.5 feet (4.5 meters) long. This gave the cavalry a reach advantage over the standard hoplite spear and did allow cavalry to force a breach in infantry formations.

The cavalry in most Greek states in the classical and early Hellenistic periods was provided from the upper echelons of society—those wealthy enough to provide their own mounts. Because of this, the cavalry was often associated with aristocratic or oligarchic government or movements. The traditional cavalry areas of Greece were in the north, especially **Thessaly** and Macedon, although the cavalry tradition was also strong in **Sicily** and **Italy**. In most of Greece proper south of Thessaly the hoplite tradition dominated, although both **Thebes** and **Athens** did field useful cavalry corps in the fifth and fourth centuries. **Xenophon**'s treatises on the duties of a cavalry commander (*Hipparchikos*) and on horsemanship (*Peri Hippikes*) are excellent sources of information on cavalry of the Classical period. *See also* ABYDUS (3); DELIUM; ELEPHANTS; HIPPARCH; *HIPPARMOSTES*; *HIPPEUS*; *HIPPOTOXOTAI*; MOUNTED INFANTRY; PHYLARCH; SHIPS; WEAPONS.

CERATA (KERATA). A range of hills (the name means the "Horns") in northwest Attica, on the border with **Megara**; site of a battle in 409 between the **Athenians** and Megarians described by **Diodorus Siculus** (13.65.1-2). The Athenians sent 1,000 **hoplites** and 400 **cavalry** against the Megarians who had recaptured Nisaea, their port. The Megarians marched out *pandemei* (in full force), supported by some troops from **Sicily**. Although they outnumbered the Athenians, the Megarians were defeated and many killed in the subsequent pursuit—one of only two recorded by the Athenian cavalry (the other was at **Abydus** in the same year).

CHABRIAS (KHABRIAS; c. 420-357). An Athenian *strategos* (general) active in the first quarter of the fourth century. An innovative and skilled commander who specialized in **naval warfare** and the use of **peltasts**, Chabrias was *strategos* at least five times from 390/89. Although apparently not from a prominent family, he became part of the aristocratic equestrian set in Athens, marrying the daughter of Menexenos, who had been a **phylarch** in 429. Chabrias also appears to have been quite wealthy, owning the winning four-horse **chariot** team at the Pythian games in 374.

Chabrias' major successes include a successful **ambush** of **Spartans** and allies on Aegina, using 800 peltasts in a hidden position and a force of **hoplites** in the open, and assisting the rebel Egyptian Akoris to resist **Persian** attacks during 385 to 383. He was so successful in **Egypt** that in 379 the Persians officially complained to Athens and he was recalled home. His swift response to the recall suggests that although commanding a **mercenary** force, he was probably at least operating with official Athenian sanction. In 378, again in command of peltasts, he was instrumental in preventing Spartan success in their invasion of **Boeotia**. In 376 he decisively defeated the Spartan fleet under Pollis at **Naxos**, breaking the Spartan blockade of grain imports. He was apparently campaigning in the Hellespont circa 375 and also served as **Iphicrates'** colleague during his very successful expedition to **Corcyra** in 373/2. Chabrias was killed fighting for Athens (but apparently not as *strategos*) at the naval battle of **Chios** in 357. His contemporary, **Xenophon**, records Chabrias' reputation "as a very good general" (*Hell.* 6.2.39). *See also* TIMOTHEUS.

CHAERONEA (CHAIRONEIA). A town northwest of **Thebes**, birthplace of **Plutarch** and the site of at least two major battles:

(1) August 338, during the war between the Greeks and **Philip II** of
Macedon. The battle was fought between a Macedonian army of 30,000
infantry and 2,000 **cavalry** and an Athenian-Theban-**Achaean** force of
about the same size or slightly larger. The coalition force was decisively
defeated, ending Greek hopes of remaining independent from Macedon.

The sources for the battle are slim, but it appears that the fighting
took place on the plain, with each flank of the two armies protected either by the Cephisus (Kephisos) River or a range of hills. Philip probably commanded the right wing of the Macedonians, stationed with its
flanks covered by the hills, while his son, **Alexander the Great**, probably commanded the cavalry on the left, next to the river. The Greeks
were deployed with the Athenians on the left, facing Philip, the Thebans
on the right, facing Alexander, and the Achaeans in the middle. Philip
drew the Athenians forward by means of a sham retreat, dislocating the
Greek line and gaining slightly higher ground for his own troops. He
then had his **phalanx** reverse direction and charge the Athenians, routing them. Alexander destroyed the Thebans opposite (every member of
the **Sacred Band** was killed) and the Achaeans in the middle of the allied Greek line caved in and were pursued from the field. About 1,000
Athenians died and 2,000 were taken prisoner and the Theban and
Achaean losses were heavier. Philip did not pursue far and imposed a
generally mild settlement on Greece. **Thebes** was one of the few places
to have a Macedonian garrison installed and **Athens** lost most of her
remaining naval empire. The lion monument, under which it is believed
the entire Theban Sacred Band was buried (Paus. 9.40.10), can still be
seen at the battle site.

(2) 86, during the First **Mithridatic War**. Fought between a **Roman** army of 30,000 men led by Lucius Cornelius Sulla and a Pontic
army of 90,000 led by **Archelaus**, **Mithridates VI (Eupator)**'s general.
Following his successful siege of **Athens**, Sulla marched north and met
the Pontic army at Chaeronea. Archelaus used his scythed **chariots** and
infantry to pin the Roman center and attempted to use his **cavalry**
against the Roman flanks. Sulla, however, made excellent use of his
mobile reserve to prevent the threat to his flanks and broke Archelaus'
left wing with a counterattack. Archelaus' army was virtually destroyed
in the ensuing rout (Plut., *Sulla*, 17-19). However, he was soon reinforced by a new army, which Sulla was forced to face at nearby **Orchomenus (Boeotia)**.

Plate 1. Hercules fighting Cycnus (500-450 B.C., Kleophrades Painter. London, British Museum, E73. Trustees of the British Museum)

Plate 2. a. Mounted (Scythian) Archer (530-500 B.C., Paseas. Oxford, Ashmolean, 1879.175 [V310]. Ashmolean Museum, Oxford). b. Thracian Cavalryman and Peltast (440-430 B.C., London, British Museum, E482. Trustees of the British Museum)

CHALCEDON (CHALKEDON/KHALKEDON; also CALCHEDON/ KALKHEDON). A **Megarian** colony in the Propontis, on the opposite shore from **Byzantium**. A wealthy trading town, Chalcedon was incorporated into the **Persian** empire around 516 (Hdt. 5.26) but little is known of its history for much of the following century, except that it became part of the **Athenian** empire. In the first stage of the **Peloponnesian War** of 431-404 Chalcedon was still loyal—in 424 the *strategos* (general) Lamachus and his army were able to seek assistance there after their ships were destroyed in a flash flood in an estuary (Thuc. 4.75.2). However, along with other cities in the area, it revolted, probably in 411, and was attacked by a force under **Alcibiades** and **Thrasyllus**.

The Athenians placed pressure on the citizens of Chalcedon by recovering all the property that they had stored for safekeeping with the neighboring **Bithynians** and then constructed a wooden stockade blocking the city off from help on the landward side. The Athenians defeated a concerted attack by Hippocrates, the **Spartan** harmost (governor), who led a **hoplite** force out from the city at the same time as the Persian satrap, **Pharnabazus**, attempted to break in from the outside with a predominantly mounted force. Thrasyllus defeated Hippocrates, with the aid of Alcibiades' **cavalry**, which tipped the balance, and Pharnabazus withdrew after failing to get through the stockade. Chalcedon subsequently came to terms, securing its safety for a cash payment and an agreement to rejoin Athens, paying the tribute at the previous rate, and making up the arrears (Xen. *Hell.* 1.3.2-12; Plut., *Alcibiades*, 29-31).

Chalcedon became part of the **Roman** empire in 74, as part of the bequest of Nicomedes IV (Philopator) of Bithynia and the same year was a focal point of resistance to **Mithridates VI (Eupator)** of Pontus. When Mithridates invaded, the Roman governor withdrew to Chalcedon and was besieged there (Plut., *Lucullus*, 8.2).

CHALCIDIAN CONFEDERACY. Also known as the Chalcidic League. A league formed in the Chalcidice in 432/1, after a revolt against **Athens**. Based on the city of **Olynthus**, the confederacy was active (with some breaks) down to 348, when Olynthus was destroyed by **Philip II** of **Macedon**. In 382 the confederated cities voted for common citizenship with the right to intermarry and own property anywhere within their combined territory. Its army was composed of contingents drawn from the member states. The army had an unknown number of **hoplites** (Xen. *Hell.* 5.2.14 states "at least 800," but the text here is almost certainly corrupt as this number is very low), numerous **peltasts**, and probably 1,000 **cavalry**.

The confederacy began its life in revolt against Athens, defeating attacks at **Spartolus** in 429 and **Amphipolis** in 422. It also had a generally hostile relationship with Macedon, although it apparently allied with Amyntas III of Macedon in 392. It was dissolved in 379 after a two-year campaign by **Sparta**, acting on an appeal from Acanthus and Apollonia concerning Olynthian domination of the league, and a request from Amyntas for the return of territory. By 375 the confederacy had reconstituted itself and joined the Second Athenian Confederacy, but left over Athenian attempts to recover Amphipolis. It allied with Macedon in 357 but was outmaneuvered by Philip II, who destroyed Olynthus in 348 and dissolved the confederacy. The league cavalry seems to have been of quite good quality, being largely responsible for defeats of Athenian expeditionary forces to the region in 429 and 422 and scoring some successes against the Spartans in 382. *See also* COMMON PEACE (1); DERDAS.

CHARES (KHARES; c. 400-325). An Athenian *strategos* (general), active in the second and third quarters of the fourth century; son of Theochares. Chares' success as a *strategos* probably elevated him to the Athenian propertied classes—he held the **trierarchy** around 349/8. In 366 he was in command of **mercenary** troops assisting Phlius against **Sicyon** and other enemies but was recalled to assist the Athenians when Oropus was occupied. He generally took a fairly hard line with **Athens'** allies in the Second Athenian Confederacy, being subject to several complaints about heavy-handed raising of money; however, the Athenians apparently preferred him as a commander because of this (Plut., *Phocion*, 14). In 361/0, he supported an oligarchic coup at **Corcyra**. This sort of action by Chares and other Athenian generals helped contribute to the outbreak of the First **Social War** in 357.

The death of **Chabrias** on the outbreak of this war left Chares as the most experienced general until the arrival of **Timotheus** and **Iphicrates**. Their failure to support Chares at the naval battle of **Embata** led to an Athenian defeat. As a result, Chares was given sole command, but insufficient resources. Short of money, he hired his army out to the satrap Artabazus, who was in revolt against the **Persian** king, Artaxerxes III. Although often seen as acting independently in this, Chares was apparently acting with Athenian knowledge (if not always before the event) and approval, and he regularly sent **booty** home to Athens. Chares was so successful against the royal forces that in 355 Artaxerxes threatened to send 300 ships to help Athens' opponents unless Chares was recalled. Athens complied, ending the Social War.

Following the war, Chares took part in Athens' activities against **Philip II** of **Macedon**, apparently assisting **Onomarchus** during the Third **Sacred War** of 355-346, during which he defeated a force of Philip's mercenaries. He fought at **Chaeronea** (1) in 338 but left Athens after 335. In the 320s he was at the great mercenary center of Taenarum in the Peloponnese but died before 324.

CHARIOT. Chariots were used in Greek warfare prior to the Dark Age, although the relative lack of flat **terrain** in Greece presumably limited their usefulness. An expensive item, they were a product of the palace societies of the Minoan and **Mycenaean** worlds and had disappeared (for military use at least) along with these societies by circa 1100. During this period, the standard chariot was a two-wheeled vehicle, drawn by two or (much more rarely) four horses and crewed by a driver and a warrior. The evidence from **Homer**, and analogy with other chariot-using societies, suggests that the warrior used either a **bow** or a spear from the chariot and may have dismounted to fight when necessary—particularly in the rougher terrain of Greece.

Following the Dark Age, chariots remained the preserve of the very wealthy and were used (except in **Cyprus**) only for transport and sport (the two- and four-horse chariot races were popular events at several Greek games). Although the chariot later reappeared in the armies of those Hellenistic monarchs based in the East, they were the preserve of local troops and did not play a decisive role. Other than this, from the Classical period on, Greek armies occasionally faced chariots fielded by enemies such as the **Carthaginians**, **Persians**, and Indians, but these almost always proved ineffective. The Persian chariots at **Cunaxa** were apparently useless (Xen. *Anab.* 1.8.18-20), and the Carthaginian chariots little better at the **Crimisus River** (Plut., *Timoleon*, 27-28). **Alexander the Great** had little trouble with the Persian chariots at **Gaugamela** or the Indian chariots at the **Hydaspes River** (Arr. *Anab.* 3.13ff. and 5.14ff.). Crimisus and the Hydaspes also demonstrate the problems chariots had on wet ground or in storms. However, they did have some successes—in 395 a party of 500 **foragers** from **Agesilaus'** army were surprised by **Pharnabazus**, who disrupted their hastily formed **phalanx** with two scythed chariots and then attacked them with **cavalry** (Xen. *Hell.* 4.1.17-19). *See also* AGATHOCLES; CHAERONEA (2); EUMENES (2); MAGNESIA.

CHIOS (KHIOS). A large island some five miles (eight kilometers) off the coast of Asia Minor. The main city on the island, which has an excellent harbor, shares the same name. An **Ionian** Greek city, Chios was not immediately incorporated into the **Persian** empire on the fall of Lydia in 546 because the Persians had no navy, but it apparently came to terms, being granted land on the mainland opposite for surrendering Pactyes, a Lydian fugitive (Hdt. 1.160). Chios joined the **Ionian Revolt** in 499, contributing 100 ships, which fought very well at **Lade** (1), but after the battle it was captured easily and sacked (Hdt. 6.8, 15, 31).

Following the Second **Persian War**, Chios joined the **Delian League**, from which it withdrew with the support of **Tissaphernes** in 412 after the failure of **Athens' Sicilian Expedition**. At that time it was one of the few league members that had maintained its original position of contributing ships rather than money to the league fleet. Chios was an important part of the empire and the Athenians reacted quickly, sending a fleet, landing troops, and **ravaging** the island—which had been untouched since the end of the Ionian Revolt—but they were unable to take the city. The Chians withstood a siege and in 411 got the better of the Athenian fleet stationed there (Thuc. 8.5-63). The city and its harbor provided valuable service to **Sparta** as a naval base for the remainder of the war (Xen. *Hell.* 1.1.32; 1.6.1, 12; 2.1.1-7).

In the fourth century, Chios joined the Second Athenian Confederacy, but fought the Athenians in the **Social War** of 357-355. In one of the first engagements of the war, the Athenians were defeated and one of their best generals, **Chabrias**, was killed in a land and sea battle near Chios. Much of the rest of Chios' history is obscure, but the Chians appear to have maintained a fairly independent position. In 208, for example, they sent envoys (along with other states) to mediate a peace between **Philip V** of **Macedon** and the **Aetolian League** (Polyb. 5.24, 5.28).

Chios had friendly relations with **Rome**, in 190 acting as a naval base for their war against **Antiochus III (the Great)** (Livy 37.27, 37.31, 38.39). The island was sacked and depopulated by **Mithridates VI (Eupator)** in 86, but two years later the population was restored under Lucius Cornelius Sulla and the island declared a free ally of Rome.

In addition to the battles at Chios which involved Chians, a major naval engagement was fought near the island in 201/0 between Philip V of Macedon and the combined fleets of Pergamum, **Rhodes**, and **Byzantium**. **Polybius** (16.2-7) provides a detailed description of the engagement, which ended rather indecisively. Philip, with a total fleet of 150, was trying to outrun the enemy fleet of 65 decked ships, nine *trihemioliae*, and three **triremes** (plus an unknown number from Byzantium). The engagement was essentially fought as two actions—one against the Rhodian ships, the other against **Attalus I (Soter)** of Pergamum. Philip had the worse of it until Attalus pursued too far and was cut off. Attalus was forced to beach his flagship and escort and narrowly escaped on foot. Seeing Philip in possession of Attalus' ship, and presuming him dead, the allies withdrew.

Philip lost several large ships (a ten, a nine, a seven, and a six), 20 other decked vessels, three *trihemioliae*, and around 65 galleys. Attalus had three fours (quadriremes) captured and two fives (quinqueremes) and one *trihemiolia* sunk. The Rhodians lost two fives and a trireme, with no ships captured. Polybius' casualty figures seem rather lopsided, given the ship casualties—130 allies dead and 600 captured, with 3,000 Macedonians and 6,000 rowers killed and 2,000 captured. Livy (32.33) records that in 197 Attalus was still seeking the return of the men and ships captured in the battle. Despite suffering the greater losses, Philip was left in possession of Attalus' flagship, and the area where the wrecks, dead, and wounded were and claimed victory. However, he refused to meet an enemy challenge the next day. *See also* WARSHIP.

CHREMONIDEAN WAR (268/7-262/1). A war fought by a coalition of Greek states led by **Athens** and **Sparta**, with support from **Ptolemy II (Philadelphus)**, against **Antigonus II (Gonatas)** of **Macedon**. It derives its name from Chremonides, the Athenian who moved the decree in autumn of 268/7 that led to war. In addition to Athens and Sparta, the Greek alliance included the **Achaean League**, **Elis**, Tegea, **Mantinea**, and various other cities, including some on Crete. The alliance consciously recalled the previous struggles of Greece, led by Athens and Sparta, against **Persia**. The motives for Ptolemy's involvement are uncertain and he provided at best half-hearted assistance to his allies—who prosecuted the war with considerable vigor. The Macedonian garrisons in **Corinth** and the Isthmus of Corinth effectively prevented Athens and Sparta from linking up and allowed Antigonus to direct attacks on Athens from both land and sea. In 265, during his third attempt to break through, **Areus I** of Sparta was killed near Corinth. Athens held out for some time, but in 262/1 was forced to surrender and accept a Macedonian garrison. Sometime toward the end of the war **Alexander II of Molossia** took the opportunity to invade Macedon while Antigonus' attention was focused elsewhere. However, Alexander was decisively defeated and temporarily went into exile. In 261 Antigonus may have sealed his success on mainland Greece with a naval victory over Ptolemy off Cos (although this could have occurred around 258, during the Second **Syrian War**). Although Antigonus' position in central and southern Greece was strengthened, the **Aetolian League** was able to take advantage of the war to extend its territory.

CIMON (KIMON; c. 510-449). An Athenian statesman and soldier, prominent in the creation of the **Delian League** and the Athenian maritime empire following the **Persian Wars**. Cimon was the son of **Miltiades** and a **Thracian** princess, Hegesipyle. As a young man in 480 he encouraged his fellow citizens to follow **Themistocles'** plan of abandoning **Athens** and fighting at sea. From 479/8 onward, Cimon was frequently elected *strategos* (general) and took part in the postwar operations in the Hellespont area. Along with **Aristides**, he apparently played a major part in encouraging the Greeks of the islands and Asia Minor to transfer their allegiance from **Sparta** to Athens, capitalizing on the arrogant behavior of **Pausanias** (1).

Cimon was very successful in the early Delian League operations to clear the **Persians** and pirates from areas in the northern Aegean (476-473). In a campaign that culminated in the battle of **Eurymedon** (c. 467/6), he destroyed Persian naval power in the Aegean and considerably extended membership of the Delian League. Shortly afterward, he cleared the Chersonese of Persians and after a two-year siege (465-463) captured Thasos and forced it back into the Delian League as a subject member.

On his return from Thasos, Cimon survived a political trial and in 462 led a force of **hoplites** to assist the Spartans in besieging the helots on Mount **Ithôme**. While he was away, his political opponents, Ephialtes and **Pericles**, began to undermine his position and, when the Spartans dismissed the Athenians from Ithôme early, his influence at Athens was effectively ended. Ostracized (exiled) in 461, Cimon was refused permission to fight on the Athenian side against Sparta at **Tanagra** in 457 but was recalled circa 450 to arrange the Five Years' Peace with Sparta. In 450/49 he led a fleet of 200 ships to a great victory over the Persians in **Cyprus** but died during the siege of Citium.

Cimon was not renowned for his intelligence and in his youth had a reputation as a hard drinker. He was known for a Spartan rather than an Athenian demeanor and maintained his influence at Athens through his very successful soldiering and a large fortune which he used to gain favor with his fellow citizens. He was the architect of the "yokefellow policy," which envisaged Sparta, the dominant land power, and Athens, the dominant sea power, as yokefellows, together defending Greece. This policy, opposed by Pericles, became increasingly tenuous with the rise in Athenian power and Spartan hostility to this and died with Cimon's dismissal from Ithôme.

CLEARCHUS (KLEARKHOS; c. 450-401). A Spartan **officer** placed in charge of the defense of **Chalcedon** and **Byzantium** in 409 during the **Peloponnesian War** (2) of 431-404. He lost Byzantium in 408 because of internal opposition arising from his unequal distribution of food during the **Athenian** siege of the city. In 403 he refused a recall to **Sparta** and went into exile under sentence of death. He took service with **Cyrus the Younger**, helping to recruit the **Ten Thousand**, the **mercenary** force Cyrus used to try to dethrone Artaxerxes II. Although not formally commander in chief of the Ten Thousand, he was regarded as the senior general. After Cyrus' death at **Cunaxa**, Clearchus was seized during a conference with **Tissaphernes** and executed.

CLEITUS (KLEITOS).

(1) Cleitus (c. 380-328). Nicknamed "the Black." A **Macedonian** noble with a distinguished military career under **Philip II**. He commanded the Royal Squadron of the Companion **cavalry** under **Alexander the Great**, saving Alexander's life at the battle of **Granicus River** (Arr. *Anab*. 1.15.8). He was still in command of the Royal Squadron at **Gaugamela** but was appointed to command half of the Companion cavalry in Alexander's reorganization after the Pages' Conspiracy and the execution of **Parmenio**. In 328 he was appointed governor of Bactria-Sogdiana, but the night before he was to leave to take up his appointment he was murdered by Alexander in a drunken quarrel. The accounts of this (Arr. *Anab*. 4.8-9; Plutarch, *Alexander*, 50-52; Curt. Ruf. 8.1.20-2.12) differ, but it appears that Cleitus objected to the increasing cult of personality around Alexander and the accompanying lack of regard for the accomplishments of Philip II and the older members of the army.

(2) Cleitus (d. 318). Nicknamed "the White." A **Macedonian** noble who served under **Alexander the Great** during the conquest of the **Persian** empire. An infantry commander at the start of the Indian campaign and at the **Hydaspes River**, he later served as a **cavalry** commander. Following Alexander's death, Cleitus was **Antipater**'s admiral during the **Lamian War** and performed well in this role. Concentrating the Macedonian fleet, in a series of battles he defeated the various Greek fleets in detail, making a major contribution to Macedonian victory in the war. In 318 he commanded Polyperchon's fleet in the Propontis during the war with **Cassander**, cutting off supplies to Cassander's fleet in the Piraeus and preventing the movement of enemy troops from Asia into Europe. Cleitus won an important naval victory over Cassander's fleet, commanded by Nicanor and supported by **Antigonus I (Monophthalmus)**, destroying at least 17 ships and capturing 40. However, the morning after the battle, Antigonus launched a surprise dawn attack on Cleitus' camp using light troops on land, followed by a naval attack, and captured or destroyed his entire fleet. Cleitus escaped in his flag-

ship, but when he landed in **Thrace** he was captured and killed by **Lysimachus'** forces (Diod. Sic. 18.72).

(3) Cleitus (fl. 335). Son of **Bardylis I**, king of the Dardanians. Cleitus was clearly a subject king when he rebelled against **Alexander the Great** in 335. Alexander penned Cleitus inside Pelium but found himself caught between Cleitus and his Taulantian ally Glaucias. Alexander overawed their troops by a precision drill display and withdrew, beating off an attack upon his rear guard by the fire from his **artillery** and **archers**. Shortly afterward, Alexander made a surprise night attack on the enemy camp, causing considerable casualties, and Cleitus sought refuge with Glaucias (Arr. *Anab.* 1.5-7). At some later date, Cleitus apparently submitted to Alexander and was left as a subject ruler in Dardania. His daughter(?) married **Pyrrhus I** of Epirus.

CLEOMENES (KLEOMENES). The name of several Spartan kings of the Agiad family. The most important militarily are:

(1) Cleomenes I (c. 550-490). King, c. 520-490. He conducted an aggressive foreign policy designed to strengthen the position of **Sparta** and the **Peloponnesian League**, principally by extending the league and weakening **Argos**. He was responsible for suggesting that **Plataea** ally with **Athens** circa 519 and in 510 expelled the tyrant **Hippias** from Athens. His attempt to interfere in Athenian affairs again two years later ended in embarrassment when he and his small number of troops were besieged on the Acropolis. They were allowed to withdraw under truce and Cleomenes made two further attempts to invade Athens (c. 506 and 504). Both of these were blocked by opposition from within the Peloponnesian League (principally from **Corinth**) and from his fellow king, **Demaratus**. In 494 Cleomenes defeated Argos in the decisive battle of **Sepia**, killing many of the Argive survivors by setting fire to the sacred grove in which they had taken sanctuary.

Cleomenes was also involved in the lead up to the **Persian** invasion of Greece in 490. In 499 he refused Spartan assistance to the Greeks of **Ionia** during the **Ionian Revolt** (on the grounds that a land campaign in Asia Minor would take the small Spartan army too far inland). However, in 491 he agreed to an Athenian request to intervene on Aegina to prevent the island from **medizing**. This failed, in part because of a further disagreement with Demaratus, whose opposition was silenced after Cleomenes and the future **Leotychides II** arranged his removal from the throne. When Cleomenes' part in this (which involved bribing a priest at **Delphi**) was discovered, he fled to **Arcadia** and began to stir up unrest against Sparta. Worried by this, the Spartans recalled him. Shortly after his recall home, Cleomenes became deranged and killed himself while being restrained in the stocks.

(2) Cleomenes III (c. 260-219). King, c. 235-222. Cleomenes was very active from 229, the year he occupied Tegea, **Mantinea**, and **Orchomenus (Arcadia)**. In 228 he built and garrisoned a fort in the territory of Megalopolis, leading to the **Cleomenic War** with the **Achaean League**. Around the same time Cleomenes revolutionized **Spartan** society by returning to the Lycurgan constitution. This involved a program similar to the one which **Agis IV** had failed to implement and included canceling debts and redistributing land. This led to a fear of social revolution elsewhere and united the propertied classes of the Peloponnese against Cleomenes.

During the Cleomenic War, Cleomenes was initially very successful against the **Achaeans**, winning several major victories. However, the tide turned against him when **Antigonus III (Doson)** allied with the Achaeans in 224 and Ptolemy III (Euergetes) withdrew essential financial support. In 222 Cleomenes was decisively defeated by the **Macedonians** and Achaeans at the battle of **Sellasia**. He fled to **Egypt** but was imprisoned on the accession of **Ptolemy IV (Philopator)**. In 219 he escaped but committed suicide when his attempt to incite a revolt in Alexandria failed.

CLEOMENIC WAR. A war fought from 228 to 222 between **Sparta**, led by **Cleomenes III** and supported by the **Aetolian League** and later Ptolemy III (Euergetes), and the **Achaean League**, led by **Aratus** and allied (from 224) with **Macedon** under **Antigonus III (Doson)**. In 229 the Aetolian League ceded several states, including Tegea, **Mantinea**, and **Orchomenus (Arcadia)** to the Spartans and seems to have formed some sort of alliance with them, although they played no real part in the war itself. Both Cleomenes and the Aetolians were acting partly from concern over **Achaean** successes in the Peloponnese, but Cleomenes also had plans to restore Sparta to its former glory, while the Aetolians also wanted to weaken a rival. In 228 Cleomenes fortified part of Megalopolis' territory and the Achaean League declared war.

Cleomenes won several fairly rapid victories over the Achaeans at the encounter battle of Lycaeum and at Ladocaea (where **Lydiadas** was killed) in 227 and at Hecatombaeum in 226. This secured the fall of a string of cities (including Mantinea, **Argos**, **Corinth**, Phlius, and Epidaurus) and gained Sparta the support of Ptolemy. However, in 224 the Achaean general Aratus (now with emergency powers and effectively in sole command of the League) secured the assistance of Antigonus by ceding him Acrocorinth (the citadel of Corinth). Cleomenes was forced to withdraw from Corinth when the Macedonian army advanced from the north and the Achaeans recaptured Argos. Even worse, Antigonus ceded some territory in Asia Minor to Ptolemy, who cut off his crucial financial support to Cleomenes. Nevertheless, in 223, despite a string of Macedonian successes which resulted in the capture of Tegea, Mantinea, and Orchomenus (Arcadia), Cleomenes succeeded in taking and destroying Megalopolis. The war ended in 222 when, desperately short of money, Cleomenes attempted to win a final decisive battle at **Sellasia** but was heavily defeated and fled to **Egypt**.

CLEON (KLEON; d. 422). An **Athenian** politician and *strategos* (general) during the **Peloponnesian War** (2) of 431-404. He became prominent after **Pericles'** death, and was associated with the change from Pericles' island policy to a more aggressive strategy. An advocate of harsh measures against disobedient members of the Athenian empire, he had motions passed to execute the males and sell the women and children as slaves when Mytilene surrendered after an unsuccessful revolt in 427 and when Scione revolted in 423. The Mytilene motion was rescinded, but the Scione motion was implemented. Cleon led light troops to **Pylos** in 425 and joined with **Demosthenes** (1) in the successful action on Sphacteria. He was also responsible for the Athenian rejection of the **Spartan** peace proposals which followed—in hindsight a mistake, but an understandable decision at the time. In 424 he was probably involved in the Athenian plan to neutralize **Boeotia** by land that failed at **Delium**. In 422 he led an expedition to the Chalcidice to recover rebellious cities there. After some minor successes he was killed in the battle of **Amphipolis**, apparently because of a basic error in handling his **hoplite phalanx**. The literary tradition is generally hostile to him, as both **Thucydides** and the comic poet Aristophanes, the two main sources, disliked Cleon. *See also* BRASIDAS.

CLEOPATRA VII (PHILOPATOR) (69-30). Last of the independent (or more accurately, semi-independent) Hellenistic monarchs, Cleopatra ruled **Egypt** from 51 to 30. This rule was initially jointly with her brother, Ptolemy XIII, but in 48 Cleopatra fled to **Syria** to avoid assassination and returned at the head of an army. It was while she was facing off against her brother's army (led by Achillas) that Julius Caesar arrived on the scene and gave his backing to Cleopatra. Cleopatra was with Caesar throughout the bitterly fought siege of Alexandria (winter 48/7) during which Ptolemy was killed. When reinforcements arrived, Caesar was quickly able to put down the opposition, placing Cleopatra firmly on the throne, with her other younger brother, Ptolemy XIV, as her husband

and joint ruler. He also restored **Cyprus** to Egyptian rule. Cleopatra joined Caesar in **Rome** in 46 but returned home after his assassination in 44, making her son by Julius Caesar coregent when Ptolemy XIV died (reputedly poisoned on her orders).

In 41 Antony (Marcus Antonius) summoned Cleopatra to Tarsus, apparently intent on boosting his treasury at Egypt's expense. However, Cleopatra headed this threat off by arriving with huge pomp and ceremony and dazzling Antony with the scale of her entertainment. Cleopatra's relationship with Antony led to large additions to Egyptian territory, including Cilicia, Coele-Syria, and parts of **Palestine** (but not Judaea). In 34 he allocated the whole of Asia Minor between her and her children.

Not surprisingly this increasingly alienated the Romans in the West, and in 33/2 led to war with Octavian (Augustus). Octavian was careful to maintain that he was at war with Egypt, not engaged in a civil war with Antony—his testament, the *Res Gestae*, mentions only his victory over Egypt. The main theater of the war was Greece. Antony had deployed most of his and Cleopatra's forces there, but late in 32 failed to prevent Agrippa, Octavian's admiral, from landing in Epirus. At the naval battle of **Actium** in late 31, both Antony and Cleopatra fled before the fighting was over. Octavian's pursuit was very leisurely—almost a year after Actium he reached Alexandria, having invaded from both Syria and Africa in the interim. Antony's troops immediately deserted to Octavian and Antony and Cleopatra committed suicide.

Our picture of Cleopatra stems largely from hostile Roman accounts, but she was extremely popular with her own people. Although Egypt became part of the Roman empire under her rule, it seems likely that this would have occurred even earlier without Cleopatra's skilled diplomacy. In terms of Hellenistic dynastic politics, what she did was eminently sensible but did not fully take into account Octavian's abilities compared to Antony's or the basic hostility of the Romans to what they regarded as unacceptable eastern and royal influence.

CLERUCHY. A colony comprised of men of military age, widely used by **Athens** during the fifth and fourth centuries to provide military security to trouble spots in areas of Athenian interest or control. The earliest was Chalcis on **Euboea** in 506, and they varied in size from the 4,000 there to the 250 established at Andros around the middle of the fifth century. Others include **Naxos**, Samos, **Potidaea**, the Chersonese, Melos, Lemnos, Imbros, and Scyrus (the last three occupied fairly continuously down to **Roman** times). Often imposed as a punishment after revolts, cleruchies were an effective method of establishing self-supporting garrisons in the empire. The settlers, or cleruchs, were Athenians of military age who retained their Athenian citizenship. They were maintained at, or elevated to, the **hoplite** class by their grant of land and were liable for both military service and the *eisphora*, or war tax. Cleruchies were extremely unpopular with members of the fifth-century empire and were forcibly disbanded at the end of the **Peloponnesian War** of 431-404. They were specifically prohibited under the terms of the Second Athenian Confederacy, but Athens did reintroduce them (in nonmember states) in the fourth century. This practice may well have helped spark the **Social War** of 357-55. Hellenistic monarchies also made use of cleruchies in Asia Minor in the second and third centuries, not only for security purposes but also to spread Greek culture and influence.

CNIDUS (KNIDOS). A Greek colony in southwest Caria. Traditionally founded by **Spartans** (Paus. 10.11), it was one of the cities that joined together to form the confederation known as the Dorian Hexapolis (Cnidus, Camirus, Cos, **Halicarnassus** [later excluded], Ialysus, and Lindus). Cnidus was strategically located on a naturally strong defensive position and had a strong maritime tradition. Following the fall of Lydia in 546, the Cnidians surrendered to the **Persians** after failing in an attempt to turn their peninsula into an island (Hdt. 1.174) and became part of their empire. After the Second **Persian War**, the city joined the **Delian League**, but revolted in 412/11 with assistance from **Tissaphernes**, during the general unrest after the failure of **Athens' Sicilian Expedition**. The Athenians made an unsuccessful attempt to recapture the city, which at that stage was unwalled, but failed (Thuc. 8.35). Cnidus was probably under the control of Ptolemaic **Egypt** in the third century, but passed under the influence of **Rhodes** at the start of the second. It assisted **Rome** against **Antiochus III (the Great)** in 190 (Livy, 37.16), later becoming a free city in the Roman province of Asia.

Cnidus was the site of at least one major sea battle, described by **Xenophon** (*Hell.* 4.3.10-12) and **Diodorus Siculus** (14.83.4-7). It was an encounter battle fought in 394 between a combined Persian-Greek fleet of more than 90 **triremes** under **Pharnabazus** and **Conon** and a Spartan fleet of 85 triremes under Peisander. Pharnabazus, leading the Phoenician ships, was supreme commander, while Conon led the Greek (predominantly Athenian) contingent, posted in front of the Phoenicians. Although outnumbered, Peisander deployed for battle against Conon's ships. His left wing, composed of allied ships, fled the battle early when the Phoenicians came up in support of Conon, and Peisander's trireme was driven ashore and he was killed. Conon captured 50 enemy ships and about 500 sailors. This battle ended Spartan naval supremacy in the

Aegean and led to the resurgence of both Athenian and Persian naval power.

COELE-SYRIA. *See* SYRIA

COMMON PEACE (*KOINE EIRENE*). An ideal of peace among all Greeks and the name given to a series of treaties made in the fourth century between the major Greek powers, often backed by **Persia**, and regarded as binding on all Greek states. In theory designed to ensure the autonomy of Greek states, they were often used by whichever power was dominant to impose its will on the other states. The treaties were:

(1) The King's Peace (Peace of **Antalcidas**), 386. Initiated by **Sparta** and supported by **Persia**, this treaty prescribed that all of Asia Minor (including the Greek cities there), Clazomenae, and **Cyprus** would belong to Persia. All other Greek states were to be autonomous, except Lemnos, Imbros, and Scyrus, which were recognized as Athenian. Persia guaranteed to enforce the peace against aggressors (Xen. *Hell.* 5.1.31). Sparta used this peace to justify action to dissolve the **Chalcidian Confederacy** and the union between **Argos** and **Corinth**, and to reduce Theban domination of **Boeotia**. *See also* CORINTHIAN WAR.

(2) Restatement of the King's Peace, 371. Initiated by **Persia**, which needed Greek assistance to quell a revolt in **Egypt**, and supported by **Sparta** and **Athens**, this agreement required that governors be withdrawn from cities, all naval and military forces be demobilized, and all Greek cities be free and allowed any Greek state to go to the aid of an injured party. **Thebes** was not allowed to sign for **Boeotia** and therefore refused to sign. This refusal led the Spartans to invade, resulting in their decisive defeat at **Leuctra** and ending their dominance over Greece. In the winter of 371/0 Athens secured general agreement to maintain the King's Peace, offering other states a general alliance with Athens, which undertook to guarantee the peace. Thebes was again excluded, but many other states (Sparta and **Elis** were exceptions) did sign.

(3) Peace of **Pelopidas**, 367 (366/5). Initiated by Spartan and Athenian embassies to **Persia** which were outmaneuvered by a **Theban** embassy led by Pelopidas in 367, this treaty recognized the independence of **Messenia** from **Sparta**, **Athens'** right to the Chersonese, and Persian control over Asia Minor; required the demobilization of the Athenian navy; and made the Greek states autonomous. Rejected by most states in 367, it may have been partially accepted in 366/5—although Sparta maintained its refusal to sign because of the Messenian autonomy clause.

(4) Peace of 362/1. This followed the **Theban** victory at **Mantinea** (2) and formed a general alliance of cities (except **Sparta**, which again refused to recognize **Messenia**'s independence). The signatories agreed to observe general peace, to defend each other against aggressors, and to settle disputes by negotiation not war.

(5) Hellenic League Peace, 337. Imposed by **Philip II** of **Macedon** after his creation of the **Hellenic League** (2) following the battle of **Chaeronea** (1). The signatories agreed to observe a general peace, use collective military force against any violator, and recognize the right of Greek states to freedom and autonomy.

CONON (KONON; c. 444-392). Athenian *strategos* (general). From an aristocratic family, Conon served as an admiral in the second half of the **Peloponnesian War** of 431-404. He first appears in charge of 18 ships at Naupactus in 414/3 (Thuc. 7.31.4-5). In 406, following the Athenian defeat at **Notium**, Conon was sent out to replace **Alcibiades** in command of the fleet. He combined the existing crews to fully man 70 **triremes** (rather than the 100 undermanned ships he found on arrival) and ravaged the **Ionian** coastline (Xen. *Hell.* 1.5.16-21). Later in the same year he was attacked by **Callicratidas** with a fleet of 170 ships and was forced to make a run for Mytilene, losing 30 of his 70 ships (but not their crews) in the harbor there. Conon was blockaded by both land and sea, but this was lifted when he managed to send for help and the Athenian relief force defeated Callicratidas at **Arginusae**. In 405, Conon manage to escape the Athenian disaster at **Aegospotami** with nine ships—all the remainder were captured. He seized **Lysander**'s cruising masts on a nearby headland (triremes went into action without their masts and sails) and sailed for **Cyprus** to join King **Evagoras** (Xen. *Hell.* 2.1.25-29)—presumably worried about his reception back home as the only surviving admiral from the defeat that cost Athens the war.

Following the Peloponnesian War, Conon served with the **Persians** and, along with **Pharnabazus**, won a decisive victory against **Sparta** at **Cnidus** in 394. This defeat, and Conon's subsequent drive up the coast and in the islands expelling the Spartan harmosts (governors), ended Sparta's naval supremacy and seriously weakened her power in Asia Minor. In 393 Conon and Pharnabazus ravaged the Spartan coast and garrisoned Cythera as a base against Sparta. Conon then took the fleet, with Pharnabazus' permission, to Athens where he began to rebuild the Long Walls and bring the islands and Ionian cities into alliance with Athens (Xen. *Hell.* 4.3.1-12, 8.1-16). Shortly afterward, Spartan pressure led to the arrest of Conon while he was on a mission to Persia and he died soon after escaping. Conon was an able admiral and made a major contribution to Athens' recovery in the early fourth century.

CORCYRA (KERKYRA). A strategically important island (modern-day Corfu) in the Ionian Sea off the west coast of Epirus; the main city on the island also bears the name Corcyra. It was colonized first by **Eretrians** from **Euboea** who were ousted and replaced in the eighth century by Corinthians. However, Corcyra fairly soon established its independence from **Corinth**—fighting in the first quarter of the seventh century what **Thucydides** (1.13.4) claimed was the earliest known naval battle between Greek states. Relations between the two cities had improved by the late seventh century, when Corcyra founded its own colony, Epidamnus, and included Corinthian colonists in the venture and a Corinthian as the *oikistes*, or head of the colony (Thuc. 1.24.1-2). Corcyra was temporarily subject to Corinth again during the reign of Periander (625-585), who treated it fairly harshly (Hdt. 3.48-53), but apparently regained its independence after his death.

Corcyra was predominantly a naval power, maintaining a strong fleet on the basis of its strategic position on the routes between Greece and **Italy** and **Sicily**. It was a relatively wealthy state because of the fertility of the island as well as the money deriving from commercial activity. It was rivalry over trade that had probably originally brought Corinth and Corcyra into conflict. Despite its apparent naval power, it never quite lived up to its potential—particularly after the internal dissension in the city in the 420s, which seems to have considerably weakened it.

During the Second **Persian War** Corcyra promised a fleet of 60 **triremes** to assist the Greek resistance, but they did not arrive in time to participate in the fighting. Whatever the reason for this—and the excuse of contrary winds may have been valid—**Herodotus** (7.168) believed that the fleet had orders to sail to Taenarum in the Peloponnese, await the results of the invasion, and then join the winning side.

Corcyra's next appearance in Greek military history was as a major participant in the outbreak of the **Peloponnesian War** of 431-404. An internal quarrel in Epidamnus led to a request for assistance to Corcyra and, when this was rejected, to Corinth. The Corinthians, apparently eager to increase their own power at the expense of Corcyra, accepted the Epidamnian offer to hand over their city to them and agreed to help. Corcyra, regarding this as interference in its sphere of influence, besieged Epidamnus. On the outbreak of this war in 435, Corcyra could muster a fleet of 120 triremes, and with 80 of them decisively defeated a 75-strong Corinthian fleet at Leucimme. On the same day Epidamnus fell, and Corcyra now held naval supremacy in the West (Thuc. 1.24-30).

The Corinthians immediately began preparations for a major naval expedition, calling on their allies in the **Peloponnesian League**, and Corcyra, which was not a member of either the Peloponnesian or the **Delian League**, in response sought an alliance with **Athens**. This was granted, somewhat reluctantly, by the Athenians on the basis that war was likely to soon come with the Peloponnesians anyway and the Corcyraean fleet would be a very useful addition to Athens' strength. The Athenian ships sent to assist Corcyra in 433 became involved in the inconclusive battle of **Sybota**, which seriously damaged Athenian relations with Corinth and contributed to the outbreak of war between Athens and the Peloponnesian League in 431 (Thuc. 1.31-55).

During the Peloponnesian War Corcyra provided 50 ships for the Athenian **raid** on the Peloponnese in 431 (Thuc. 2.25) but, rather ironically, was unable to provide much other assistance to her new ally because of a civil war between aristocrats and democrats that broke out in 427 (Thuc. 3.69-85). It was able to supply 15 triremes to assist **Demosthenes** (1) in Acarnania but otherwise was generally limited to providing occasional reinforcements and a safe harbor and support base to Athenian fleets—even after the destruction of the aristocratic party in 422 (Thuc. 3.94; 4.46-48; 6.27, 32; 7.26). In 410 the internal strife broke out again, with the upper classes favoring **Sparta** and the democrats Athens. The latter gained the upper hand, with Athenian support, and staged a massacre of their opponents, although the survivors later reconciled (Diod. Sic. 13.48).

Corcyra's history immediately after the Peloponnesian War is unknown, but in 375 **Timotheus** restored Athenian domination over the island, which joined the Second Athenian Confederacy (Xen. *Hell.* 5.4.64). The following year the island and the city suffered very badly from a Spartan-led attack and siege. However, the arrival of 600 **peltasts** under Athenian command to assist and the overconfidence of the Spartan commander (and his failure to treat his **mercenaries** properly) allowed the desperate Corcyraeans to launch a successful attack. They defeated the enemy army and almost captured the camp, which was soon evacuated when news reached the island of an Athenian relief expedition under **Iphicrates**. The Athenians had reacted swiftly to the Spartan attack, persuaded by the Corcyraean argument that the island was still the second strongest Greek naval power after Athens and a very valuable ally (Xen. *Hell.* 6.2.1-26; Diod. Sic. 15.47).

From around 360 Corcyra appears to have been independent from Athens, but in 340 joined the Athenians in an (unsuccessful) attempt to prevent **Macedonian** expansion into its region. After **Alexander the Great**'s death, Corcyra enjoyed brief periods of independence, interspersed by longer periods under the control of various rulers. In 312 it managed to eject the Macedonians from Epidamnus and Apollonia, but nine years later the island fell under the temporary control of the Spartan royal adventurer, Cleonymus (Diod. Sic. 19.78.1, 89.3; 20.104-105). In 300, apparently independent again, Corcyra was attacked by **Cassander**, who was only beaten off with the aid of **Agathocles**, tyrant of **Syracuse**, who promptly seized the island for himself. Around 295 Agathocles gave Corcyra as a dowry to **Pyrrhus I** of Epirus when he married Agathocles' daughter, Lanassa. A few years later in 290, Lanassa invited in **Demetrius I (Poliorcetes)**, who occupied Corcyra with her consent (Diod. Sic. 21.2-4; Plut., *Pyrrhus*, 9-10).

Around 274 Pyrrhus recaptured Corcyra and held it until his death, when it again became independent. This period of independence lasted until 229 when the Illyrians under Queen Teuta seized the island—despite the intervention of an **Achaean** naval relief force, which was defeated by Illyrian tactics and the superiority of the Illyrian marines. However, the same year, her garrison commander, Demetrius of Pharos, surrendered it without a fight to a **Roman** fleet (Polyb. 2.9-11). From this point on, Corcyra was part of the Roman empire and was used first as a naval station and then also as a base for the Roman operations against Greece in the second century (Diod. Sic. 31.11.1; Polyb. 21.32).

CORINTH (KORINTHOS). A large Dorian city, located at the southern end of the Isthmus of Corinth, controlling the land route between the Peloponnese and the rest of Greece. Although its agricultural land was poor, Corinth was from early times a powerful maritime trading state and is referred to in both **Homer** and Pindar as "wealthy Corinth." Corinth possessed two ports: **Lechaeum** on the Gulf of Corinth and Cenchreae on the Saronic Gulf. Corinth was a major colonizing power in the eighth and seventh centuries, founding **Corcyra**, **Potidaea**, and **Syracuse** and a string of other colonies at strategic locations on the sea routes to the west. The first **pentecontors** and **triremes** built in Greece were constructed at Corinth, which was the major commercial and naval power in Greece until eclipsed by **Athens** in the fifth century.

Corinth was a member of the **Peloponnesian League** from the sixth century, although occasionally taking an independent line—around 506 it withdrew its forces from a **Spartan** expedition against Athens. In the early fifth century Corinth continued its generally pro-Athenian policy, supplying ships to Athens during its war with Aegina. In 481, Corinth hosted the meeting of Greek states to plan resistance to the forthcoming **Persian** invasion and provided 40 triremes to the Greek fleet during the Second **Persian War**. According to **Herodotus** (8.1, 44-48), this was twice the size of the next largest contingent, although less than one-third of Athens' contribution.

Following the Persian Wars, Corinth became increasingly concerned with the growth in Athenian naval and commercial power, and hostilities erupted in 459/8 when Corinth's neighbor **Megara** defected from the Peloponnesian League and allied with Athens. This led to the First **Peloponnesian War**, in which Corinth played a major role—despite a poor start in 458 when a Corinthian invasion of Megara was defeated by a scratch Athenian army of veterans and teenagers under Myronides. Corinth was one of the allies that helped pressure Sparta into a declaration of war in 431—largely because of Athenian support to Corcyra in 433. Corinth provided the backbone of the small and outclassed Peloponnesian flcct during the Peloponnesian War (2) until Persian funding allowed for a massive expansion during the Decelean War. Corinth, along with **Thebes**, refused to accept the Peace of **Nicias**, instead signing a truce with Athens that had to be renewed every seven days. At the end of the war, Corinth and Thebes unsuccessfully demanded the destruction of Athens.

Early in the fourth century, Corinth joined the anti-Spartan movement and during the **Corinthian War** of 395-387 fought against its old ally. Much of this war revolved around the attempts of both sides to secure control of the Isthmus of Corinth and several engagements occurred in the area, including the destruction of a Spartan infantry *mora* at Lechaeum in 390. During this war Corinth became a democracy and temporarily merged with **Argos**—although this union was dissolved by Sparta under the provisions of the King's Peace (**Common Peace** [1]) of 386 and Corinth rejoined the Peloponnesian League.

Corinth remained loyal to Sparta during its struggle against Thebes but was allowed to make a separate peace with Thebes in 366 because of the losses it had suffered (Xen. *Hell.* 7.4.6ff.). This event effectively marked the end of the Peloponnesian League. Corinth generally supported the anti-Macedonian cause in Greece during the middle of the century, although distracted by providing assistance to allies in **Sicily** under **Timoleon** from 344 onward. Following the Greek defeat at **Chaeronea** in 338, Corinth's citadel, the Acrocorinth, was garrisoned by **Macedon**. From this point on, the Acrocorinth was regularly garrisoned by whichever power was dominant, earning it the dubious title of one of the "fetters of Greece."

From 243 to 223 Corinth was a member of the **Achaean League** and had a league garrison in the Acrocorinth. However, the league handed the Acrocorinth over to **Antigonus III (Doson)** to gain his support during the **Cleomenic War**. It was retained by his successor **Philip V**. After **Cynoscephalae** (2) in 197, the Romans declared Corinth free and it rejoined the Achaean League. In 146, during **Rome**'s war with the Achaean League, the city was sacked and destroyed by the Romans. It was subsequently rebuilt by Julius Caesar and later became the capital of the Roman province of Achaea.

CORINTHIAN WAR. A war fought from 395 to 387 between the **Spartan** alliance and a coalition that included **Athens**, **Argos**, **Boeotia**, and **Corinth**, attempting to end Spartan hegemony over Greece. The war began when the **Boeotian League**, led by **Thebes**, took advantage of Sparta's involvement against **Persia** in Asia Minor to ally with **Locris** in its struggle against **Phocis**. A Phocian appeal to Sparta led to an ultimatum to Thebes not to interfere and a declaration of war when this was ignored. Athens allied with Boeotia.

Late in 395 a two-pronged Spartan attack on Boeotia went badly wrong. In a supposedly coordinated attack, **Lysander** was defeated and killed at **Haliartus** in the north before **Pausanias** (2) arrived from the south. When Pausanias did arrive he found the Athenians there in support of the Boeotians and instead of fighting agreed to withdraw in exchange for Thebes returning the bodies of Lysander and his men. Over the winter of 395 this led to Argos, Corinth, Ambracia, Acarnania, **Euboea**, and the **Chalcidian Confederacy** joining the anti-Spartan coalition. Although many states desired to be free from the heavy-handed Spartan domination, the Persians, wanting to end the Spartan campaign in Asia Minor, had helped the outbreak and scale of the war by providing money to various anti-Spartan individuals and groups.

Following the defeat at Haliartus, the Spartans sentenced Pausanias to death (although he escaped into exile) and recalled **Agesilaus II** and his army from Asia Minor in 394. The anti-Spartan alliance had mixed success prior to Agesilaus' return—they lost a battle at **Nemea** but at sea **Conon** decisively defeated the Spartan navy off **Cnidus**, ending Spartan naval superiority. Shortly after this battle, Agesilaus led his experienced army from Asia Minor to victory over the allied forces at **Coronea**, although his losses meant he was unable to fully exploit this success.

From this point on, Sparta focused on the Isthmus of Corinth and opposing Athens in the Aegean and Asia Minor. In the first theater, during internal strife in Corinth which saw the city merge with Argos, Sparta temporarily occupied Corinth's Long Walls in 392. Continual attacks were mounted on Argive territory, and in 391 Sparta secured naval control of the Corinthian Gulf, blocking the route to Corinth. This allowed Agesilaus to attack Acarnania in 388, forcing it to come to terms.

In the naval war, Athens made a slow start, needing Persian money to rebuild its Long Walls and fleet. In 391 Sparta took advantage of this breathing space to reopen the naval war against Persia, thereby securing the goodwill of many **Ionian** Greek cities. In 389 the Athenian fleet under **Thrasybulus** had considerable success in the northeast Aegean, although he was unable to dislodge the Spartans from the Hellespont. Unfortunately, lack of resources meant that much of Thrasybulus' good work was undone by his need to extract operating money from the local inhabitants. During 389-388 the naval war continued indecisively but in 387 the Spartan **Antalcidas** gained Persian support and, with the additional assistance of **Dionysius I** of **Syracuse**, established an 80-strong fleet in the Hellespont, cutting Athens off from its grain supply.

The war effectively ended in 387—Athens was in trouble from Antalcidas' interdiction of the grain supply through the Hellespont, Corinth was affected by the Spartan blockade of the Corinthian Gulf, Argos was exhausted, and Thebes was increasingly isolated. Sparta too was short of money and needed a respite. In 386 the belligerents signed the King's Peace (**Common Peace** [1])—guaranteed by Persia, which undertook to use force against any state not abiding by the terms. Persia was keen for peace in order to curb Athenian action against Persian interests in Asia Minor and free up **mercenaries** to suppress revolts in **Cyprus** and **Egypt**. Spartan power was certainly weakened by the war, but it was able to maintain its domination of Greece (albeit less firmly) until 371—although Athens' strength had increased and Thebes, which had reluctantly signed the peace under threat, was clearly still a danger.

CORONEA (KORONEIA). A district in **Boeotia**, west of **Thebes**, which was the site of several battles.

(1) A battle fought in late 447 (or early 446) between 1,000 Athenian **hoplites** under Tolmides, supported by some allies, and a combined force of Boeotian, Locrian, and Euboean exiles. Tolmides had marched north to restore Athenian influence in **Orchomenus (Boeotia)**, **Chaeronea**, and other cities where exiled oligarchs had overthrown the democracies established by **Athens**. Tolmides recaptured Chaeronea and installed a garrison. However, on its return the Athenian army was attacked at Coronea and decisively defeated. Tolmides was killed, along with a substantial number of his men, and the rest were captured. Athens agreed to evacuate the whole of Boeotia in exchange for the return of the prisoners. This battle reversed the result of **Oenophyta**, cut Athens' communications with **Phocis** and **Locris**, and led to the revolt of **Euboea** against Athens.

(2) A battle fought on 14 August 394 during the **Corinthian War**. The Spartan force (which included troops from **Sparta, Orchomenus [Boeotia], Phocis,** and Asia Minor) was led by **Agesilaus II**. The opposing army included troops from Boeotia, **Argos, Athens, Corinth, Euboea,** and **Locris**. According to **Xenophon** (*Hell.* 4.3.15) the numbers of cavalry on each side were about equal but Agesilaus had a marked superiority in **peltasts**; the **hoplite** numbers are unknown. Agesilaus' right wing was victorious, but on the left wing the **Thebans** defeated the contingent from Orchomenus. When the Thebans attempted to rejoin the remainder of the army, instead of letting them past and attacking their rear, Agesilaus met them head on with his **phalanx**—in a move which Xenophon states with some irony "can only be described as courageous" (*Hell.* 4.3.19). At least some, and perhaps many, of the Thebans broke through and Agesilaus was wounded. The final result was a technical victory for Sparta, but Agesilaus, unable to follow up, evacuated Boeotia.

CORUPEDIUM (KOROUPEDION). A battle fought to the west of Sardis in February 281 between **Seleucus I (Nicator)** and **Lysimachus**. Almost nothing is known of the battle itself, except that Lysimachus was defeated and killed, which made Seleucus the most powerful monarch—until his assassination very shortly afterward.

CRANNON (KRANNON). A town near the Peneius River in **Thessaly**, site in late August 322 of the final battle of the **Lamian War**. The **Macedonian** army of 40,000 heavy infantry, 3,000 **archers** and **slingers**, and 5,000 **cavalry**, led by **Antipater**, defeated a depleted Greek army of 25,000 infantry and 3,500 cavalry, led by **Antiphilus**. Many Greeks had returned home after their previous success against Leonnatus at **Lamia** earlier in the year, and despite the excellent performance of their cavalry (the bulk of whom were Thessalian), the Greek infantry were decisively defeated by the Macedonian **phalanx**. Antipater avoided the mistake made by Leonnatus earlier in the year and engaged the Greek infantry before the cavalry action was complete. When the Greek infantry retired, their cavalry abandoned the fight and went with them. The loss of this battle led to the Greek surrender. *See also* MELITAEA.

CRATERUS (KRATEROS; c. 370-320). A **Macedonian officer** who rose to the position of **Alexander the Great**'s second-in-command. Craterus commanded an infantry brigade at the **Granicus River** (Arr. *Anab.* 1.14.3) and the entire infantry on the left wing at **Issus** (Curt. Ruf. 3.9.8; Arr. *Anab.* 2.8.4) and **Gaugamela** (Arr. *Anab.* 3.11.10;

Curtius Rufus [4.13.29] is almost certainly wrong in giving him command of "the Peloponnesian **cavalry**" here). He also commanded the left wing of the fleet during part of the operations at **Tyre** (Curt. Ruf. 4.3.11; Arr. *Anab.* 2.20.6). When **Parmenio** was left behind in Ecbatana, Craterus assumed a more prominent role and was essentially Alexander's second-in-command—unequivocally so after Parmenio's execution in 330. Craterus played an active part in the campaign in Bactria and Sogdiana in 329-328, often being entrusted with independent command. He also played an important part in the defeat of **Porus** at the **Hydaspes River**, where he was a senior **cavalry** commander (Arr. *Anab.* 5.11.3-4, 5.15.3-4, 5.18.1).

In 324 Alexander married Craterus to **Darius III**'s niece and then sent him home with the discharged Macedonian veterans under instructions to replace **Antipater** as regent in Macedon (Diod. Sic. 18.3.4; Arr. *Anab.* 7.4.5, 7.12.1-4). On Alexander's death Craterus was well positioned to be a major force among the *diadochoi*, although he voluntarily subordinated himself to Antipater—whose daughter he married in 322 (Diod. Sic. 18.5.7). During the **Lamian War** Craterus brought large reinforcements to Antipater at **Crannon** in 322—although there is no record of his role in the battle itself (Diod. Sic. 18.16.4-5, 18.1). Following this, he was active in the campaign against the **Aetolian League** and was having considerable success until he was forced to allow the Aetolians fairly easy terms in order to meet a new threat from **Perdiccas** (2) (Diod. Sic. 18.24-25, 18.25.1-2). In 320 Craterus crossed to Asia Minor to fight Perdiccas' general, **Eumenes of Cardia**, but was killed in the first battle in Cappadocia. According to **Diodorus Siculus**, Craterus was killed when his horse stumbled while leading a cavalry charge and Craterus was trampled to death. Eumenes returned Craterus' bones to his wife for burial (Diod. Sic. 18.29-30, 19.59.3).

CRIMISUS RIVER (KRIMISOS also CRIMESUS/KRIMESOS). Probably the modern Belice River, in the west of **Sicily**, site of a major **Carthaginian** defeat by the Sicilian Greeks in May/June 341 (or 339). The Carthaginians, preparing for a major attack on Greek Sicily, landed at Lilybaeum with an army, reputedly of 70,000 men and including four-horse **chariots** (Plut., *Timoleon*, 25.1; Diod. Sic. 16.77.4). **Timoleon** led the Greek army across Sicily to meet the enemy on their own territory. **Plutarch** (*Timoleon* 25.2-3) states that his force, comprising **Syracusans** and **mercenaries** (although some deserted just before the battle) was only 6,000 strong (5,000 infantry and 1,000 **cavalry**). **Diodorus Siculus** (16.78.2), more plausibly, puts it at around 12,000 men—including troops supplied by Hicetas, the tyrant of Leontini, with whom Timoleon had been at war.

Timoleon deployed his men on a line of hills and attacked the Carthaginian force as it was crossing the Crimisus—nicely judging when his force could deal with those who had already crossed. When the Carthaginian chariots reduced the impact of Timoleon's initial cavalry charge, Timoleon personally led the infantry into the attack, ordering his cavalry against the enemy flanks. Initially successful against the first part of the Carthaginian force, Timoleon's men got into some difficulties when more began to cross. However, a major storm arose, blowing up from behind the Greeks and into the faces of the Carthaginians. More heavily equipped than the Greeks, the Carthaginians soon had problems in the swollen river and mud and broke and ran. Timoleon instituted a vigorous pursuit, his *psiloi* (light troops) in particular inflicting major casualties on the fleeing enemy.

The reports of the Carthaginian casualty figures vary from 10,000-12,500 dead, including 2,500-3,000 native Carthaginians—an unusually high figure as they generally relied on their subject peoples and mercenaries in battle. A large quantity of **booty** and between 5,000 and 15,000 **prisoners** were taken. The magnitude of the defeat and the number of casualties from Carthage itself led the Carthaginians to make peace. The terms were that the Greek cities were to be free, Carthage would not aid any tyrants at war with Syracuse, and the boundary between the Greeks and Carthaginians on the island was fixed at the Halycus River (giving the Greeks about two-thirds of the island). This victory considerably boosted Timoleon's support, both in Sicily and in Greece, and helped speed up his settlement of Sicily.

CROCUS PLAIN. A battle fought in 352, at an unknown location in southeast **Thessaly**, between the **Macedonians** and the **Phocians** during the Third **Sacred War** of 355-346. **Philip II** of Macedon had been defeated twice by **Onomarchus** and the Phocians in 353 but defeated and killed him in this battle. One of the reasons for the Macedonian victory was that the 500 Phocian **cavalry**, who were generally of good quality, were overwhelmed by the 3,000 Macedonian and Thessalian cavalry serving with Philip. Around 6,000 of the Phocians died and 3,000 were captured. An ambiguous reference in **Diodorus Siculus** (16.35.6) has usually been taken to mean that the **prisoners** were flung into the sea and drowned, but this may refer only to the corpses of the dead. Because it was a Sacred War and the enemy were regarded as temple robbers, denial of burial (which meant their spirits would wander the earth eternally) was regarded as an appropriate punishment. This victory established Philip as the dominant force in Thessalian affairs, although the Phocians did continue to resist under Onomarchus' brother Phayllus.

CUNAXA (KOUNAXA). Site in Babylonia of a battle in 401 between **Cyrus the Younger** and his brother, Artaxerxes II, king of **Persia**. Cyrus was attempting to overthrow his brother and marched against him with an army of 100,000 local troops, including 20 scythed **chariots** and a force of Greek **mercenaries**. According to **Xenophon** (*Anab.* 1.7.10), these included around 10,400 **hoplites** and 2,500 **peltasts**—although the figures he gives earlier for the individual contingents add up to around 12,600 hoplites. Alerted by scouts of the enemy's approach about mid-morning, it took several hours for Cyrus' column to deploy in battle line. Artaxerxes, reputedly leading a force of 900,000 (*Anab.* 1.7.12, but surely exaggerated), advanced with his 200 scythed chariots leading.

Cyrus ordered **Clearchus**, the senior mercenary general, to concentrate his attack on the Persian center, where the king was stationed. However, Clearchus was worried that he might be outflanked by the more numerous enemy and kept his right wing on a river, opposite the enemy's left wing. The Greeks were victorious, quickly driving the opposing enemy from the field, but were in danger of being outflanked and taken in the rear by the center of the opposing force. When Cyrus saw this, he led his personal bodyguard of 600 cavalry directly against the enemy center and, despite initial success, was killed trying to attack his brother. Cyrus' Persian troops fled and the Greeks were left victorious on their side of the battlefield, unaware of the death of their employer. Following the battle, the Greek mercenaries (later known as the **Ten Thousand**) staged a fighting withdrawal in an epic march recorded in Xenophon's *Anabasis*.

CURTIUS RUFUS (d. A.D. 53). Author of the only surviving history in Latin of **Alexander the Great**, probably to be identified with the Quintus Curtius Rufus who served as praetor under Tiberius, became consul circa A.D. 43, later served as legate of Upper Germany, and died in A.D. 53 while proconsul of Africa. He based his account on the work of earlier writers, including some—such as Callisthenes—who were contemporaries of Alexander. His work is not as good as **Arrian**'s history and is marred by geographical errors, internal contradictions, and a tendency to moralize. Unfortunately his military narrative is also frequently poor. However, Curtius Rufus' work is valuable in preserving an alternative and less favorable tradition of Alexander than Arrian's—probably based heavily on Cleitarchus. Cleitarchus' work is now lost but he wrote shortly after Alexander's death, was not involved in his expedition, and apparently tended to exaggerate Alexander's vices and believe the incredible.

CYNOSCEPHALAE (KYNOSKEPHALAI). A ridge line (literally "Dog's Heads") overlooking a plain on the Pharsalus-Larissa road in **Thessaly**; site of two major battles:

(1) 364, fought between a **Thessalian-Boeotian** army led by **Pelopidas** and a Thessalian and **mercenary** army under **Alexander of Pherae**. Pelopidas, who harbored a personal grudge against Alexander, led a small Boeotian contingent to help the Thessalians end Alexander's interference in their cities. The two armies met at Cynoscephalae. Pelopidas' **cavalry** routed the enemy cavalry in the plain before the two **phalanxes** met and advanced on Alexander's infantry. However, Alexander managed to secure the high ground first and repelled the cavalry, which was advancing uphill. Pelopidas detached the cavalry to operate against those enemy soldiers still on the plain, and attacked with his own infantry. Alexander was slowly forced back and his troops began to waver when Pelopidas' cavalry returned to threaten them. At this point Pelopidas was killed by **javelins** while trying to kill Alexander. A Thessalian cavalry charge then routed Alexander's army, which fled leaving around 3,000 dead. Although an important victory, the loss of Pelopidas helped to hasten **Thebes'** military decline.

(2) 197, during the Second **Macedonian War**, fought between the Macedonians under **Philip V** and the **Romans** under Flamininus. Philip's army consisted of 18,000 heavy infantry, 2,000 **cavalry**, and 5,500 light troops (which included 1,500 **mercenaries**). Flamininus' army probably consisted of around 22,000 heavy infantry, 2,500 cavalry, and 8,000 light troops. Philip was attempting to reach Demetrias, but ran into Flamininus just south of Pherae. Short of supplies, he withdrew west toward Scotussa, while Flamininus moved parallel and to the north. Delayed by rain and low clouds that considerably reduced visibility, Philip sent forward an advance guard, which clashed at Cynoscephalae with a force of 1,300 Romans sent by Flamininus to locate the Macedonians.

Both commanders deployed additional forces to assist. Philip, encouraged by an outdated report that his advance guard had the Romans on the run, quickly moved forward with part of his **phalanx**, the rest following behind. The impact of Philip's phalanx moving downhill against the first Roman legion drove it back with some loss. The other half of the phalanx, arriving slightly later in some disorder, was routed by Flamininus' other legion and his **elephants**. A Roman officer in this legion detached around 2,000 men from the pursuit and swung them into the rear of Philip's phalanx, rapidly destroying it. The Macedonian *sarissa* was very cumbersome under these circumstances, and once it was discarded, the small Macedonian **shield** and dagger were no match for the Roman sword and large shield in close-quarter combat. Around 8,000 Macedonians died and 5,000 were captured, and Philip fled; the Romans lost 700 men. Philip's defeat (along with reverses on other fronts) forced him to come to terms with Rome, ending the war. This battle, confirmed by **Pydna** (1) in 168, clearly demonstrated the superiority of the more flexible Roman maniple system over the Macedonian phalanx.

CYNOSSEMA (KYNOSSEMA). See ABYDUS (1)

CYPRUS. An island off the coast of **Syria**, first colonized by Greeks in the 15th century, during the **Mycenaean** period, largely submerging the original culture. Further waves of immigrants in the 11th and ninth centuries introduced Syrian and then Phoenician elements to the population. Cyprus fell under Assyrian control, was apparently independent for about 50 years, and then passed to Egyptian control around 570, during the reign of Amasis (Hdt. 2.182). For most of the rest of its history, Cyprus was ruled by **Egypt**, or whichever power controlled Egypt.

Cyprus joined **Persia** around 525, providing ships to assist Cambyses' conquest of Egypt and subsequently paying tribute to Persia (Hdt. 3.19, 91). All of Cyprus except Amathus joined in the **Ionian Revolt** against Persia, led by Onesilus of Salamis (on Cyprus). After a year of Cypriot independence, the Persians launched a major effort to recover the island. A Persian army was landed, supported by the Phoenician fleet. Cyprus' **Ionian** allies defeated the Phoenicians at sea, and on land the battle initially went in favor of the Cypriot forces. Onesilus killed the Persian commander in single combat—despite the latter's specially trained warhorse—but then the troops from Curium deserted, followed by the war **chariots** from Salamis. The remaining Cypriots were routed, Onesilus was killed, and the cities were besieged and captured one after the other (Hdt. 5.104, 5.108-115).

Following the Persian defeat at **Salamis** (in Greece) in 480, the **Hellenic League** (1) fleet liberated a large part of Cyprus. However, a later **Athenian** attempt in 450 to 449 to clear the Persians from the island failed because of a shortage of food and the death of **Cimon**, the expedition's commander—although the Athenians won a major battle on both land and sea near (Cypriot) Salamis (Thuc. 1.94, 112). At the end of the fifth century, Cyprus enjoyed a resurgence under **Evagoras**, king of Salamis, who embarked on a deliberate program of hellenization on the island and temporarily established its independence from Persia around 389. During this period Athens was on friendly terms with Evagoras and at least twice sent military assistance to him against Persia (Xen. *Hell.* 4.8.24, 5.1.10). However, Persia's right to the island was recognized by the Greeks under the terms of the King's Peace, the **Common Peace** (1) of 386 (Xen. *Hell.* 5.1.31), and they managed to re-establish control by about 380.

In 351 the whole island revolted against Persia (at the same time as Egypt and Phoenicia), but the island was fairly quickly recovered by a Persian expeditionary force led by the Athenian **Phocion** and Evagoras, the grandson of the earlier Evagoras, both in **mercenary** service with Persia (Diod. Sic. 16.42, 46). After the battle of **Issus** in 333 Cyprus joined **Alexander the Great** and sent 120 ships, including some fives (quinqueremes), to assist in the siege of **Tyre**. These provided valuable service during the siege (along with Cypriot engineers) and Cypriot ships were among the 100 then dispatched to the Peloponnese to counter **Agis III** of **Sparta**'s revolt (Arr. *Anab.* 2.20-24, 3.6.2; Curt. Ruf. 4.3-4).

Following the death of Alexander, Cyprus became part of Ptolemaic Egypt but was shortly afterward captured by **Demetrius I (Poliorcetes)**, who besieged Salamis in 307 and inflicted a decisive defeat on the Ptolemaic relief fleet near there in 306 (Plut., *Demetrius*, 15-18; Diod. Sic. 20.47-53). However, **Ptolemy I (Soter)** recovered Cyprus in 295, and from this time, on it remained part of Ptolemaic Egypt (although occasionally attacked by Seleucid Syria) and often served as the personal possession of younger members of the royal family. In 58 **Rome** simply annexed Cyprus and incorporated it into the Roman empire.

CYRUS THE YOUNGER (KYROS). Son of **Darius II**, king of **Persia**, Cyrus was appointed overall commander of coastal Asia Minor in 408, with instructions to assist **Sparta** against **Athens** in the **Peloponnesian War** of 431-404. He played a major role in supporting the Spartan war effort, increasing the **pay** of the Peloponnesian fleet at the request of **Lysander**. Although less amenable to Lysander's successor, **Callicratidas**, Cyrus rapidly resumed his friendship with, and support of, Lysander on his return to naval command in 405. Cyrus was accused by **Tissaphernes** of plotting against Cyrus' elder brother, Artaxerxes II, who had succeeded Darius as king in 404. Artaxerxes imprisoned Cyrus but was persuaded by their mother, Parysatis, to release him and restore him to his position in Asia Minor.

Incensed at his treatment Cyrus secretly raised several forces of Greek **mercenaries**. In 401 he brought them together with his local troops and, with Spartan support, rebelled against his brother. Despite his careful preparations and the success of his mercenaries (the **Ten Thousand**), Cyrus' revolt failed with his death at **Cunaxa**. **Xenophon**, who personally knew Cyrus, portrays him in the *Anabasis* as a charming and gifted prince with excellent leadership skills. Xenophon states that Cyrus had a wide reputation as the Persian who was "of all those after Cyrus the Elder most kingly and the most deserving to rule" (*Anab.* 1.9.1).

CYZICUS (KYZIKOS). A Milesian colony founded in the Propontis in the mid-eighth century. Cyzicus was strategically located at a favored harbor on the main sea route between the Aegean and the Black Sea. The prosperity generated by this location is indicated by the fact that, as a member of the **Delian League**, it paid the largest *phoros* (tribute) of any city in the Hellespont. The city was incorporated into the **Persian** empire in the second half of the sixth century but joined in the **Ionian Revolt**. It made terms with the Persians in 493 after the fall of **Miletus** (Hdt. 6.33).

Cyzicus joined the Delian League after the Second **Persian War**. With the support of **Pharnabazus** it revolted against **Athens** in 411, during the general unrest which followed the failure of the **Sicilian Expedition**. The Athenians easily recaptured the city, which was unwalled, and imposed a fine on it (Thuc. 8.6.1, 8.107.1). The following year the city was again in Peloponnesian hands, but **Alcibiades** surprised the Spartan admiral, **Mindarus**, driving his 60 **triremes** ashore and capturing them. He then reoccupied Cyzicus and extracted large sums of money from the population (Xen. *Hell.* 1.1.14-20). **Agesilaus II** of **Sparta** used Cyzicus as a base in his Asian campaign of 396/5 (Xen. *Hell.* 3.4.10), but the city was presumably Persian again after the King's Peace (**Common Peace [1]**) of 386.

In the Hellenistic period, Cyzicus appears to have maintained good relations with the Attalids of Pergamum (Polyb. 22.20, 33.13). It was given the status of a free city by **Rome** after its sustained opposition to **Mithridates VI (Eupator)** in 74.

- D -

DARIUS. The name of several Achaemenid rulers of Persia:

(1) Darius I (d. 486). King of **Persia**, 521-486. He spent the early years of his reign quelling numerous revolts and instituting administra-

tive reforms (including the establishment of provinces or satrapies). In 512 he extended Persian control into Europe with a punitive expedition against the **Scythians** that led to the conquest of most of the **Thracian** coastline and the Chalcidice. He suppressed the **Ionian Revolt** of 499-493 and subsequently sent two invasion forces against Greece. The first was wrecked off Mount Athos in 492; the second, which properly began the **Persian Wars**, was defeated at **Marathon** in 480. His preparations for a further attack on Greece were delayed by a revolt in **Egypt** and ended by his death in 486. Darius' son, **Xerxes I**, implemented the planned invasion in 480, initiating the Second Persian War.

(2) Darius II (d. 405). King of **Persia**, 424-405. Although distracted by a series of revolts (apparently caused by his own mismanagement) in **Syria**, Lydia, and Media and the successful revolt of **Egypt** in 410, Darius played a major role in the latter part of the **Peloponnesian War** (2). The diplomatic, financial, and military support provided by his satraps **Tissaphernes** and **Pharnabazus** initially helped keep **Athens** and **Sparta** evenly balanced. However, the dispatch of his son **Cyrus** to **Ionia** to provide full support to Sparta played a crucial role in the defeat of Athens.

(3) Darius III (c. 380-330). King of **Persia**, 336-330. He had the misfortune to be in power when **Alexander the Great** invaded. Although the size of the empire, which diffused his initial response, and the high quality of Alexander's generalship and army played a major role in Darius' defeats, he was not a skilled general. At **Issus** he deployed his troops in an area that reduced his major advantage—overwhelming numbers. He avoided this mistake at **Gaugamela**, but his deployment and keeping his troops at arms overnight contributed to his defeat. A further factor in the Persian loss of both battles was Darius' headlong flight when his personal safety was threatened. After Gaugamela he was arrested by one of his own satraps, Bessus, and stabbed to death just before Alexander caught up with him.

DASCYLIUM (DASKYLION). Site of a meeting in **Ionia** in 398/7 between the army led by the Spartan **Dercylidas**, which included both Peloponnesians and Ionians, and the army of **Tissaphernes** and **Pharnabazus**, which included Carians, **Persians**, and Greek **mercenaries**. Tissaphernes and Pharnabazus had crossed into Ionia in response to Dercylidas' foray, supported by the Peloponnesian fleet under Pharax, against Tissaphernes' territory in Caria. Dercylidas returned to Ionia to face the threat, and the two armies met at Dascylium. A significant proportion of Dercylidas' Ionian troops fled, but the steadiness of his Peloponnesians and the memory of **Cunaxa** caused the Persians to have second thoughts about the engagement. The danger of battle was ended by negotiation, followed by a truce the next day, to be ratified by their home governments.

In 396 Dascylium was also the site of an inconclusive cavalry action between **Agesilaus II** and Pharnabazus.

DATIS. A Mede, appointed co-commander (with Artaphernes, son of **Artaphernes**) of the Persian expedition against Greece in 491. He is supposed to have made a generous sacrifice at Delos en route to **Marathon** and, on his return trip to **Persia**, to have deposited at Delos for safe return to Greece a stolen statue which he had discovered in his ships. His two sons were **cavalry** commanders in the 480/479 invasion of Greece. *See also* PERSIAN WARS (1).

DECADARCH (*DEKADARCHOS*, pl. *DEKADARCHOI*). A file-leader, literally "leader of 10"; a rank in the **Spartan** infantry. **Xenophon** recommended in his **cavalry** general's manual, the *Hipparchikos*, that decadarchs should be used as subordinate commanders in the cavalry squadrons at **Athens**.

DECELEA (DEKELEIA). An Athenian *deme* (administrative district), located northeast of **Athens** on the eastern pass through Mount Parnes, on one of the main routes to **Boeotia**. **Mardonius** withdrew from Attica via this pass prior to the battle of **Plataea** during the Second **Persian War**. During the **Peloponnesian War** of 431-404 it was the main overland route for grain imported via **Euboea** and Oropus. In 413 it was occupied and fortified by the **Spartans** (on **Alcibiades**' advice, according to **Thucydides**, 6.91). This occupation not only meant that food imports had to come all the way around Cape Sunium (famous for its contrary winds and shipping delays) but opened the whole of the Attic Plain to **ravaging** the whole year round. Other effects were interference in the mining of silver at Laurium and an increasing loss of **slaves**, who now had a focal point for desertion. The effect of this occupation was serious enough for the second phase of the Peloponnesian War to be termed the Decelean War.

DECELEAN WAR. *See* PELOPONNESIAN WAR (2)

DELIAN LEAGUE. A naval alliance set up by **Athens** in 478/7, following Greek victory in the Second **Persian War** of 480-479. The name is a modern one (coined because the original league treasury was on Delos)—contemporary inscriptions refer to it as "the Athenians and their allies." The league was set up after the arrogant behavior of the **Spartan** regent, **Pausanias** (1), alienated most of the Greeks campaigning against **Persia** in the Hellespont region. Impressed by the contrasting behavior of **Aristides**, the Greeks requested that Athens take over leadership. Essentially uninterested in a naval campaign so far from home, and regarding the Athenians as both competent and friendly, the Spartans agreed—although this did contribute later to growing Spartan resentment of Athens.

According to **Thucydides** (1.96), the Delian League was originally set up to **ravage** Persian territory to compensate the members for the damage done to them during the recent war. However, Mytilene later stated that the original purpose of the league "was to free the Hellenes from Persia" (Thuc. 3.10). Although the Mytilenians had a vested interest in claiming this (they were trying to persuade the Spartans to support their revolt against Athens), it seems highly likely that many states joined with the aim of throwing off Persian domination. The early membership included the **Ionian** states of Asia Minor, most of the Aegean islands, and cities in the Hellespont. Each member swore an oath of loyalty and agreed to provide either **triremes** and men to the league fleet or a monetary contribution. Under Aristides and **Cimon**, Athens encouraged the payment of money rather than the provision of ships, and this was used to expand the Athenian navy.

The league was successful in driving the remaining Persians from **Thrace** (including the Chersonese) and removing Persian control from most of the west coast of Asia Minor and from Caria. In about 467/6, the league fleet under Cimon won a resounding victory over the Persian fleet at the **Eurymedon River**. However, by this time the league had developed some authoritarian tendencies. Around 472 **Euboea** had been forced to join, **Naxos** was forcibly prevented from leaving circa 468/7, and around 465 Thasos revolted when Athens started impinging on its economic interests on the mainland opposite. Around 463/2 Thasos was finally forced to surrender and became a subject ally, stripped of her mainland possessions. The treatment for states like this normally involved paying money instead of providing ships and the loss of autonomy in foreign policy and in legal cases involving serious charges. To ensure Athenian control, sometimes the Athenians sent a political officer and/or garrison, or confiscated land and set up a **cleruchy**.

Despite a reverse in 454, when a large allied and Athenian fleet was destroyed by the Persians in **Egypt**, the Delian League continued its success, having around 200 members at its peak. From the 450s, Athens became increasingly dominant, to the point where the league is more accurately described as the Athenian empire. By the outbreak of the **Pe-**

loponnesian War of 431-404, Athens had imposed its coinage and a uniform set of weights and measures on the league, was trying in Athens all legal cases involving Athenians and in some cases capital charges in member states, had reduced several of the larger members to the status of subject allies, and was receiving tribute rather than ships from all the members except Lesbos and **Chios**. The tribute not expended on the fleet or league activities was now regarded by the Athenians as legitimately theirs and was used for domestic purposes. **Pericles**, for example, used tribute to pay for the construction of the Parthenon. The income generated by the league/empire also allowed Athens to institute **pay** for political, jury, and military service. Although the league was valuable in suppressing piracy, aiding commerce, and freeing the Greeks in the Aegean and coastal Asia Minor from Persian control, it was very unpopular. During the Peloponnesian War of 431-404, many allies took the first opportunity to revolt. At the end of this war, the league was dissolved and the Athenians were careful to avoid the Delian League's most hated features when establishing the Second Athenian Confederacy in 378/7.

DELIUM (DELION). An area on the coast in the territory of Tanagra in **Boeotia** and site of a well-known temple of Apollo. The **Athenians** and Boeotians fought a major battle nearby in 424, during the **Peloponnesian War** (2). The Athenians, as part of a more aggressive prosecution of the war, had fortified the temple with a ditch, earthen ramparts, and a wooden stockade and towers and left a garrison. The remainder of the army, led by Hippocrates, consisted of about 7,000 **hoplites**, a few poorly equipped *psiloi* (light troops), and perhaps the full, 1,000-strong **cavalry** corps. This force was only a short distance from Delium on its way home when it was challenged by the Boeotians. The Boeotian force comprised 7,000 hoplites, 10,000 *psiloi*, 500 **peltasts**, and 1,000 cavalry with Pagondas, one of the two boeotarchs from **Thebes**, in supreme command.

Hippocrates deployed his hoplites eight deep, with cavalry on each wing and a further 300 left at the temple to assist in its defense and to intervene in the battle if the opportunity arose. The Boeotians deployed with the Theban contingent 25 deep (the other contingents were of varying depths), the cavalry and *psiloi* on the wings, and a detached force to prevent the 300 Athenian cavalry at the temple from deploying. The battle was initially fairly even, with the Theban hoplites on the right slowly pushing the Athenians back while the Boeotian left wing was defeated by the Athenians, with considerable loss. At this point Pagondas dispatched two squadrons of cavalry to assist his left wing. They moved under cover of hills and when they emerged into the open the Athenian right wing fled, causing the collapse of their entire line. **Casualties** were caused by the Boeotian cavalry but night prevented a full pursuit. **Alcibiades**, serving in the Athenian cavalry, assisted in the resolute withdrawal of Socrates, who was serving as a hoplite. Following the battle the Thebans refused to return the Athenian corpses until the temple was evacuated. The Athenians refused, but 17 days later the Boeotians took the temple anyway, using an ingenious flamethrower they constructed on site. **Thucydides** 4.89-101 gives a full account of the action. *See also* LAWS OF WAR.

DELPHI. A city and cult center on the north coast of the Corinthian Gulf. Delphi played little direct part in Greek military activities—for most of its history it was under the protection of other states. However, as the most famous Greek center of prophecy, Delphi exercised considerable influence through the oracle. The Pythia (who delivered Apollo's oracles) was regularly consulted on major decisions, including questions of war and peace. According to **Herodotus** (1.53-56, 90-91), even foreigners such as Croesus of Lydia consulted it about the chances of success in war. In the sixth century Delphi appears to have had a close relationship with **Sparta**, although at the end of the century the Pythia was bribed by Athenian exiles to persuade Sparta to overthrow the Pisistratid tyranny in **Athens**.

During the **Persian Wars** Delphi, through the Pythia, appears to have encouraged several Greek states, including **Argos** and Crete, not to resist **Persia** (Hdt. 7.148-149, 169). Although this may have caused some damage to its reputation, Delphi continued to take sides in various conflicts. It seems to have favored the Spartans in the **Peloponnesian War** of 431-404 and later (when not occupied by **Phocis**) supported **Philip II** of **Macedon**. Although its influence declined in the Hellenistic period, Delphi remained a place for **sacrifice** by leaders prior to campaigns or after them. Examples in the early second century include the **Roman** general Flamininus and **Antiochus III (the Great)**.

Delphi was attacked by the Persians in 480 and the **Gauls** in 279, although apparently suffering little damage. It was the site of one major battle, which drove the Gauls from Greece in 279. The Gauls were attacked from two sides and, although they maintained their cohesion, they spent a miserable night in snow and were then forced to retire under harassing attacks from the Phocians. *See also* SACRED WARS.

DELPHIC AMPHICTIONY. *See* AMPHICTIONIC LEAGUE

DEMARATUS (DEMARATOS; fl. c. 500). Spartan king of the Eurypontid family who reigned c. 510-491. Around 506 he sided with disgruntled members of the **Peloponnesian League** against his fellow king, **Cleomenes I**, causing the invasion of Attica to be abandoned. He opposed Cleomenes again in 491 over the **Athenian** request to **Sparta** to arrest pro-Persians on Aegina and was dethroned on a false charge of illegitimacy organized by Cleomenes and the future **Leotychides II**. Demaratus left Sparta and sought refuge in **Persia**. According to **Herodotus**, Demaratus provided useful advice to **Xerxes I** both before and during the invasion of Greece in 480-479—although Xerxes often failed to follow it. Following the Persian defeat, Demaratus was given estates in Persia, where he ended his days. *See also* PERSIAN WARS (2).

DEMETRIUS I (POLIORCETES) (DEMETRIOS POLIORKETES; 336-283). Son of **Antigonus I (Monopthalmus)**. Demetrius played a major role in the military struggles among the *diadochoi*. He was entrusted with an independent command in **Syria** and **Palestine** but was heavily defeated by **Ptolemy I (Soter)** at the battle of **Gaza** in 312. Subsequent to this he proved a valuable asset to Antigonus' plans. One of his greatest successes was the Cypriot campaign of 306. Following a decisive defeat of Ptolemy's navy at Salamis (**Cyprus**), Demetrius brought the island under Antigonid control. Antigonus chose this occasion to have himself and Demetrius proclaimed kings (Plut., *Demetrius*, 15-18; Diod. Sic. 20.47-53). The following year, although ultimately unable to capture **Rhodes**, the scale of Demetrius' activities and his persistence during this siege earned him the nickname Poliorcetes ("Besieger"). Demetrius was then unleashed upon Greece, where he had considerable success in reducing the influence of **Cassander** and Ptolemy by both political and military means. In 302 he successfully capitalized on the perennial desire of Greek *poleis* for autonomy by refounding the **Hellenic League (2)**, based, like **Philip II of Macedon**'s earlier version, at **Corinth**.

However, in 301 Demetrius directly contributed to his father's death at the battle of **Ipsus** by pursuing the enemy cavalry too far and leaving Antigonus' **phalanx** exposed to a flank attack by **Lysimachus** and **Seleucus I (Nicator)**. Demetrius escaped with only 8,000 troops. His main asset now was the loyalty of the islands, especially Cyprus, and some coastal towns in Greece, notably Corinth. The Antigonid fleet was intact and Demetrius retained naval superiority. In his attempt to restore Antigonid power, Demetrius first allied with Seleucus against Lysimachus and Ptolemy but the next 15 years saw frequent and rapid shifts in his alliances.

In 296, following the death of Cassander, Demetrius launched a major campaign in Greece aimed at securing the throne of Macedon. **Athens** surrendered in 294 after a siege and later that year Demetrius seized Macedon and had himself proclaimed king. However, this success was balanced by the loss of key eastern possessions, especially Cyprus and Cilicia (to Ptolemy and Seleucus, respectively). An invasion of Lysimachus' territory the following year was aborted because of a revolt against him in Greece backed by **Pyrrhus I** of Epirus. Although initially successful here, Demetrius was drawn into several years of debilitating warfare with Pyrrhus and the **Aetolians** and forced to exert increasing brutality to maintain his precarious hold on Greece. At the same time Ptolemy was successfully extending his control over Demetrius' island possessions and allies. When Pyrrhus and Lysimachus simultaneously invaded Macedon in 288, Demetrius' army deserted. Abandoning Macedon, he invaded Asia Minor with a small force, attacking the territory of Lysimachus and then of Seleucus. He was forced to surrender to Seleucus in Cilicia in 285 and drank himself to death in captivity. His son, **Antigonus II (Gonatas)** was left holding a few possessions in Greece. Demetrius possessed undoubted talent as a general, particularly when following his father's direction. However, after the death of Antigonus I (Monophthalmus), Demetrius proved less than adequate in coping with the complex political and strategic issues of the time.

DEMOSTHENES.

(1) An **Athenian** general (d. 413) active during the **Peloponnesian War** of 431-404. He was a very innovative commander who appeared to have learned valuable lessons from his defeat in 426 at the hands of **Aetolian** *psiloi* (light troops) while trying to protect Naupactus and threaten **Boeotia** from the west. He redeemed himself by a decisive victory at **Olpae**, destroying two Ambraciot armies in successive days. In 425 he suggested the extremely successful tactic of fortifying the headland of **Pylos** on the Peloponnesian coast to act as a base for **ravaging Spartan** territory and causing unrest among the helots. In 424 he surprised and took Nisaea, the port of **Megara**, but failed to capture Megara itself. The same year he also took part in (and, along with **Cleon**, was perhaps a prime instigator of) the Athenian attack on Boeotia that ended in failure at **Delium**. He was responsible for all of the recorded Athenian **ambushes** and surprise and night attacks during the Peloponnesian War. In 413 he was captured and executed after the destruction of the Athenian force during the **Sicilian Expedition**. *See also* BRASIDAS; ELLOMENOS.

(2) Athenian orator and statesman (384-322), who played a leading role in **Athens'** opposition to Macedonian expansion under **Philip II**. In early 351 he delivered the first of his *Philippics*, major speeches attacking Philip. In 349 he delivered three speeches (the *Olynthiacs*) urging intervention against Philip in support of **Olynthus**. He apparently performed poorly in the embassies to **Macedon** which led to the Peace of Philocrates and shortly afterward began open moves to overturn it. In both 344 and 342 Demosthenes toured the Peloponnese trying to win support against Philip. By 341 Demosthenes was exerting the major influence on Athenian policy, bringing about an alliance with **Byzantium** and, after Philip's declaration of war in 340, with **Thebes**. He played a large part in the events leading to the war and was present at the battle of **Chaeronea** (1), which ended it in Philip's favor.

Despite this defeat, Demosthenes continued to play a leading role in Athenian politics, generally continuing his anti-Macedonian stance whenever he could. He was involved in creating unrest in Greece on Philip's death, although **Alexander the Great** very quickly restored the situation. He was also implicated in the Theban revolt of 335—although Alexander chose not to punish him for it. He went into exile after being found guilty of embezzlement. However, he was recalled and assisted in Athenian preparations for the **Lamian War**. He committed suicide in 322 while on the run from **Antipater** after the Macedonian victory at **Crannon**. Although Athenian speeches have to be considered critically, Demosthenes' oratory is a valuable source for the history of Greece and Macedon in the third quarter of the fourth century.

DERCYLIDAS (DERKYLIDAS also **DERCYLLIDAS/DERKYLLI-DAS; fl. c. 400).** A **Spartan officer** who secured the revolt of **Abydus** and Lampsacus against **Athens** in 411, during the **Peloponnesian War** of 431-404. He succeeded **Thibron** in command in **Ionia** in 399/8. **Xenophon** describes him as "a man considered to be extremely cunning; he was also nicknamed Sisyphus" (*Hell.* 3.1.8). He made a truce with **Tissaphernes** and **ravaged** the territory of **Pharnabazus**, being careful not to repeat Thibron's mistake of harming allied territory. He rapidly secured the area of Aeolis and fortified the Chersonese to protect the Greek cities from attack by **Thracians**. In 396 he was replaced by **Agesilaus II**, who sent Dercylidas as one of three commissioners to negotiate with Tissaphernes. He fought in the battle of **Nemea** in 394, taking the news of victory to Agesilaus and the cities in the Hellespont. He was then reappointed harmost (governor) of **Abydus**. After the battle of **Cnidus** in 394, Dercylidas persuaded Abydus and Sestus to stay loyal to Sparta and generally stiffened Greek resistance to Pharnabazus in the region. *See also* DASCYLIUM; DISCIPLINE.

DERDAS (fl. 382). King of Elimia in Upper **Macedon** who led the 400-strong Elimian **cavalry** contingent serving with the Spartan-led force under Teleutias at **Olynthus** in 382. He proved the outstanding cavalry commander of the Olynthian campaign of 382-381. *See also* CHALCIDIAN CONFEDERACY.

DIADOCHOI. Literally "successors." The men who took over and ruled the various parts of **Alexander the Great**'s empire after his death in 323. *See also* ANTIGONUS (1); ANTIPATER; CRATERUS; EUMENES OF CARDIA; LYSIMACHUS; PTOLEMY (1); SELEUCUS (1).

DIENECES (DIENEKES; d. 480). A **Spartan** killed at **Thermopylae** (1); author of one of the most famous sayings of the **Persian Wars**. On being told that their **Persian** opponents were so numerous that their arrows would hide the sun, Dieneces replied that this was "very good news because if the sun is hidden by the Medes the battle will be in the shade and not in the heat of the sun." (Hdt. 7.226).

DIODORUS SICULUS (c. 90-20). Historian, born in Agyrium in **Sicily** in the first century. He wrote (c. 60-30) a massive general history of the known world, covering from the earliest times to Julius Caesar's first consulship in 59. Only 14 of the original 40 books survive, but these include the **Persian** and **Peloponnesian Wars** and the campaigns of **Alexander the Great**. Diodorus essentially compiled his history from earlier authors and adopted an annalistic or chronological framework, dealing with the events of each year before moving to the next. Although he preserves some useful accounts from earlier historians whose works are now lost, his accuracy is often suspect.

DION (c. 408-354). Brother-in-law of **Dionysius I** and uncle of **Dionysius II**; ruler of **Syracuse** for parts of the period 357-354. He was a prominent member of Dionysius I's court and attempted to have the future Dionysius II educated by Plato as the model philosopher-king. Dion was exiled by Dionysius II on the grounds of treasonable correspondence with the **Carthaginians** during the peace negotiations of 366. He went to **Athens**, where he was closely associated with Plato's Academy. Provoked by Dionysius II's confiscation of his property, Dion raised a small force of 800 men (**mercenaries**, according to Diod. Sic. 16.6.5) and sailed to Syracuse in 357. Dionysius was away in **Italy** and Dion eluded the force sent to block him when a storm blew his small force to the south, ensuring that he approached Syracuse from an unexpected direction. Aided by Dionysius' unpopularity, Dion quickly captured the city (but was unable to take its island fortress, Ortygia). Another exile, Heraclides, arrived with reinforcements and defeated Dionysius' fleet.

Rather aloof in character and tainted by his previous association with Dionysius I's tyranny, Dion soon lost favor with the people, particularly as Heraclides was advocating a program of popular reform. In 356 Dion was forced to leave Syracuse and went to Leontini. However, he was soon recalled to meet the very threatening attack of Dionysius' Campanian commander Nisypius. Dion and his followers played a major part in beating Nisypius back and in capturing Ortygia in 354. Dion then had Heraclides assassinated and became *strategos autokrator* (sole general with full powers). In effect, he was now tyrant of Syracuse and his rule appears to have become increasingly autocratic, to the point that he was assassinated by another follower of Plato, Callippus (who seized power in his own right).

Dion was an able military leader (if rather lucky on occasion) but was unable to consolidate his power in the fluid situation after his initial capture of Syracuse. His personality, combined with his background (highlighted in Dionysius II's propaganda in Syracuse), played a major part in this failure. His links with Plato have ensured him a more favorable reputation in the main sources (**Diodorus Siculus** and **Plutarch**'s *Dion*) than he perhaps deserves.

DIONYSIUS (DIONYSIOS). The name (not to be confused with the Greek god Dionysus [Dionysos]) of two tyrants of **Syracuse** in **Sicily**, in the fifth and fourth centuries.

(1) Dionysius I (c. 430-367). Tyrant of **Syracuse** from 406/5 to 367. Associated with the democrat **Hermocrates**, as a young man Dionysius came to power in Syracuse in 406/5 by attacking the performance of the city's generals against the Carthaginians at Gela. He soon had himself made *strategos autokrator* (sole general with full powers) and set himself up as tyrant—despite opposition, in part caused by his own poor military performance against the Carthaginians. However, in return for very favorable peace terms, **Carthage** recognized Dionysius as ruler of Syracuse. Shortly afterward, using Campanian **mercenaries**, he suppressed a serious uprising. From this point on, Dionysius exercised full power in the city and gradually extended his dominion over most of eastern **Sicily** and much of southern **Italy**.

In the first years after his accession Dionysius built up a large army of mercenaries, and by 401 he had taken over every large Greek city in eastern Sicily except **Messana**. His methods were designed to strike fear into the hearts of opponents. Naxos was razed and Catana settled with mercenaries—after the inhabitants of both cities were sold into **slavery**. Leontini made terms and its inhabitants were spared but were moved to Syracuse. Dionysius also enlisted military engineers from all over Greece. It was this team which invented the first **artillery** in 399, to be used in a new war with Carthage that began with his attack on Motya in 398.

The Carthaginians seem to have been taken by surprise and Dionysius had considerable initial success—Eryx surrendered and he captured Motya, a naturally very strong defensive position, after a siege in which artillery played an important role. However, the next year Carthage recaptured Motya, took Messana, and founded a new city, Tauromenium, near the old site of Naxos, specifically as a base to besiege Syracuse. Dionysius was saved by a plague which decimated the Carthaginian force. Dionysius defeated and enslaved the remnants but released the surviving Carthaginian citizens. This sparked a serious revolt in Libya, which distracted Carthage from Sicily for several years. Dionysius took advantage of this lull to attack (unsuccessfully) Tauromenium, resettle Messana, and found the city of Tyndaris. When the Carthaginians attacked again in 392, Dionysius was victorious and made an advantageous peace (which placed Tauromenium under his control).

From 391-387 Dionysius campaigned in southern Italy. Allied with Locri, he first moved against Rhegium but was prevented from taking it by a coalition of south Italian states (the Italiote League). Turning his attention to them, he inflicted a major defeat on the league at **Elleporus** in 388. Splitting the league by his selective release of **prisoners**, he made terms with several cities and took Caulonia and Hipponium. Rhegium, now isolated, fell in 387.

The history of the rest of Dionysius' reign is rather uncertain, as the main surviving source, **Diodorus Siculus**, is much less detailed for this part of his rule. However, it was marked by continuing struggles against Carthage and further expansion in southern Italy. He had mixed success against Carthage—winning a major victory at Cabala in 383 and capturing Croton, but losing heavily at Cronium in 379 (the location of both sites and details of the battles are unknown). He was forced to surrender some territory after Cronium, but appears to have renewed the war in 368/7, with some success before his death in 367. During the course of his wars with Carthage, Dionysius became the dominant power in southern Italy and even extended his influence into the Adriatic. He also supplied military aid to **Sparta** in its struggle to retain hegemony in Greece—his troops and/or ships were particularly helpful in 387 and 373 and were instrumental in the Spartan victory at the **Tearless Battle** in 368. His alliance with **Athens** in 368/7 marks an interesting change in policy, but he died shortly afterward from natural causes.

Dionysius was a successful ruler of Syracuse, ushering in a period of great success for the city. His foreign relations were marked by a preference for cities with the same Dorian background as Syracuse (e.g., Locri, **Tarentum** [Taras], and Sparta). However, his rule was financially very expensive for Syracuse, marked by a considerable amount of repression, particularly in the early period, and by an inability to decisively remove the Carthaginian threat.

(2) Dionysius II (b. c. 397). Son of **Dionysius I**. Tyrant of **Syracuse**, 367-357 and 347-344 (although not in control of the whole city in the latter period), of Rhegium until 351, and of Locri until 346. In 366, during peace negotiations with **Carthage** which restored peace to Syracuse for around 10 years, he exiled his chief adviser, **Dion**. When Dionysius later confiscated Dion's property, Dion raised troops in Greece and attacked while Dionysius was absent in Italy in 357/6. With some good fortune and aided by Dionysius' unpopularity, Dion's small force captured Syracuse—except for the island fortress of Ortygia (which held out until 354).

Dionysius based himself in Locri and Rhegium (until its loss in 351). From Locri he recaptured Ortygia in 347. However, Dionysius was soon left with Ortygia as his sole possession when the Locrians took advantage of his absence to overthrow his rule there—and to kill his wife and children. Dionysius surrendered Ortygia to **Timoleon** circa 344 and retired to **Corinth** as a private citizen. Not as able as his father, he suffered from a bad press in antiquity, although regarded as having some wisdom and dignity in retirement—partly because of Plato's involvement in his education (Plut., *Timoleon*, 15).

DIPAEA (DIPAIA). An Arcadian town, site of an important but obscure battle between **Sparta** and all of **Arcadia**, except **Mantinea**. The battle took place sometime between 479 and 464, at a time when Sparta's position as dominant power in the Peloponnese was under serious challenge. Around 471 **Themistocles** was in the Peloponnese stirring up anti-Spartan activity. **Messenian** helots revolted in 469, occupying Mount **Ithôme** and were reinforced in 464 after a serious earthquake precipitated a new revolt. Around 468 a coalition of **Argos**, Tegea, and Cleonae destroyed **Mycenae**. The most likely dates for the battle are

473-470 or 466, depending on whether the helot defeat in **Herodotus** (9.35; cf. Paus. 3.11.7, 8.8.6) was in 469 or 464.

The cause of the battle was Tegea's wish to secede from the **Peloponnesian League**, and perhaps a general desire for a more independent Arcadia (coins, dating from 490, have been found with the wording "of the Arcadians"). Sparta's victory was an important step in bringing the general situation under control. However, if Sparta had lost at Dipaea, there is a good chance that its position in the Peloponnese would have been seriously weakened, at least in the short term.

DISCIPLINE. Discipline was particularly important in a volunteer citizen army of the type dominant in Greece down to the mid-fourth century and which survived in most *poleis* (city-states) for many years after that. Greek states did not have codified bodies of military law to aid discipline, which was generally maintained by commitment to the cause or the *polis*, the relative shortness of most campaigns, and pressure from one's fellow citizens to perform.

In **early Greek warfare** discipline was probably aided by the hierarchical nature of society, reinforced by the threat of violence. When Thersites spoke out of turn at an assembly of the Greeks during the **Trojan War**, **Odysseus** verbally abused him and then beat him (Homer, *Iliad* 2.211-277). Discipline and performance could also be influenced by the use of rewards such as **booty**, land, and an increase in status and prestige levels or by the use of shame or reduction in status as a punishment. The wartime power of leaders presumably closely reflected their level of power in the community, although limited by the need not to alienate the group that was currently doing the fighting.

In the Classical period the level of discipline varied from state to state and from time to time. This is not surprising, as good discipline is often a product of good leadership, which also varies. **Sparta**, though, had a comparatively high minimum standard of discipline, deriving from the *agoge* and the generally military tenor of life there. This sometimes caused problems for Spartans who were unable to adjust their behavior when commanding allied troops. **Pausanias** (1), for example, alienated many by his harsh and arrogant actions at **Byzantium** after the Second **Persian War** (Thuc. 1.95; Plut., *Aristides*, 23; and *Cimon*, 6; see also Thuc. 8.84.2).

The Spartan system made considerable use of shame to produce the desired results. Aristodemus, the sole survivor of the Spartans at **Thermopylae**, was disgraced on his return and called "the coward" by his fellow citizens (Hdt. 7.231). Demeaning punishments were also used, even on fairly prominent men. When he was harmost (governor) of **Abydus**,

Dercylidas was forced by **Lysander** to stand guard carrying his **shield**, which **Xenophon** states "is considered by distinguished Spartans to be a disgrace because it is the punishment for ill-discipline" (*Hell.* 3.1.9). However, Spartan discipline did not necessarily produce unthinking automatons. In 418, when **Agis II**, driven by a slight upon his reputation from a previous failure, was about to attack the Argives and their allies under very unfavorable circumstances, he was dissuaded by the shouted comment from an old soldier that "he was planning to cure one evil with another" (Thuc. 5.64.5-65.3).

The Spartan army had a higher level of discipline than the armies of other *poleis* and discipline in a democracy such as **Athens** was much less easy to maintain. Despite this and the lack of a corpus of military law and regulations, there were some mechanisms for aiding obedience (other than good leadership and the soldiers' acceptance of the need to be on campaign). Athenian *strategoi* (generals) seem to have had the right to employ summary punishment in the field. The Athenian orator Lysias preserves two fifth-century examples—a **hoplite** cashiered for insubordination involving an assault on a **taxiarch**, and a soldier executed for signaling to the enemy (Lysias, 3 [*Simon*] 45; 13 [*Agoratus*] 67); another punishment was fining. However, the power of the *strategoi* to impose summary punishment seems to have been curtailed in the fourth and later centuries.

Subordinate **officers**, such as taxiarchs and **phylarchs**, probably did not have the power of summary punishment in the field, but like any other citizen, they could use the civil courts on return from campaign. Some specifically military offenses were recognized by Athenian courts, including desertion, failure to report for service, cowardice, and throwing away one's shield. Punishment for these could involve loss of citizen rights (*atimia*). **Argos** apparently had a formal procedure whereby a returning army was subject to a public court to try any offenses committed while on campaign (Thuc. 5.72). Officers could also curb problem soldiers by assigning them unpleasant or dangerous duties, or by using peer pressure and shame, and, at Athens at any rate, may have derived some authority from the fact that they were elected officials of the state.

The problem of maintaining discipline was generally even worse in **mercenary** forces. Here, discipline was less likely to be sustained by a belief in the cause or city for which the soldiers were fighting and the officers were often elected by the army (not by the entire citizen body, as was the case at Athens). Xenophon records some real problems with discipline among the **Ten Thousand**, including an attempt to stone **Clearchus** (*Anab.* 1.5.11-17), insubordination (*Anab.* 3.4.47), general lack of discipline, and stoning traders and officers (*Anab.* 5.7.13-23,

6.65-68). Conversely, mercenary troops could at times be more tightly controlled than citizen-soldiers, as demonstrated by **Iphicrates'** strict training and discipline regime in the early fourth century (cf. Frontinus, *Strategemata*, 3.12.2-3).

Philip II and **Alexander the Great** seem to have had few disciplinary problems in their armies, which were rather different in character from contemporary Greek citizen armies. Although Alexander faced some internal opposition during his **Persian** campaign, this was mainly from his court and concerned his adoption of Persian dress and customs and the use of Persians in the army. Alexander's troops, though, did mutiny in India when they realized that Alexander intended to extend his conquests even further.

During the Hellenistic period the **Macedonian** elements of the various successor armies were the source of royal legitimacy that gave them some power over their leaders. **Demetrius I (Poliorcetes)** lost control of Macedon in 287 when he lost the support of his army. Desertions also occasionally occurred, as at **Gabiene** in 316, when the *argyraspides* handed over their general, **Eumenes of Cardia**, to preserve their baggage which was in enemy hands. The increasing use of mercenaries in the Hellenistic kingdoms in the East led to an additional aid to discipline—the drawing up of very specific contracts, and the withholding of **pay** when required.

- **E** -

EARLY GREEK WARFARE. Greek warfare prior to the rise of the **hoplite** in the late eighth/early seventh centuries is difficult to reconstruct. **Homer** provides some clues, but raises just as many problems, while archeological remains are not extensive. However, there is some agreement between archaeology and Homer. Prior to the Dark Age, the status of the warrior was high and considerable resources were expended on arms and **armor.** The palace records in Crete record expenditure on **chariots** (and spare parts) and other military equipment. The finds at **Mycenae** include some beautifully inlaid daggers and gold-handled swords which must have been quite expensive.

However, the Homeric picture of warfare fought largely between aristocratic and royal heroes is unlikely to be correct. There are references in Homer to formations of warriors (e.g., *Iliad* 4.419-432), and the "Warrior Vase" (Athens, National Museum, 1426) from Mycenae, dated to circa 1200, shows a remarkably uniform body of infantry departing for war, each carrying a bag of rations tied to his spear. Forces of this size and apparent rough uniformity must have included more than just the aristocratic heroes of Homer. The primary **weapon** (according to Homer) was the spear, although this was very commonly thrown as well as used for thrusting. In conjunction with the other evidence, this suggests that when clashes occurred, they involved a fairly loose, skirmishing style of fighting.

Sites such as Mycenae were heavily fortified and probably extremely difficult to attack given the lack of siege **artillery**. Partly because of this, and despite the epic scale and treatment of war in Homer, the most common form of warfare prior to the eighth century was probably the **raid**, aimed at stealing women and livestock and/or paying back an enemy for a previous slight. **Ambushes** must also have been common.

Dark Age warfare is even more difficult to reconstruct, although the status of the warrior is unlikely to have diminished much during what seems to have been a fairly troubled period. Weapons, although not armor, were buried with dead warriors. By the early eighth century, the start of the Archaic period, the practice of interring weapons died out, although there are a couple of burials which include armor, including a very well-preserved **helmet** and breastplate from **Argos**, dating to around 750 (figure 2). The archeological and artistic evidence suggests an increasing importance of infantry, especially heavy infantry—which implies some sort of formation—as well as an apparent preference for the sword. **Archers, mounted infantry**, and to a much lesser extent, **cavalry**, are also depicted in contemporary vase paintings.

EGYPT. Greece had contacts, principally trade, with Egypt from at least the Mycenaean period, but its first recorded military contact was the provision of **mercenaries**. According to **Herodotus** (2.152-154), Carian and **Ionian** Greek soldiers served the Saite kings, notably Psammetichus I, in the middle of the seventh century, and this is confirmed by graffiti on the temple of Abu Simbel. These mercenaries assisted Psammetichus in freeing Egypt from Assyria. Apries, who had 3,000, used them unsuccessfully against internal rebels (Hdt. 2.163, 169)

During the fifth century, the **Athenians** saw Egypt, by then part of the Persian empire but frequently in revolt, as a natural ally against **Persia**. A desire to damage Persia, as well as the long-standing contacts with Egypt, its wealth, and its importance as a potential source of grain supply, all caused the Athenians to assist an Egyptian revolt under Inaros in 460. Inaros, using mercenaries as well as locals, had been quite successful in the opening engagements of the rebellion and Athens dispatched the 200-strong **Delian League** fleet operating in **Cyprus** to assist him. It sailed up the Nile, joined Inaros, and inflicted a major defeat on the Persians. A portion of the fleet remained to assist Inaros against the remaining Persians and any future reinforcements, but in 456 it was caught at Prosopitis (an island between a canal and the Nile) and besieged for 18 months. This ended in 454 when the Persians diverted the canal and attacked by land. A relief force of 50 ships was soon afterward surprised and destroyed, unaware that Prosopitis had fallen. This defeat was a major blow to Athens and her allies and played a major part in their decision to end the First **Peloponnesian War**.

The Egyptians staged a series of revolts against Persia in the fourth century, and this was often a factor in Persian involvement in various **Common Peace** treaties in Greece. While Greece was at peace, Persia had a better chance of employing the Greek mercenaries it needed to stiffen its own army against rebels in Cyprus and Egypt. Not surprisingly, Egypt surrendered without a fight to **Alexander the Great**. In 332 he restored religious freedom, was crowned pharaoh, founded Alexandria, and left six governors in charge (two **Macedonians**, with military powers and command of the garrisons, and two Greek and two Egyptian governors with civil powers). The foundation of Alexandria as a Greek city, as well as the garrison and Macedonian/Greek governors, began a process of hellenization. On Alexander's death, Ptolemy (later **Ptolemy I [Soter]**) seized control in Egypt, which by 321 was capable of resisting the other *diadochoi*.

From this point onward, Egypt became a Hellenistic monarchy, with a Macedonian ruler, a Greco-Macedonian administration, and an increasingly Greek culture. However the Greco-Macedonian ruling class was generally very careful to cater for local religion and customs and the monarchy had a definite Egyptian flavor to it. Egypt was arguably the most successful of the Hellenistic monarchies—although far from the most powerful. It survived a series of wars against the other monarchies, particularly Seleucid **Syria**, as well as native Egyptian revolts (the most serious creating a separate native Egyptian kingdom in the Thebaid about 205-185—perhaps influenced by the use of Egyptians in the **phalanx** at **Raphia**). From 321 to 169 no foreign army successfully penetrated Egypt and the occupation of the country by **Antiochus IV (Epiphanes)** in 169-168 was ended by Roman edict. After that, Ptolemaic Egypt was increasingly influenced by **Rome** and its wishes. Egypt retained its formal independence and was the last Hellenistic monarchy to be annexed by Rome after **Cleopatra VII (Philopator)**'s alliance with Marcus Antonius led to the battle of **Actium** in 31.

EKDROMOS (pl. EKDROMOI). A soldier who charged (as part of a group) out of a formation to drive off an enemy who was harassing it.

Ekdromoi were normally chosen from the youngest and fittest soldiers. *See also* LECHAEUM.

ELEPHANTS. The first recorded contacts of a Greek or **Macedonian** army with war elephants were at **Gaugamela** in 331, when **Darius III** fielded about 15, and at the **Hydaspes River** in 326. The numbers at the Hydaspes are variously given in the sources as 85 (Curt. Ruf. 8.13.6), 130 (Diod. Sic. 17.87.2) and 200 (Arr. *Anab.* 5.15.4). **Alexander the Great**'s troops handled them relatively easily, directing missile fire against the drivers and the elephants themselves. Boxed in by the pressure of the Macedonians, riderless and wounded elephants did as much damage to their own troops as to the enemy. This illustrates the main disadvantage of elephants in battle—they were spectacular and frightening (particularly to horses and troops which had never seen them before), but relatively vulnerable and hard to control. Although the crew, equipped with **javelins** or **bows**, were protected by a crenellated tower, the driver was completely unprotected sitting on the elephant's neck. Because of these vulnerabilities, elephants were usually protected by detachments of light troops.

Despite these problems, war elephants became a fairly regular feature of Hellenistic warfare, particularly in Seleucid armies. **Syria** from the time of **Seleucus I (Nicator)** had the closest contacts with India and eventually gained a monopoly over the supply of Indian elephants. Seleucus received elephants from the Indian king Chandragupta, and the 75 he deployed at **Ipsus** in 301 were instrumental in the victory over **Antigonus I (Monophthalmus)**. Later on, **Antiochus III (the Great)** reestablished links with Bactria and Gandhara, receiving war elephants in return for recognition and renewal of friendship. However, Ptolemaic **Egypt** also imported elephants and when the Indian supply was closed to them, Ptolemy III (Euergetes) and **Ptolemy IV (Philopator)** sent expeditions to what is now Somalia to capture "Ethiopian" elephants. These were African forest elephants, smaller than their Indian equivalents.

At the battle of **Raphia** in 217, the Egyptians fielded 73 elephants, the Seleucids 102. **Polybius** (5.84) provides an interesting account of their use in this battle, along with a general description of elephant-to-elephant combat. Antiochus took only six elephants to Greece with him when he invaded in 192, but deployed 54 at **Magnesia** in 189. **Ptolemy I (Soter)** successfully used a system of chains and spikes against the 43 elephants deployed by **Demetrius I (Poliorcetes)** at **Gaza** in 312. **Pyrrhus I** of Epirus used elephants in his war with the **Romans**, achieving success on the first occasion at the **Siris River** in 280, but the

Romans had learned to deal with them by **Beneventum** in 275. The contribution of elephants to **Antiochus I (Soter)**'s victory over the Galatians **(Gauls)** in 275(?) was such that it was called the "Elephant Battle." Elephants were occasionally even used in sieges, with limited success. Polyperchon used them unsuccessfully at Megalopolis in 318, where they fell victim to spiked boards hidden in shallow trenches (Diod. Sic. 18.71); their last recorded employment at a siege was by **Antigonus II (Gonatas)** circa 270.

ELIS. The major city in the region of the same name to the north of **Messenia** on the west coast of the Peloponnese. The city of Elis seems to have been dominant over the other parts of the region by about the eighth century, although clashes, particularly over the right to conduct the Olympic games, continued intermittently through the Classical period. Elis was apparently a loyal ally of **Sparta** from the Second **Messenian War** (when other parts of the region, Pisatis and Triphylia, supported the Messenians) down to the Peace of Nicias during the **Peloponnesian War** of 431-404.

In 421 Elis (along with **Boeotia**, **Corinth**, and **Megara**) voted against the Peace of Nicias. In retaliation Sparta backed the claim to independence of Lepreum, in the Triphylia region of Elis, and placed a garrison of **hoplites** to protect it. Elis joined **Argos**, **Mantinea**, and Corinth in an anti-Spartan alliance (Thuc. 5.17.2, 31). This alliance was broken up after its defeat at the battle of Mantinea in 418 (where the Elean **cavalry** saved the **Athenian** hoplites from disaster) and at the end of the Peloponnesian War Sparta sought revenge against Elis. The war was caused by Spartan demands that Elis give independence to the cities under its control. It lasted from 402 to 400 and the Eleans were forced to come to terms after a comprehensive **ravaging** of their territory by Sparta, assisted rather opportunistically by **Arcadia** and **Achaea**. The unfortified city of Elis was directly threatened but not attacked, and internal violence erupted between a party of wealthy pro-Spartans and the democrats. The Eleans were forced to surrender control over Triphylia and destroy the fortifications of their harbor of Cyllene. They also lost territory to Arcadia (Xen. *Hell.* 3.2.21-31). This effectively destroyed Elis' regional power base.

An attempt to retrieve their former possessions after the Spartan defeat at **Leuctra** in 371 failed when Triphylia joined the newly formed **Arcadian League**. Rather ironically, the Eleans then allied with Sparta and on the basis of this went to war with the Arcadians in 366. After some initial success, the Eleans were forced back to their own city, where internal strife again broke out. The democratic party almost took control but was forced out and occupied **Pylos**. The Arcadians had also garrisoned **Olympia**. In 365 the Arcadians invaded again and defeated the Eleans in battle, but withdrew when Sparta created a diversion by invading Arcadia. Pylos was taken and the defenders massacred. The Elean cavalry and a body called the Three Hundred (apparently the Elean *epilektoi*) figured prominently in these actions. The following year the Arcadians held the Olympic games under the presidency of the Pisatans and narrowly beat off an attack by Elis within the Olympic precinct. Although the Arcadians were clearly winning the war, they lost support when they stole treasure from the temples at Olympia. The outcry, especially within the Arcadian League, which began to fragment over the issue, was such that peace was made with Elis, which regained the presidency of the Olympic games for 362 (Xen. *Hell.* 7.4.12-35).

Later in the century Elis backed **Philip II** of **Macedon**—partly because of his bribery, but also because of support he gave to the antidemocratic party that seized power in 343. Elis sent 150 soldiers (probably cavalry) to join **Alexander the Great** after **Granicus River**, but after Alexander's death fought on the Greek side against Macedon in both the **Lamian War** (323-321) and the **Chremonidean War** (268-261). In the middle of the third century Elis had regained Triphylia and was generally hostile to the **Achaean League** and, in concert with the **Aetolian League** and at times Sparta, engaged in fairly regular warfare with the Achaeans. In 219/18 Elis was ravaged by **Philip V** of Macedon (an Achaean ally), who garrisoned Triphylia. The Eleans riposted with a successful invasion of Achaea in 218/7. In 192 the Eleans requested assistance from **Antiochus III (the Great)**, who sent 1,000 men to help against the Achaean League, but details of this period are generally obscure. Elis was ultimately incorporated into the **Roman** empire along with the rest of Greece.

ELLEPORUS RIVER. Also known as the Helleporus River. Site of a battle, fought in **Italy** in 389, between the forces of **Dionysius I** of **Syracuse**, allied with Locri and the Lucanians, against the Italiote League (which consisted predominantly of Greek cities in the southwest of Italy). While trying to neutralize the league, in order to isolate and defeat Rhegium, Dionysius besieged one of its members, Caulonia. He then **ambushed** the Italiote League relief force of 15,000 infantry and 2,000 **cavalry**, led by a Syracusan exile, Heloris, at the Elleporus River. Dionysius located the enemy first and deployed his 20,000 infantry and 3,000 cavalry to meet the enemy's unsuspecting advance guard. He defeated this, killing Heloris in the process, and destroyed the rest of the army as it raced to the attack in separate bodies. The 10,000 survivors were trapped on a hill and forced to surrender unconditionally. However, Dionysius released them without ransom, enabling him to make favorable agreements with many league members. He then defeated the two which held out (Caulonia and Hipponium) and besieged and captured Rhegium in 387, effectively ending the Italiote League and making Dionysius the dominant force in southern Italy.

ELLOMENOS. A town on the island of Leucas. A contingent of its garrison troops was successfully **ambushed** by **Demosthenes** (1) in 426 on his way to attack the island's main city (also called Leucas) during the **Peloponnesian War** (2). *See also* OLPAE.

EMBATA (also EMBATUM). A naval battle fought in autumn 356 in the straits between **Chios** and the **Ionian** mainland during the **Social War** of 357-355. The **Athenians** had 120 **triremes**, the rebels 100, but two of the Athenian generals, **Timotheus** and **Iphicrates**, refused to attack because the weather was bad and **Chares**, attacking alone with only part of the fleet, was defeated. This was the second Athenian naval defeat of the war and, together with serious problems in equipping and manning ships, seriously weakened Athens' chances of victory. Public outrage was such that Iphicrates and Timotheus were prosecuted on their return—Iphicrates was acquitted but Timotheus was fined 100 talents. Chares was given sole command, but without sufficient resources to really achieve anything.

EPAMINONDAS (EPAMEINONDAS; d. 362). Theban general and political leader. Epaminondas first drew attention to himself by his personal bravery in battle at **Mantinea** in 385. He stayed in **Thebes** in 382 when **Phoebidas** occupied the Cadmea, but actively worked against the pro-**Spartan** group. He played a major role in the coup of 379, during which the Cadmea was retaken and Thebes liberated from Spartan control. In partnership with **Pelopidas**, Epaminondas perhaps exerted a major influence on Thebes' rise to hegemony in Greece over the following years, although his role is obscure until his election as boeotarch in 371.

As boeotarch, Epaminondas led the **Boeotians** at **Leuctra** in 371 and was responsible for the oblique formation that destroyed both the Spartan army and Spartan military power. This victory made his reputation as a general. He was joint commander, perhaps the more influential, with Pelopidas in the first invasion of the Peloponnese in 370/69 which established **Arcadian** independence from Sparta. He led two subsequent invasions of the Peloponnese in 369 and 366.

Apparently temporarily out of favor, in 368 Epaminondas took part as a private soldier in the first expedition to **Thessaly** to free Pelopidas, but rallied the army when it was in great danger during a withdrawal. In 367 he commanded the expedition that succeeded in freeing Pelopidas. After this, he developed a navy and challenged **Athens** at sea—with some initial success.

Epaminondas was killed in 362 at the battle of Mantinea (2), during another invasion of the Peloponnese, although his army was victorious. **Xenophon**, not noted for a pro-Theban attitude, comments (*Hell.* 7.5.8) that, although it was an unlucky campaign, Epaminondas "was beyond reproach in terms of planning and daring" (*Hell.* 7.5.19-20). He later adds that Epaminondas had trained his army to a very high level and the troops (including his allies) were prepared to follow him anywhere. An innovative and highly successful military leader, Epaminondas arguably brought the **hoplite phalanx** to its peak of perfection. His death (along with Pelopidas') contributed to the military decline of Thebes. *See also* BOEOTIAN LEAGUE.

EPHEBEIA. The state youth-training system at **Athens**, probably founded as a formal training institution in the second half of the fourth century. It involved two years of training/garrison duty for youths (*epheboi*) between the ages of 18 and 20, who served as light-armed frontier guards. In the Hellenistic period a greater variety of training, including equestrian and **artillery**, was provided. It continued into the **Roman** period, although taking on an increasingly less military character.

EPHESUS (EPHESOS). Ionian Greek city in Lydia, the closest port to Sardis, and therefore an important entry point into **Ionia**. It was also a major cult center for the goddess Artemis. Croesus besieged the city in 555 and apparently conquered it despite its countermeasure of linking itself to the neighboring temple of Artemis with a rope to ensure divine protection. In 499 the **Athenian** and **Eretrian** contingents sent to help against **Persia** in the **Ionian Revolt** landed at Ephesus. They marched inland from there with the rebels and fired Sardis. However, during their withdrawal the Persians caught up with them just outside Ephesus and defeated them in battle. The commander of the Eretrian contingent was killed and the Athenians refused to contribute any further help to the revolt (Hdt. 5.99-103). Ephesus seems to have taken no active part in the revolt, and its temple, unlike many others, was spared at the end of the war.

During the **Peloponnesian War** of 431-404 Ephesus was the site of another engagement, when an Athenian force under **Thrasyllus** was defeated by a force of Ephesians, **Syracusans**, and Persians under **Tissaphernes** (Xen. *Hell.* 1.2.6-10). **Lysander** used the city as his base in 407 and from there defeated the Athenian fleet at **Notium**. **Agesilaus II** also used Ephesus as his initial base for the start of his Persian campaign in 396.

The city was occupied without opposition by **Alexander the Great**, and was subsequently controlled at various times by **Lysimachus**, the Antigonids, the **Ptolemies**, and the Seleucids, who used it as one of their principal naval bases. Sometime during the Second **Syrian War** the Ptolemaic navy was apparently defeated off Ephesus by the **Rhodians** and their allies. In 189 **Antiochus III (the Great)** used the city as his headquarters during his war with **Rome**. Under the Treaty of **Apamea** Ephesus was given to **Eumenes II (Soter) of Pergamum**. The city became the principal one of the Roman province of Asia, formed in 129, and was the usual seat of the governor. Despite this it sided with **Mithridates VI (Eupator)** in the First **Mithridatic War** and was punished with a heavy fine by Lucius Cornelius Sulla.

EPHIALTES (d. c. 470). A Trachinian who showed **Xerxes I** the Anopaea path around the Greek flank at **Thermopylae** (1) which allowed Xerxes to destroy the Greek force there. The **Amphictionic League** placed a price on his head and around 10 years later he was killed by a fellow Trachinian motivated by a personal grudge.

EPIBATES **(pl.** *EPIBATAI***).** A soldier who fought from a platform such as a **chariot, elephant**, or **warship**; at **Sparta**, a subordinate naval **officer**. In the Classical period, the word is commonly used to denote a **hoplite** serving as a marine on a warship. The function of *epibatai* in a sea battle was to defend their ship against boarders and, in turn, to board enemy ships. Other duties included taking part in amphibious **raids** and probably the provision of security to the ship when it was beached. An Athenian **trireme** normally carried around 10 *epibatai* (supplemented by four **archers**). These *epibatai* were designated part of the ship's complement for that campaign and were distinct from any extra troops who might be carried for particular missions or ordinary hoplites carried on warships converted to troop carriers. As far as we know, *epibatai* were not specialist troops—at **Athens** they were usually drawn from the normal hoplite lists, although there were exceptions to this. According to **Thucydides** (6.43), the Athenians (probably still suffering from the demographic effects of the plague) deployed 700 *thetes* as marines in the **Sicilian Expedition** in 415. As the *thetes* comprised the nonhoplite class they must have been armed by the state. Like many Greek terms, *epibates* is often used loosely and can mean any troops on a ship who could engage in combat.

EPILEKTOI. Selected troops who formed a state-financed permanent or semipermanent force in many Greek states in the fourth century. Their creation was motivated by an acceptance that part-time **hoplites** could not compete with the **Spartans** nor operate as effectively in the more intensive conditions of fourth-century warfare. The Theban "**Sacred Band**" was one of the first but others include the "**Three Hundred**" of Elis and the "**Thousand**" of Argos.

EQUESTRIAN EQUIPMENT. Throughout antiquity Greek **cavalry** operated without the stirrup, which did not come into widespread use in Europe until the fifth century A.D. Horseshoes were also unknown. The lack of this equipment affected the ability of cavalry to operate on hilly and/or broken and rough **terrain**. However, as **hoplites** too could not maintain formation in this type of country, it was not a barrier to the military use of the horse for much of Greek history.

Bits and bridles were in common use from **Mycenaean** times. Examples of bronze bits survive from this period (case 16, Archeological Museum, **Thebes**); later bits were made of iron. These were often "severe" bits, designed to cause discomfort to the horse when the reins were pulled. This type of bit (recommended by **Xenophon**, *On Horsemanship*, 10.6-12) would have aided in control of the horse and was necessary given the lack of stirrups.

Most riders rode bareback—saddlecloths and saddles are attested, but are rare in the pictorial and literary evidence. When saddles were used, it was apparently for comfort rather than to make the rider's seat more secure. **Demosthenes** (2) certainly criticizes Meidias for taking luxury items, including a padded saddle, on campaign in **Euboea** (21 [*Against Meidias*] 133-134).

ERETRIA. One of the two major cities located on the western coast of **Euboea**. It was an early maritime and colonizing power in Greece. Prior to the **Lelantine War** it controlled Andros, Tenos, and Ceos, as well as other islands. At around this time it could apparently field 3,000 **hoplites**, 600 **cavalry**, and 60 **chariots**. Although its power was severely curtailed by the loss of this war to Chalcis, in recognition of Milesian assistance during this war, Eretria sent five **triremes** to support **Miletus** at the start of the **Ionian Revolt** in 499. In revenge, the **Persians** sacked the city in 490 and removed the surviving citizens to Asia Minor, settling them near Susa.

Eretria was rebuilt slightly south of the original site and fairly quickly regained its previous size. It became part of the **Delian League**, and then the Athenian empire during the course of the fifth century. Eretria initiated a revolt against **Athens** in 411 by massacring the Athenian sailors who had sought refuge there after an Athenian fleet was defeated by the **Spartans** off the city (Thuc. 8.95.6).

In the first half of the fourth century the city was ruled by a series of tyrants who generally took an anti-Athenian line. However, in 348 the tyrant Plutarchus appealed to Athens for aid against Chalcis and **Phocion** was sent with an army. After winning the battle of **Tamynae**, Phocion then expelled Plutarchus from Eretria and established a democracy. Shortly afterward, **Philip II** of **Macedon** sacked Eretria's harbor and installed a pro-Macedonian tyrant, who ruled until expelled by Phocion in 341. Following the battle of **Chaeronea** in 338, Eretria became subject to Macedon and remained so until captured by the **Romans** and their allies in the Second **Macedonian War** and subsequently declared free.

EUBOEA (EUBOIA). The largest Aegean island, off the east coast of **Boeotia**, the southeast coast of **Thessaly**, and the northeast coast of Attica. The two main cities on it were Chalcis and **Eretria**, both wealthy and influential in early Greece and major colonizing powers in the eighth and seventh centuries. At the end of the eighth century the two fought the **Lelantine War**, which left Chalcis the dominant power on the island and Euboea enjoyed considerable prosperity for around 200 years. However, at the end of the sixth century Chalcis joined with Boeotia against **Athens**, which defeated it around 506 and settled 4,000 **cleruchs** on territory confiscated from the city. Eretria was destroyed by the **Persians** in 490 during the First **Persian War**—as a punishment for having sent five **triremes** to assist the rebels in the **Ionian Revolt**.

The island was subject to Athens by the mid-fifth century and formed a particularly valuable part of the empire. Chalcis was already under Athenian domination prior to the Persian Wars, Carystus (at the southern end of the island) was early on forced into the **Delian League**, and there is evidence of more cleruchs established on the island around 450. Euboea revolted in 446 (at the same time as **Megara**) but was defeated by an Athenian expedition under **Pericles**. He restored Athenian control and expelled the population of Hestiaea, establishing Athenian cleruchs on their land.

During the **Peloponnesian War** of 431-404, the island played an important part in Pericles' strategy of avoiding a land battle with the Peloponnesians. One of the ways he was able to achieve this was by evacuating the livestock from Attica and storing it on Euboea during the annual Peloponnesian invasions. This considerably reduced the damage that could be inflicted on Athenian agriculture. Imports also came by sea to Euboea and were then brought overland via Rhamnus, avoiding the voyage around Attica, which was often difficult because of contrary winds at Cape Sunium. Because of this, the revolt of Euboea in 411 was a serious blow to the Athenian war effort and, according to **Thucydides** (8.96.1), caused the greatest panic of all time at Athens—greater even than the news of the destruction of the **Sicilian Expedition**.

In the fourth century a resurgent Athens reestablished its influence over Euboea, but never to the same extent as before. **Jason of Pherae** had unsuccessfully tried to interfere there in the 380s and tyrannies arose in many of the cities in the 370s and 360s. Euboea was also subject to **Theban** influence and seceded from the Second Athenian Confederacy shortly after the battle of **Leuctra** in 371 but Athens had restored the situation on the island by 357. In 349-348 Euboea was in revolt against Athens, perhaps with encouragement or assistance from **Philip II** of **Macedon**. This is not proven, but it would certainly have been in his interests to distract Athens from his activities in the north against **Olynthus**. Athens recognized Euboean independence in 348 after the generally unsuccessful campaign of **Tamynae**. In 346 a proposal was put forward by Chalcis for a Euboean League. Philip sent 1,000 **mercenaries** to Euboea in 343 (or possibly 342) to intervene in internal strife in Eretria and in 342 Macedonian troops assisted in the overthrow of the democracy at Oreus. The following year Callias, the Chalcidian leader, negotiated a defensive alliance with Athens and, with additional help from Megara, liberated Oreus and then Eretria. A general alliance between Athens and the Euboean cities was followed by a visit by Callias and **Demosthenes** (2) to the Peloponnese to create more support for the anti-Macedonian cause. This policy ended in failure at **Chaeronea** in 338.

From this point on, Euboea was under Macedonian influence, although which particular Macedonian group was in control varied. For example, around 249 (or possibly 252) the son of **Craterus**, Alexander, revolted against **Antigonus II (Gonatas)** and established an independent kingdom based on **Corinth**. He removed the Macedonian garrisons from Euboea and revived the Euboean League, but Antigonus recovered the island in 245 and reinstalled the garrisons, especially at Chalcis. In 194 the Romans freed the island from **Philip V** of Macedon and it subsequently supported **Rome** against the **Aetolians**. Euboea was taken by **Antiochus III (the Great)** in 192 when he invaded Greece and subsequently became part of the Roman province of Achaea.

EUMENES OF CARDIA (c. 362-316). One of the *diadochoi*. A Greek from Cardia (Kardia) in the Peloponnese, he was **Alexander the Great's** principal secretary and was appointed satrap of Cappadocia on Alexander's death. He fought for **Perdiccas** (2) against several of Alexander's generals who were trying to establish separate kingdoms, defeating and killing **Craterus** in 320. In the same year, Perdiccas' murder in **Egypt** removed Eumenes' legitimacy as a commander and he was treated as a rebel by his opponents. An able leader, Eumenes was executed in 316 by **Antigonus I (Monophthalmus)** when his *argyraspides* deserted, causing his defeat at the battle of **Gabiene**. *See also* PARAETACENE.

EUMENES OF PERGAMUM (PERGAMON).
    (1) Eumenes I (d. 241). Ruler of Pergamum from 263 to 241. He played a major part in continuing the movement of Pergamum toward independence from the Seleucids that was started by his uncle Philetaerus and formalized by Eumenes' adopted son, **Attalus I (Soter)**, the first of the dynasty to assume the title of king. Eumenes, with assistance from **Egypt** that allowed him to hire a large force of **mercenaries**, defeated **Antiochus I (Soter)** at Sardis in 262 and then expanded Pergamene territory, incorporating the neighboring coastal areas as well as Mount Ida (which produced timber and pitch important for ship-building and maintenance). However, Antiochus II (Theos), seems to have recovered some of the lost Seleucid possessions from Eumenes during the Second **Syrian War** (260-253). Despite this setback, Eumenes managed to retain the core of his possessions (even extending them again slightly during the Third Syrian War) and Pergamene independence, leaving a solid base to his successor, Attalus.
    (2) Eumenes II (Soter) (d. 160). Son of **Attalus I (Soter)** and king of Pergamum, 197-160. Throughout his reign he followed a policy of friendship with **Rome** and of using Rome to weaken or neutralize his enemies. His complaints to Rome in 196 helped persuade it to go to war with **Antiochus III (the Great)** and his visit to Rome and complaints against **Perseus** in 172 helped spark the Third **Macedonian War**.
    During the war with Antiochus, Eumenes (along with **Rhodes**) supplied naval support to Rome and engaged Antiochus on land in Asia Minor—initially without success. Pergamum itself was besieged for a short time until relieved by Roman and Rhodian naval successes and the arrival of a relief force from the **Achaean League**. In 190 Eumenes commanded the right wing of the Roman-allied army at **Magnesia**, using his **slingers** to defeat Antiochus' **chariots** and then leading a well-timed attack that materially contributed to victory. The Treaty of **Apamea** which formally ended this war forced Antiochus to repay 400

talents to Eumenes and hand over to him his war **elephants** and large sections of western Asia Minor (including Lydia and Phrygia on the Hellespont). Cappadocia also made peace with Eumenes, who was now arguably the most powerful Hellenistic ruler in Asia Minor. His close links with Rome helped him against the expansionist Prusias I of **Bithynia** and against Pharnaces of Pontus. Rome ordered an end to the war of 186-183 and ruled in Eumenes' favor against Pharnaces during the war of 183-179. However, Rome provided no material assistance and Eumenes was forced to win the war against Pharnaces on his own—which he did by an invasion of Pontus in 179.

Perseus of **Macedon**'s dynastic marriage alliances with Bithynia and the Seleucids early in his reign aroused Eumenes' suspicions and he worked tirelessly to convince Rome that Perseus was a threat. During the ensuing Third Macedonian War (171-167), Eumenes provided military and naval support to Rome, but for some reason fell rather out of favor. Stories circulated that he intended to act as a mediator between Rome and Perseus, or even to switch sides—both highly unlikely. However, according to **Polybius** (29.4-6) some sort of negotiation did occur between Eumenes and Perseus. The Romans seem to have transferred their favor to Eumenes' brother Attalus, although they did support Pergamum against the Galatians, who rose in revolt in 168/7. Polybius (32.22) provides a very favorable assessment of Eumenes, pointing out that he inherited a small state and bequeathed the largest kingdom of his time.

EURYBIADES (fl. c. 480). A Spartan officer during the **Persian Wars**; son of Eurycleides. He commanded the combined Greek fleet against the **Persians** in 480, although his level of naval experience was probably fairly low as Sparta only contributed 16 ships to the fleet. His appointment illustrates the confidence the Greeks had in Spartan military leadership at the time, as well as their suspicion of **Athens**—if **Herodotus** 8.2-3 is correct. In Herodotus' account of the 480 campaign, the credit for the Greek naval decision-making and success is given to **Themistocles**, probably correctly. *See also* ARTEMISIUM; SALAMIS.

EURYMEDON RIVER. A river near Phaselis in Pamphylia (southern Asia Minor), site of a major victory circa 467/6 by the **Delian League** fleet, led by **Cimon**, over the **Persians**. Cimon's fleet consisted of 300 **triremes** (200 provided by **Athens**, 100 by her allies), while the Persians had around 350, with an additional 80 Phoenician **warships** en route from **Cyprus**. The Persians appear to have mustered their fleet with a view to challenging Athenian naval supremacy in the Aegean but

Cimon learned of this and attacked first. Cimon had modified his triremes, originally built by **Themistocles**, by building bridges between
the side decks. This allowed the additional **hoplites** he had for the expedition to move more easily from one side of the boat to the other and
suggests that ramming was less important than at **Salamis**.

Striking before the ships from Cyprus could arrive, Cimon sailed
into the mouth of the Eurymedon River but the Persians withdrew up-
stream. He pursued them and forced an engagement in which the Persian
fleet was routed, with the loss of 200 ships captured and many others
destroyed. Cimon then landed his hoplites and defeated the Persian army
on land. **Plutarch** (*Cimon*, 13) states that Cimon surpassed Salamis and
Plataea by achieving dual victories on one day—like a champion ath-
lete. Cimon then followed this up by destroying the 80 Phoenician
ships which had not yet reached Eurymedon. The victory ended any
hopes of a Persian naval recovery and the Persians kept their naval forces
out of the Aegean—whether under the terms of the Peace of **Callias**, or
de facto observing similar conditions. Eurymedon River secured Athenian naval supremacy for most of the rest of the century and paved the
way for an increasing hold over her allies.

EVAGORAS (EUAGORAS; c. 435-374). A member of the Greek Teucrid
dynasty that had ruled Salamis in **Cyprus**. He returned to Salamis from
exile in 411/10 and with only 50 followers seized power from the Phoenicians. As king, Evagoras fostered Hellenism and forged close links
with **Athens**, in particular with **Conon**, who had sought refuge with
him after **Aegospotami**. He assisted in the naval victory over **Sparta** at
Cnidus in 394 but by 389 was in revolt against **Persia**—with Athenian
support. Evagoras (operating in concert with **Egypt**, which was also in
revolt) was initially successful. His navy reached a peak of 200 **triremes**
and he secured temporary control over parts of Cilicia and Phoenicia.
However, with the King's Peace of 386 (**Common Peace** [1]), Persia
was free to deal with Egypt and Cyprus and had a ready supply of Greek
mercenaries. Following an unsuccessful campaign against Egypt, a
large force was raised and Cyprus was attacked in 382. Evagoras beat
this off but the loss of the naval battle at Citium (381?) allowed the Persians to besiege Salamis. Around 380 Evagoras was forced to come to
terms, but dissension between the Persian army commander and the Persian admiral led to the imposition of a relatively mild settlement.
Evagoras was left as king but was assassinated in 374, having made a
major contribution to restoring Cyprus' strength, reputation, and Greek
character.

- F -

FOOT COMPANIONS (*PEZHETAIROI*). The Foot Companions were
the core of the **Macedonian** infantry, comprising the six **phalanx**
groups in the Macedonian army under **Philip II** and **Alexander the
Great**. This corps consisted of 9,000 men, grouped into three brigades
(*taxeis*) of 1,500 men (each comprising three *lochoi* of 500 men). All of
the *taxeis* may have been recruited regionally—at least three came from
areas of Upper Macedonia. They were probably instituted in the reign of
Alexander II of Macedon (*FGrH* 72 fr. 4). *See also* HYPASPISTS.

FORAGING. As Greek armies of the Classical period generally lacked an
organized commissariat, foraging was one of the main methods Greek
soldiers used to feed themselves. It was often combined with **ravaging**
and/or plundering. However, troops engaged in these activities usually
scattered in small groups across the countryside and were vulnerable to
cavalry or *psiloi* (light troops). For example, **Persian** cavalry under
Pharnabazus killed 500 men from a large foraging party of the **Ten
Thousand** and in 395 he repeated the success against a 700-man party
from **Agesilaus II**'s army, killing 100 of them (Xen. *Anab.* 6.4.24;
Hell. 4.1.17-19). The first general attested as regularly using other
troops to protect his foragers is **Alexander the Great**. This practice may
have become more common in later Greek armies, though, as
Onasander certainly advocated protecting authorized parties and ban-
ning unauthorized foraging.

FORMATIONS. From the time the **hoplite phalanx** was introduced,
Greek heavy infantry fought in close formation. **Peltasts** equipped with
the thrusting spear and **cavalry** also operated in formation, while most
psiloi (light troops) did so only in very loose order at best. The most
common fighting formation for heavy infantry (whether hoplites or *sarissa*-armed **Macedonians**) was the phalanx. However, marching troops
usually moved in column (unless under immediate threat, in which case
careful generals moved them in fighting order). The hollow square was
also used, particularly when heavy infantry were threatened by *psiloi*.
With professional troops such as the **Spartans**, the *epilektoi* of various
states, and the **Ten Thousand**, reasonably complex maneuver in formation was possible. During their withdrawal from **Cunaxa**, for example,
the Ten Thousand are recorded as having marched in a square formation
that was contracted by a set drill when traversing a defile and expanded
again afterward (Xen. *Anab.* 3.4.19-23). However, this was beyond the
average citizen-soldier army.

The most common cavalry formation was rectangular. However, the **Thessalian** cavalry used a rhomboidal formation and **Philip II** of Macedon is credited with introducing the wedge or triangular formation. The advantage of these over the rectangular formation was that they were easier to maneuver, particularly in terms of changing direction, and gave cavalry equipped with suitable **armor** and a thrusting spear or *sarissa* the opportunity to penetrate an infantry formation. Under the Hellenistic monarchies, most cavalry seems to have reverted to the traditional rectangular formation.

FORTIFICATIONS. Although the palaces of the Minoan world were not protected by large walls, from the Mycenaean period onward the Greeks made extensive use of fortifications to protect themselves. While many of the early fortifications were designed to protect against **raids**, especially from pirates, some of the fortified sites from the Mycenaean period were on a massive scale. The walls at **Mycenae** and Tiryns were roughly 23 feet (7 meters) thick and at least 30 feet (9 meters) high. As in later periods, the walls enhanced natural slopes and considerable attention was paid to the gates—the weakest point in a wall. Measures to defend the entrances included designing ramps to force attackers to move uphill with their unshielded side to the walls, projecting bastions (but probably not towers at this date) to allow fire to be brought to bear from two directions, and narrow and/or winding exits from the gate into the city.

During the Dark Age the Mycenaean techniques were lost and nothing to compare with them was constructed until the sixth and fifth centuries; the earlier Mycenaean sites were often repaired and reused down to the third century. Iron-age Greek fortifications were influenced by Near Eastern techniques, and the most elaborate defensive walls down to the late sixth century were in **Ionia** and the islands, where the Greek inhabitants had to face the more sophisticated siege techniques of the Lydians and **Persians**. Under the tyrant Polycrates, Samos was protected by walls with towers and a ditch. To ensure a water supply during a siege (and as an additional place of refuge), a tunnel almost a mile long was cut through a hill behind the city. This feat is even more impressive considering that the tunnel was dug from both sides of the hill and met in the middle. In the west, where the Greeks in **Sicily** faced the **Carthaginians**, the evidence suggests that defensive systems were more advanced than in mainland Greece.

Many Greek cities possessed a fortified citadel (e.g., the Acropolis at **Athens**, the Cadmea at **Thebes**, Acrocorinth at **Corinth**, and the island citadel of Ortygia at **Syracuse**) as the last line of defense. The city itself (and sometimes sections of agricultural land) was surrounded by a wall, which down to the fourth century was often of brickwork on top of a stone base of varying height. In the fourth century solid stone walls became much more common (although mudbrick continued in use because it was cheap and also had good shock-absorbing qualities, particularly as a facing material). It was fairly common from the early fifth century for maritime cities to link with their ports using "Long Walls." The most famous connected Athens to the Piraeus and rendered the city immune to siege except by a strong land and naval power; other examples include Corinth and **Megara**. Long Walls, though, had fallen into disuse by the end of the third century at the latest because of improved siege techniques. One exception to this general pattern of defensive fortification was **Sparta**, which throughout the Classical period had no walls, relying instead on the prowess of its **hoplites**.

In the fourth century, with the development of **artillery** and the experience of several sieges during the **Peloponnesian War** of 431-404, Greek fortifications became much more elaborate. For example, Athens seems to have developed a series of walls and forts throughout Attica to aid in its defense. These served several purposes, including allowing signals to be sent from location to location, providing local rural refuges, and facilitating sorties. Larger fortresses (such as Eleusis and Oenoe) allowed invading forces to be harassed on the way both in and out of Attica. A good example, probably dating from the first half of the fourth century, is at Eleutherai (see plate 3). Although the Athenian system is a particularly sophisticated example, other states, such as Thebes, also made use of walls and forts to help protect their territory.

As siege techniques and artillery improved, fortifications kept pace, with larger numbers of bigger towers mounting artillery to cover the walls and to suppress the artillery and siege machines of besiegers. Ditches and stockades were also employed to keep the enemy at a distance from the walls. Occasionally temporary structures were constructed. At Torone, for example, a wall of sand-filled baskets was constructed when the city was surprised by **Timotheus** (who cut through the baskets with scythes mounted on poles or perhaps his [detached] ships' masts; Polyaenus, *Strategemata*, 3.10.15). Although attested earlier, sally ports became increasingly common in the third century to allow the defenders to make sorties against the enemy siege engines. At around the same time, large bastions were built to mount batteries of catapults—one of the biggest examples being the Euryalus fort at Syracuse (unfinished when the **Romans** captured the city in 211 after a hard siege). Despite these advances, by the third century the balance of power had swung to the besieger, who was much more likely to have a professional army well versed in siege techniques. The citizens of Greek city-states were usually unable to match the well-organized forces of Hellenistic monarchs and later the Romans. *See also* DELIUM; HALICARNASSUS; SIEGE WARFARE; TYRE.

- G -

GABIENE. A battle fought northwest of Persepolis in early 316 between **Eumenes of Cardia** and **Antigonus I (Monophthalmus)**. Eumenes had around 35,000 infantry and 6,000 **cavalry**, while Antigonus led 22,000 foot and 9,000 cavalry. The battle was fairly even, with Eumenes' *argyraspides* ("Silvershields") victorious on their right but Antigonus' cavalry victorious on the left. This cavalry captured Eumenes' camp and his *argyraspides* mutinied and handed him over in exchange for their baggage and families. Eumenes was subsequently executed. This battle reversed the result of **Paraetacene** and ended the chance of centralized **Macedonian** control of the territories in the East.

GALATIA. *See* GAULS

GASTRAPHETES **(pl.** *GASTRAPHETAE***).** Almost certainly invented in **Syracuse** in 399, the *gastraphetes* was essentially a crossbow. It consisted of a larger-than-usual composite **bow** attached to a stock, to which a sliding mechanism was attached. The rear of the stock was braced against the stomach (hence its name, literally "belly-bow") and the bowstring drawn using the slider attached to the stock. Although still limited by the strength of the operator, the mechanical assistance allowed the construction of bigger bows than would otherwise have been possible. This probably allowed an arrow to be fired as far as 250 yards (230 meters), an increase of around 50 yards (46 meters) over the range of a normal bow. Later versions were fitted with a winch to assist in drawing the string, further reducing the limitations of musclepower and allowing even bigger bows to be constructed. Overcoming the next limitation, the size of bow a man could hold and aim, probably led to the development of nontorsion **artillery**.

GAUGAMELA. A plain in northeastern Mesopotamia (modern-day Iraq), site of the final battle in **Alexander the Great**'s conquest of **Persia**. Also known as the battle of Arbela, after the neighboring town, it was fought on 1 October 331. The ancient sources give unbelievable totals for the Persian numbers (up to 1,000,000 infantry and 400,000 **cavalry**), but it is safe to say that **Darius III**'s army, reinforced by troops under the command of Bessus, the satrap of Bactria, heavily outnumbered Alexander and that its main strength was in its cavalry. Determined not to repeat his mistake at **Issus**, Darius chose the broad plain of Gaugamela for the battle because it was "level on all sides" and the Persians further worked on any uneven places to make "them convenient for chariots and cavalry to traverse" (Arr. *Anab.* 3.8.7). On arrival, Alexander sensibly stopped, made a fortified camp, and rested his army. The Persians, however, were afraid of a night attack and remained awake the entire night before the battle. Consequently they began the battle the next day tired and in generally low morale.

Alexander's careful battlefield **reconnaissance** identified the Persian dispositions: Darius in the center with his royal guard, 50 scythed **chariots**, and Greek **mercenary** infantry. His best cavalry and 100 scythed chariots were on the left, and to the right were the **Syrian** and Mesopotamian troops, flanked by more cavalry and 50 scythed chariots. To counter this, Alexander posted a reserve force to protect his rear and deployed with his **phalanx** in the center, the **Thessalian** and allied Greek cavalry on the left, and on the right his **hypaspists** and then the Companion cavalry, under his command, on the right wing. Light infantry and **archers** were also posted on the flanks, linking the front rank with the reserve line.

In the prebattle maneuvering, Alexander continually moved his right wing further to the right to prevent the Persians from encircling him. He then precipitated the Persian attack by reaching uneven ground which the Persians wanted to avoid because it was unsuitable for their chariots. Darius' chariot attack was destroyed when specially tasked light troops targeted the horses and drivers with their **javelins**. A few chariots got past this barrage, but the main body of Macedonians opened up gaps and allowed them through to the rear where they were easily mopped up. When the Persians created a gap in their line by sending off some cavalry to help turn the Macedonian right wing, Alexander seized the critical moment. Although his left wing was under dangerous pressure, he formed a wedge with his cavalry and that part of the phalanx which was nearby and charged into the gap straight at Darius. In the ensuing hand-to-hand combat, the Macedonians succeeded in putting Darius to flight. The reserve force partly relieved the pressure on the Macedonian left wing and the battle ended when Alexander led the Companion cavalry back to assist it. This battle did not cause the same number of Persian casualties as at Issus, but it effectively ended centralized Persian resistance to Alexander. Darius was murdered shortly afterward by Bessus, who had escaped the battlefield with his large force of cavalry virtually intact. *See also* GRANICUS RIVER.

Figure 4. Greek Helmets

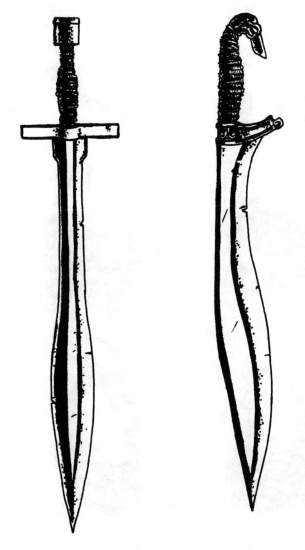

Figure 5. Greek Swords

GAULS. A Celtic people which first appeared in small bands on the frontiers of Greece at the start of the fourth century. Much larger groups migrated through Greece in the early part of the third century before settling permanently in Asia Minor and becoming known as the Galatians or Gallogrecians.

In 281-280 the death of **Lysimachus** and Ptolemy Ceraunos' seizure of power in **Macedon** created the opportunity for a major incursion by three large groups of Gauls into **Thrace**, Paeonia, and Macedon. Each of these groups may have consisted of around 20,000 fighting men, along with families, **slaves**, and camp followers. Ceraunos was decisively defeated and killed and the Gauls devastated large areas of Macedon. One group, under Brennus, **ravaged** as far south as **Delphi** after being temporarily held up by a coalition of states at **Thermopylae**. Delphi was saved by **Phocis** and the **Aetolians**. The **Aetolian League** in particular made a major contribution to Greek success, harrying the Gauls with guerrilla tactics and destroying large parts of their force. The Gallic band that had invaded Thrace was decisively defeated in 277 by **Antigonus II (Gonatas)** at Lysimacheia in the Thracian Chersonese. However, one group remained in Thrace and founded the kingdom of Tylus, which survived until the end of the third century and exacted tribute from Greek cities such as **Byzantium** and **Chalcedon**.

The Gallic migration and attendant damage and destruction caused fear and panic in many parts of Greece. The Greeks regarded the Gauls as uncivilized barbarians, savage and bloodthirsty warriors, who treated their **prisoners** brutally (cf. Paus. 10.22.2). Those Greek states and monarchs that successfully opposed them gained considerable fame. Antigonus' victory at Lysimacheia was the first Gallic defeat in a pitched battle and the prestige he gained reversed his fortunes and secured him the throne of Macedon. However, he soon incurred considerable unpopularity as the first monarch to employ Gauls as **mercenaries**.

Despite the fear and loathing in which the Gauls were held, their reputation as effective soldiers soon led other Hellenistic kings to overcome their scruples and employ Gallic mercenaries. For example, as early as 274, Gauls served under **Pyrrhus I** of Epirus in his invasion of Macedon. Pyrrhus' defeat of Antigonus' Gallic rear guard (and the capture of his **elephants**) spread consternation through Antigonus' main force and caused his **phalanx** to desert to Pyrrhus. Pyrrhus was particularly pleased with this success over Gauls and dedicated spoils from their arms and **armor** at the temple of Athena Itonis. He had at least 2,000 Gauls with him when he attacked **Sparta** in 272 (Plut., *Pyrrhus*, 26.2-5, 28.1-3). Around 4,000 Gauls fought for **Egypt** at **Raphia** in 217.

In 278 several groups of Gauls crossed over into Asia Minor, at least one band invited by Nicomedes I of **Bithynia**, who soon lost control of them. Faced by resistance weaker than in Greece, the Gauls initially ravaged the country with impunity, causing damage and fear out of all proportion to their relatively small numbers. Although walled cities were generally safe because of the Gauls' inability to conduct **siege warfare**, many chose to buy off the Gauls to avoid damage to their land. Others, such as **Miletus**, chose to resist but were captured. This first incursion was eventually defeated at the "Elephant Battle," perhaps in 275, by **Antiochus I (Soter)**, until now distracted by both internal and external enemies. Nicomedes and **Mithridates** I of Pontus settled the survivors in northern Phrygia as a buffer against Antiochus and this region became known as Galatia.

Attalus I (Soter) of Pergamum defeated the Gauls sometime before 230 and drove them from the coastal regions (Paus. 1.8.2). From this point on, they were firmly restricted to the area of Galatia and exerted much less influence on the region, although they were still troublesome to neighboring states, especially Pergamum. However, later in the third century Attalus brought more Gauls across to Asia Minor to help him in his war against **Achaeus**. They soon left his employ and began ravaging the area of the Hellespont until they were destroyed by Prusias I of Bithynia around 218/7 (Polyb. 5.78, 111). **Rome** subdued Galatia in 189 and insisted the Galatians respect the border with Pergamum. However, Rome may subsequently have used the Galatians as a check on Pergamum. Although Galatia became part of the Roman empire in 25, it retained its strong Celtic character for many years.

GAZA. A Philistine city in southern **Palestine** strategically placed at the southern end of the main route between **Egypt** and **Syria**. It was part of the **Persian** empire when captured by **Alexander the Great** after a siege vigorously fought over several months in late 332. Alexander's engineers constructed large earthworks to enable their **artillery** to engage the city, which was built on high ground, but Alexander was unable to take it until the heavy artillery used at **Tyre** was brought up. Although large sections of the strong walls had been destroyed by artillery and sapping and the **Macedonians** attacked under a hail of missile fire, the defenders beat off three assaults. The fourth succeeded. The defenders fought to the last man, and the women and children were sold as **slaves**. Alexander subsequently repopulated the city, using it as a fort to secure the north-south road (Arr. *Anab.* 2.26-27).

Because of its location, Gaza was constantly fought over in the wars of the *diadochoi*—especially the **Syrian Wars**. **Ptolemy I (Soter)** annexed it in 320 but lost it to **Antigonus I (Monophthalmus)** in 315. Ptolemy retrieved the city in 312 but lost it to Antigonus again the following year. Antigonus used Gaza as the forward base for his unsuccessful attack on Egypt in 306, but Ptolemy regained the city during the campaign leading up to **Ipsus** in 301. In 218 **Ptolemy IV (Philopator)** stocked the city with supplies against the threat from **Antiochus III (the Great)**, which was temporarily ended by Antiochus' defeat at **Raphia**, just south of Gaza. In 202/1 Antiochus suffered a further reverse at the city in the campaign leading up to his decisive victory over Ptolemy at **Panion**. Gaza was left isolated but refused to surrender. When it fell to him in 198 Antiochus destroyed the city because of its faithful support of the Ptolemies. It was later repopulated and at some stage it was apparently colonized by Greeks (perhaps as early as Antigonus' control during 311-301)—it was certainly regarded as a Greek city in the early **Roman** empire.

Excluding sieges, Gaza was the site of at least one major battle (Diod. Sic. 19.80.3-86.5). This was fought in 312 between an Antigonid army led by **Demetrius I (Poliorcetes)** and Ptolemy I (Soter) of Egypt. Ptolemy had 18,000 infantry and 4,000 **cavalry**; his forces included Macedonians, **mercenaries**, and native Egyptians. Demetrius fielded 12,500 infantry (including 2,000 Macedonians and a high proportion of mercenaries), 4,600 cavalry, and 43 **elephants**. Ptolemy employed a movable device of chains and spikes to neutralize Demetrius' elephants and won a decisive victory, killing 500 of Demetrius' men (5,000 according to Plut., *Demetrius*, 5) and capturing 8,000. This battle was a major setback to Antigonid plans. Ptolemy recovered Gaza and (temporarily) Phoenicia and Syria, and **Seleucus I (Nicator)** recaptured Babylon and then Media, the first steps in creating the Seleucid empire.

GELON (c. 540-478). An aristocrat and **hipparch**, with a distinguished military reputation at Gela in **Sicily**, who became tyrant circa 490. Over the next few years he extended his control over considerable parts of the south coast of Sicily. Around 485 he captured **Syracuse** and installed himself as tyrant there, leaving his brother in charge in Gela. He considerably strengthened Syracuse's power, conquering Camarina and Megara Hyblaea and removing their upper classes to Syracuse—the poor were sold into **slavery** (Hdt. 7.154-156).

According to **Herodotus** (7.157-158) in 481 the Greeks asked Gelon to help against the forthcoming **Persian** invasion. He offered 20,000 **hoplites**, 4,000 **cavalry** (half heavy and half light), 2,000 **archers**, 2,000 **slingers**, and 200 **triremes**, but this was rejected when he insisted he must command the **Hellenic League** forces. However, it seems more

likely that his failure to commit troops to Greece during the Second
Persian War was influenced by an impending **Carthaginian** invasion
of Sicily.

In a major expedition, perhaps involving 100,000 troops, the Carthaginians, led by Hamilcar, began their campaign by besieging Gelon's
ally Theron at Himera in 480. Gelon advanced to Himera with an army
around half the size of Hamilcar's, but stronger in cavalry. Once there,
he used his superior cavalry arm against the Carthaginian **foragers** and
penned them up in their camp. Gelon then breached the Carthaginian defenses by a ruse: learning from a captured courier that the Carthaginians
were expecting cavalry reinforcements from Selinus, Gelon sent his own
cavalry in their place on the appointed day. They probably had orders to
kill Hamilcar and secure the way for a general attack. Once inside the
Carthaginian camp the cavalry fired the enemy fleet on the beach and Gelon followed this up with an infantry attack. The Carthaginian army was
destroyed and Carthage forced to pay a large indemnity to secure peace
and the return of **prisoners**. The victory was so complete that it was another 70 years before Carthage attempted major operations in Sicily. Gelon died shortly afterward.

GRANICUS RIVER (GRANIKOS). The site of the first battle in **Alexander the Great**'s conquest of the **Persian** empire, fought in May 334
between the **Macedonian**-Greek army and a Persian-Greek army. The
Macedonian army was perhaps as large as 50,000 men (Alexander
brought 32,000 infantry and 5,000 **cavalry** over to join with the advance
party already there); the Persian army consisted of 20,000 cavalry and
about 20,000 Greek **mercenary** infantry. The Persians occupied the river
bank, blocking Alexander's eastward movement toward **Cyzicus**. According to **Arrian** (1.14.4), they drew up their 20,000-strong cavalry,
their best troops, along the river bank and posted their infantry behind
the horse. Alexander led with his cavalry, but included one brigade of
infantry, with a squadron of cavalry in front of them, in the initial
thrust. He employed his cavalry to dislodge the Persian horse from the
bank and win a bridgehead for the infantry. Arrian describes this initial
action as "a cavalry fight, but on infantry lines" (1.15.4-5). Alexander
also had the following infantry cross the river at an oblique angle so that
they would not be forced to engage the Persians on an extended front.
During the action Alexander, leading from the front, was almost killed
by the Persians. After a hard engagement the Macedonian cavalry, aided
by their longer spears, drove the Persian cavalry off the field, killing
about 1,000 of them. The 20,000 Greek mercenaries were surrounded
and, despite vigorous resistance, massacred. Only around 2,000 **prison-**

ers were taken. According to Arrian Alexander lost 115 men, but this total seems very low in the light of the fierce resistance by the Greek mercenaries. *See also* GAUGAMELA; ISSUS.

GREEK LEAGUE. *See* HELLENIC LEAGUE

GYLIPPUS (GYLIPPOS; fl. c. 445-400). A **Spartan officer** sent, on **Alcibiades**' advice, to assist **Syracuse** against **Athens' Sicilian Expedition**. Gylippus arrived in Syracuse in 414, passing through the Athenian siege lines and entering the city when the Syracusans were seriously considering coming to terms with the Athenians. His first attempt to lead the Syracusans against the Athenians failed when he deployed the Syracusan army in a narrow area where it was unable to take advantage of its superior **cavalry**. Admitting his mistake, he led the Syracusans to victory in the next battle. Under his leadership Syracusan resistance stiffened and the defenders rapidly gained confidence. His leadership was a major factor in the destruction of the Athenian force, a turning point in the Spartan struggle against Athens. Gylippus returned home but in 405 was convicted of embezzling public monies and went into exile. *See also* NICIAS; PELOPONNESIAN WAR (2).

- **H** -

HALIARTUS (HALIARTOS). One of the smaller *poleis* (cities) in **Boeotia**, to the northwest of **Thebes**. Around 395, along with **Coronea** and Lebadaea, it constituted one of the 11 wards of the **Boeotian League** and presumably therefore contributed one-third of the ward's normal contribution of 1,000 **hoplites** and 100 **cavalry** to the league army. In 395 it was the site of the opening battle between **Sparta** and Thebes during the **Corinthian War**. **Xenophon** preserves the most detailed account of the engagement (*Hell.* 3.5.6-7, 17-25), although unfortunately without specifying the size or composition of the armies involved.

The Spartan plan was for **Lysander** to raise an army from Sparta's allies in **Phocis** and neighboring areas; King **Pausanias** (2) was to meet him at Haliartus when he arrived with the main **Peloponnesian League** army. Lysander arrived first, having secured the revolt of **Orchomenus** from the Boeotian League (currently dominated by Thebes) and decided to try to detach Haliartus without waiting for Pausanias. He was surprised and killed outside the walls by Theban hoplites and cavalry who had rushed to assist Haliartus. Lysander's army was scattered but the hoplites re-formed on reaching nearby broken high ground and staged a successful counterattack, which killed more than 200 Thebans.

Pausanias arrived the next day (as did **Athenian** reinforcements sent to help the Boeotians), but after deploying for battle, he elected not to fight. According to Xenophon this was partly because he was discouraged by the death of Lysander but also because of the enemy's cavalry superiority and doubts about his allies (**Corinth** had refused to send a contingent and other states were there reluctantly). Pausanias also reasoned that even if the Spartans won, they still could not recover the bodies of Lysander and those killed with him without a truce, because they were right next to the city walls. Using the bodies of the enemy dead as a bargaining chip (as they had at **Delium**), the Thebans refused to return them unless the Spartans agreed to withdraw. Pausanias accepted and left Boeotia without engaging the enemy. **Plutarch** (*Lysander*, 28) preserves an alternative version (perhaps based on local Boeotian tradition) that Pausanias failed to arrive in time because the letter from Lysander arranging the date of their meeting was captured by the Thebans, who deliberately ambushed him. He also records the Theban losses as 300. Both accounts, though, suggest that Lysander got himself into trouble by acting before Pausanias arrived.

On his return to Sparta, Pausanias was put on trial but fled into exile in Tegea. The victory at Haliartus gave considerable impetus to the Boeotian-Athenian cause and in conjunction with **Persian** gold, persuaded other Greek states, including **Argos** and Corinth to join against Sparta.

HALICARNASSUS (HALIKARNASSOS). A Greek colony on the Carian coast in southwest Asia Minor; birthplace of **Herodotus**. A naval contingent from Halicarnassus fought for **Xerxes I** in the Second **Persian War** (480-479), under the command of **Artemisia**. After the **Persian** defeat, Halicarnassus joined the **Delian League**. In the 370s Mausolus greatly expanded the city and made it the capital of Caria.

In 334 **Alexander the Great** besieged the city on his march south toward **Syria**, Phoenicia, and **Egypt**. The siege was a hard-fought one and is notable for the first attested use of stone-throwing (as opposed to bolt-shooting) **artillery**; however, these were apparently smaller machines used against personnel rather than the walls. Alexander first had to fill in the moat, which was about 50 feet (15 meters) wide and 25 feet (7.5 meters) deep, in order to get his mobile towers and battering rams up to the wall. This was completed under the protection of his stone-throwing and bolt-shooting catapults. The defenders made several sorties from the city and succeeded in firing some of the siege engines. Even when Alexander succeeded in knocking down a section of wall, the defenders countered by building a crescent-shaped wall inside the perimeter wall to contain the breach. Despite the effectiveness of the defensive fire (including bolts fired from numerous catapults on three sides), the **Macedonians** were able to demolish more of the wall. At this point, the Persians thought the capture of the city was inevitable, and after setting fire to it, they retired to two strongholds. Alexander rescued the inhabitants, razed the town, and moved on. He simply ignored the two strongholds, which could do little against the troops he had left in the region.

The city was soon restored, under local dynastic control appointed by Alexander. After Alexander's death, the city was subject at different times to control by the various Hellenistic monarchies. In 129 Halicarnassus became part of the **Roman** empire. *See also* TYRE.

HAMIPPOI (sing. *HAMIPPOS*). Infantry (usually light) that operated closely with **cavalry**—and, ideally, were trained to do so. *Hamippoi* were intermingled with the cavalry and moved with them. At least one stone relief (Louvre, 744) depicts them keeping up with advancing cavalry by holding on to the horses' tails. The use of *hamippoi* enhanced the combat power of cavalry, not only by the addition of extra firepower (*hamippoi* were often equipped with **javelins**), but also by allowing a measure of surprise. **Xenophon**, for example, suggests (*Hipparch.* 8.19) that infantry can be hidden from the enemy by deploying them within a cavalry **formation**. *Hamippoi* are attested in **Sicily** as early as 480 and in **Boeotia** by 419/8. They are attested from around 365 as part of the **Athenian** army and were annually assessed by the council or *boule* for physical fitness. **Agesilaus II** of **Sparta** deployed *hamippoi* with his cavalry on the plain of Sardis in 395 and **Thebes** had them at **Mantinea** in 362. The cavalry of **Perseus** of **Macedon** (179-167) also regularly operated with *hamippoi*.

HECTOR (HEKTOR). Trojan prince and hero during the **Trojan War**. **Homer**'s *Iliad* portrays Hector as a devoted family man and humane soldier, although not as skilled a warrior as **Achilles**. Hector forms a strong contrast to Achilles and the more brutal and savage style of warfare the latter represents. This is particularly shown in the difference between Hector's treatment of fallen enemies and Achilles' savage treatment of Hector's body.

HELLENIC LEAGUE. The name adopted for several panhellenic leagues. The most important are:

(1) Founded in 481. The league formed at the Isthmus of **Corinth** to resist the impending **Persian** invasion. **Herodotus** is our main source of information on the league and its activities. Each state was repre-

sented by elected delegates and had one vote. A majority vote was binding. The members chose **Sparta** to exercise command on land and sea and took an oath to punish **medizers**—as well as several practical measures for the impending war. The number of states involved was not large—31 are recorded on the victory monument after **Plataea** (1): Sparta, **Athens**, **Corinth**, Tegea, **Sicyon**, Aegina, **Megara**, Epidaurus, **Orchomenus**, Phlius, Troezen, Hermione, Tiryns, **Plataea**, Thespiae, **Mycenae**, Ceius, Malis, Tenos, **Naxos**, **Eretria**, Chalcis, Styra, Halieis, **Potidaea**, Leucas, Anactorium, Cythnus, Siphnos, Ambracia, and Lepreum. This, however, does not acknowledge all contributions, especially those of states that resisted initially but were forced to medize. There was some attempt to make this league permanent after Plataea but while it did oversee the initial conduct of the follow-up campaign against Persia, it was soon overtaken by the creation of the **Delian League**. *See also* PERSIAN WAR (2).

(2) Established in 337 by **Philip II** of **Macedon** after **Chaeronea** (1). The League included most mainland Greek states from Macedon south, except for **Sparta**. The members swore an oath of allegiance to Philip and his heirs and agreed to maintain (by force if necessary) a general peace (*koine eirene*) in Greece. Members also agreed to maintain their existing laws and not undertake subversive measures such as land redistribution. They were to refrain from executions and undertook to suppress pirates and brigands. The league had an assembly (*synedrion*), with representation according to each member's size or military strength, and a majority decision was binding on its members. Philip was elected *hegemon* and **Alexander the Great** succeeded him in this position, using the league to provide troops for his invasion of **Persia**. There is considerable continuity between this league and the one established by **Demetrius I (Poliorcetes)** in 302.

(3) Founded in 224/3 by **Antigonus III (Doson)**; on his death in 221, **Philip V** became *hegemon*. The league had a *boule* (council) and an *ekklesia* (assembly). It initially consisted of **Macedon** and the **Achaean League**, but soon extended membership to much of the Peloponnese. The league was used by Philip V to mobilize military support against the **Aetolian League** in the **Social War** (2). The league ended when Philip V was defeated by **Rome**.

HELLENIC WAR. *See* LAMIAN WAR

HELMETS. Greek **cavalry** and heavy infantry regularly wore helmets from Mycenaean times on. The earliest helmets were made of leather, leather and boar's tusk (figure 1), or bronze. Later helmets were made of iron. Although leather continued throughout antiquity as a cheap alternative to metal, it was generally worn by light rather than heavy infantry in the Archaic, Classical, and Hellenistic periods. However, there is evidence that the *pilos* helmet (figure 4) was worn in a felt (or leather) version by **hoplites**, especially **Spartans**, from the latter half of the fifth century. Early metal helmets were constructed from as many as five pieces, later ones from two pieces or even one piece. Metal helmets were padded inside (with the lining secured by stitches and/or glue) and were regularly surmounted with horsehair crests, dyed in a variety of colors.

The infantry helmet of the Classical period seems to have developed from two types, the *Kegelhelm* (cone-shaped helmet) and the Corinthian. The Corinthian helmet (figure 4) was widely used by hoplites, particularly in the sixth and fifth centuries—although it survived for some time afterward. It gave considerable protection, although limiting hearing and peripheral vision. The *Kegelhelm* developed into the Illyrian and the Chalcidian helmets (figure 4). These supplanted the Corinthian as the most popular helmet, probably because they were lighter and did not restrict hearing and vision as much. (For a useful description of the chronological development of Greek infantry helmets from the eighth to the fifth centuries, see P. Connolly, *Greece and Rome at War*, 60-62.)

Given the different function and characteristics of cavalry, helmets for mounted troops were always designed to maximize the wearer's ability to see. Some, particularly the **Boeotian**-style helmet (figure 4), seem to have been modeled on hats designed to provide shade while riding. Cavalry helmets also tended to be more elaborate than infantry ones because of the aristocratic origins of the arm and the practice of drawing the cavalry from the wealthier classes, which could afford to provide their own horses.

HERACLEA. *See* SIRIS RIVER

HERMOCRATES (HERMOKRATES; d. 407). A **Syracusan** general and democratic politician active during the **Peloponnesian War** (2) of 431-404. According to **Thucydides** (4.58-65), in 424 Hermocrates was pivotal in persuading the **Sicilians** to resolve their differences and unite against **Athenian** interference on the island. He also led resistance to the Athenian attack on Syracuse in 415. In 412, after the failure of the Athenian **Sicilian Expedition**, Hermocrates commanded the Syracusan naval contingent operating with **Sparta** in Asia Minor but was exiled in absentia (along with the other generals). He remained with the Spartan force until 408, when he returned to Sicily. He was killed in 407, during an attempt to seize control of Syracuse. **Dionysius I** married his daughter. Both Thucydides and **Xenophon** portray him favorably in their works.

HERODOTUS (HERODOTOS; c. 484-420). The main ancient source for information about the **Persian Wars** of 490 and 480-479. Born of a good family in **Halicarnassus** in Asia Minor, Herodotus was forced into exile after civil unrest against the tyrant Lygdamis and traveled extensively, giving public performances of his work. He joined the **Athenian** colony at Thurii in **Italy** and died there sometime before 420. His only known work is a history of the Persian Wars. He is the first Western historian and preserves a strong epic tradition in his writings as well as introducing new techniques of analysis and explanation. Because his history was delivered orally to public audiences, Herodotus included some stories purely for their entertainment value. However, as he states at 2.123 and 7.152, he did not necessarily believe these stories (some of which are plainly incredible), but saw his duty as an historian to record them. Although in places there are problems with his accuracy, Herodotus often tells us the source of his information for a specific event and, within the limitations of his information and methodology, is often a careful and analytical historian. *See also* ARTEMISIA; EURYBIADES; MARATHON; PLATAEA (1); SALAMIS.

HIERON II (c. 306-215). Tyrant of **Syracuse**, 275-215. Hieron served under **Pyrrhus I** of Epirus and then in 275/4 was illegally elected general by the troops in the field, although his successes caused the Syracusans at home to ratify this. He was apparently defeated by Campanian **mercenaries**, known as the Mamertines, in 270 (although **Polybius** [1.9] claims Hieron deliberately planned the loss of a contingent of disaffected mercenaries while securing the safety of his citizen-troops). Two years later he decisively defeated the Mamertines at the River Longanus near Mylae and was proclaimed king on his return to Syracuse.

In 264 Hieron allied with **Carthage** to attack the **Roman** expeditionary force that had landed to aid the Mamertines in Messana (the start of the First Punic War). The Romans, however, sallied out in strength and defeated him in a hard-fought battle. Impressed by this, and by the scale of the subsequent Roman reinforcements, Hieron switched his allegiance to the Romans in 263. The use of Syracuse as a firm base and major supply depot was an important factor in the Roman success on land during the First Punic War. Hieron remained a loyal ally of Rome until his death, and under him Syracuse increased its territory and generally prospered. He also strengthened the city militarily by employing

Archimedes to develop the defenses of the city itself and by expanding the navy. Polybius 1.8-18 preserves a favorable account of his rule.

HIPPARCH (*HIPPARCHOS,* pl. *HIPPARCHOI*). A **cavalry** commander, at **Athens** and elsewhere. At Athens there were two *hipparchoi*, elected by the citizens and holding joint command for one year. There was no limit, though, on the number of times a man could hold the office. When all the cavalry was deployed, each hipparch probably commanded a contingent of 500. In the mid-fourth century an additional Athenian hipparch appears, the *"hipparchos eis Lemnon."* This **officer**, probably slightly junior in status to the other two hipparchs, commanded the cavalry contingent sent annually to the island of Lemnos. The details of the hipparchs of other regions are often obscure and the word is sometimes used to mean any cavalry commander. **Xenophon** calls the **Spartan** cavalry officer killed at **Olynthus** in 381 a hipparch (*Hell.* 5.2.41). *See also* HIPPARMOSTES.

HIPPARMOSTES. A **cavalry officer** at **Sparta**, apparently equivalent to a **phylarch** (or squadron commander) at **Athens**, as they led a *mora* of cavalry (Xen. *Hell.* 4.5.12). Their method of appointment and duration of office are unknown. *See also* HIPPARCH; PASIMACHUS.

HIPPEIS. See HIPPEUS

HIPPEUS (pl. *HIPPEIS*). A cavalryman; the word can also simply mean "rider." In the plural it means "the **cavalry**" (or the title of an aristocratic group in several Greek cities). At **Athens** it identifies members of both the cavalry and the second highest of the census classes established by Solon. At **Sparta** it was the name of the Spartan kings' 300-strong bodyguard. These may originally have been mounted but in the sixth to the fourth centuries the force was composed of **hoplites**. They are first mentioned in connection with the recovery of the bones of Orestes in the second half of the sixth century and may have been the 300-strong force who died at **Thermopylae (1)** with **Leonidas I** (Hdt. 1.67, 7.205).

HIPPIAS (fl. 530-490). Eldest son of Pisistratus; tyrant of **Athens** 527-510. His forces, assisted by **Thessalian cavalry**, defeated a **Spartan** amphibious landing at Phalerum in 511/0. In 510, **Cleomenes I**, operating with the assistance of Hippias' Athenian enemies, defeated Hippias' army and expelled him from Athens. Hippias sought refuge in **Persia** and three subsequent military attempts by Cleomenes to interfere in Athens on his, or his supporters', behalf failed. **Artaphernes'** subsequent insistence that Athens restore Hippias helped influence Athenian participation in the **Ionian Revolt**. Hippias accompanied the Persian expedition against Greece and in 490 advised that the plain of **Marathon** would be a good site for the Persian landing in Attica.

HIPPOTOXOTAI (sing. *HIPPOTOXOTES*). Mounted **archers**. Although firing a **bow** from a horse without stirrups was a difficult skill to acquire, mounted archers were used in ancient Greek warfare. The best were foreigners, especially the **Scythians**, who spent a considerable part of their lives on horseback.

Scythian mounted archers are attested in **Athens** from the sixth century (see M. F. Vos, *Scythian Archers in Archaic Attic Vase Painting*) and the city had a corps of 200 throughout the fifth century. Its status and composition are a little uncertain—Athens had a contingent of publicly owned Scythian archers who acted as police, but the mounted archers were probably a separate group and **mercenaries** rather than **slaves**. They were certainly paid, as around 403/2 their daily pay was increased from two obols to eight (Lysias, fr. 6, *Against Theozotides*). However, citizens could also serve—in 395 Alcibiades the Younger, the son of **Alcibiades**, served with the *hippotoxotai* when he was expelled from the **cavalry** *phyle* (squadron) in which he was enrolled (Lysias, 15 [*Against Alcibiades 2*], 6). Sometime between 395 and 365 this 200-strong corps was disbanded, probably under the financial constraints imposed by the loss of empire after the **Peloponnesian War** of 431-404. It was replaced by *hamippoi*, infantry trained to operate with the cavalry.

Tactically, the Athenian *hippotoxotai* seem to have been used to ride ahead of the rest of the cavalry to soften the enemy up with their arrows (cf. Xen., *Memorabilia*, 3.3.1). In addition, the dispatch of 20 to Melos in 416 and 30 to **Sicily** in 415 (Thuc. 5.84.1, 6.94.4) suggests that they may also have been used for small-scale **raids**. However, horse archers were never a major arm at Athens or in any other part of the Greek world.

Nevertheless, with **Alexander the Great**'s conquest of the **Persian** empire and his acquisition of Persian military resources, mounted archers did become a more prominent feature of Greek armies and warfare—particularly in the East. Alexander apparently had no mounted archers with him during his campaign against **Darius III** but had them in India. He used them to good effect against **Porus**' son in the preliminaries to the battle at **Hydaspes River** and against Porus' left wing during the battle (Arr. *Anab.* 5.14-18). In both cases, the *hippotoxotai* were used to disrupt the enemy formation with their arrows. The use of mounted archers continued under Alexander's successors. For example,

Antiochus III (the Great) fielded more than 1,200 mercenary horse archers at **Magnesia** in 190 (Livy, 37.40).

HOMER. Two long epic poems, the *Iliad* and the *Odyssey*, are attributed to the poet Homer. The *Iliad* covers a short series of events during the **Trojan War**. It focuses on **Achilles'** quarrel with **Agamemnon** and his withdrawal and return to battle and climaxes in the death of the Trojan warrior **Hector**. The *Odyssey* covers the wanderings of the warrior **Odysseus** at the end of the Trojan War and his final return home to Ithaca. Both these poems provide us with considerable material on **early Greek warfare**, particularly arms and **armor**. However, they have to be treated carefully, as they were written down several hundred years after the events they describe and contain various anachronistic elements.

There is no reliable evidence for Homer's date, or even of details of his life. In fact, many scholars (from ancient times onward) argued that the *Iliad* and the *Odyssey* were composed by different poets or even that they were composites, put together from the works of many authors. However, there is sufficient unity in the poems to suggest that they were produced by a single writer, even if perhaps based on earlier oral epic. Both Homeric poems were probably in circulation by about 700 and had assumed roughly the form in which we know them between 750 and 700—the period during which writing was reintroduced into Greece after the Dark Age.

HOPLITE (*HOPLITES*, pl. *HOPLITAI*). A heavy infantryman equipped with **shield**, spear, and sword and perhaps (but not always) breastplate, **helmet**, and greaves. The central items of the hoplite panoply were the large 3-foot (95-centimeter) diameter shield (*aspis*) with central arm-ring (*porpax*) and handle (*antilabe*) inside the offset rim and the hoplite spear of 8-9 feet (2.5 meters) in length. The sword was used as a secondary **weapon** if the spear broke (which must have occurred fairly frequently). Hoplites fought in a close **formation**, the **phalanx**. This allowed full protection across the front, as the left-hand half of a hoplite's shield, which extended left from the wearer's elbow, provided protection for the right-hand side of the hoplite to his left. The hoplite equipment was probably introduced in the late eighth century, and the first half of the seventh century saw the introduction of hoplite tactics (the first known artistic depiction of the hoplite phalanx is the "Chigi Vase" of c. 650).

The Greeks themselves thought that the hoplite equipment was originally imported from Caria, but there has been considerable debate among modern scholars over whether the equipment was introduced first, leading to the creation of the phalanx formation, or the Greeks already used a phalanx and adopted the hoplite panoply for it. It seems perhaps more likely that the equipment led to the development of the phalanx in its traditional form, but the Greeks may already have been used to fighting in some sort of formation, probably looser than the hoplite phalanx, in **Homeric** or **early Greek warfare**.

The hoplite was the mainstay of Greek armies down to the late fourth century, when it proved inferior in battle to the **Macedonian** phalanx. Although vulnerable to harassment by **cavalry** and missile-armed *psiloi* (light troops), the hoplite's close formation must have provided considerable psychological reassurance, as well as a high degree of physical protection. It also required little formal **training** to deploy. Hoplite equipment could only be afforded by those of moderate or better means and in most states the backbone of the hoplite arm was the small farmer. The military dominance of the hoplite for several hundred years aided in the social dominance of the hoplite class. Military service often brought with it the right to vote, and by the fourth century most states which used hoplites were limited democracies.

Although by the end of the fourth century the hoplite was no longer the dominant arm in Greek warfare, it continued in use for much longer than that. For example, archaeological evidence shows that the Macedonians, who were not traditionally a hoplite power, continued to employ this equipment continuously from **Philip II** in 359 down to their last king **Perseus**, defeated by the **Romans** at **Pydna** in 168. *See also* HYPASPISTS; *ZEUGITAI*.

HYDASPES RIVER. The modern-day Jhelum River, site of a battle in 326 during **Alexander the Great**'s Indian campaign. Alexander was faced with an opposed river crossing against the Indian ruler **Porus**. By a series of troop maneuvers, he managed to deceive Porus into thinking he would wait for winter, when the water levels would fall, and then ferried around 6,000 infantry and 5,000 **cavalry** across in boats and rafts under cover of a storm. Alexander used his cavalry and **archers** to destroy an attack by Porus' son and a force of **chariots**. In the main engagement, Porus was outnumbered, but possessed 200 **elephants**. These were deployed at intervals along the line, separated by infantry and with the cavalry and chariots on the wings.

Alexander initially made an attack on the Indian left wing with mounted archers and cavalry. When the Indians moved cavalry from the right wing to assist, **Macedonian** cavalry attacked the depleted Indian right wing. Alexander's infantry then advanced against the elephants and Indian infantry. The Indians had no answer to the Macedonian *sarissa* and Alexander had ordered detachments armed with **javelins**, axes, and sickles to attack the elephants. The Indian line was thrown into confusion and Porus was left virtually alone, surrendering only when overcome by his **wounds**. A vigorous pursuit launched by **Craterus**, who crossed the river elsewhere, destroyed the Indian survivors as they fled. Alexander confirmed Porus as a vassal king.

HYPASPISTS (*HYPASPITAI*, sing. *HYPASPISTES*). Macedonian infantry, probably armed as **hoplites** (although perhaps without a cuirass) and perhaps also trained to use the *sarissa* when necessary. **Arrian** (2.4.3) demonstrates they were less encumbered than the *sarissa*-equipped **Foot Companions** and elsewhere they were chosen by **Alexander the Great** to take part in pursuits. The hypaspists under Alexander formed a 3,000-strong corps (divided into six *lochoi*, or companies) and were often stationed on the right wing of the Foot Companions' **phalanx** in order to protect their vulnerable right flank and to link them with the **cavalry**. Hammond, and others, believe they were the royal bodyguard under Alexander (Hammond, *A History of Macedonia*, vol. 2, 414). However, although the hypaspists were probably descended from the royal bodyguard, Sekunda plausibly suggests they were not the bodyguard under Alexander (Sekunda and Warry, *Alexander the Great: His Armies and Campaigns, 334-323*, 13). At the end of his reign Alexander equipped the hypaspists with silver-colored shields and termed them the *argyraspides* ("Silvershields"). After his death this term was kept by his hypaspists to differentiate them from the hypaspist forces formed by the *diadochoi*.

The term is also used in Classical **Sparta**, most probably to denote helots who carried the **shields** for their Spartan masters when they were not in combat. *See also* MACEDONIAN ARMY.

- I -

INFANTRY. *See* ARCHERS; FOOT COMPANIONS; *HAMIPPOI*; HYPASPISTS; MOUNTED INFANTRY; PELTAST; *PSILOI*; SLINGERS

INTELLIGENCE. Although they generally had no formal intelligence organization (unlike the **Persians**), the ancient Greeks gained military information in a variety of ways. The main methods were spying and **reconnaissance**. Prior to the development of reconnaissance skills in the late fifth and early fourth centuries, the prime source of military information was from spies. The literary prototype was Sinon, a Greek who penetrated Troy during the **Trojan War** by pretending he had been maltreated and deserted. More historically, in his treatise on the duties of a **hipparch (cavalry** general), **Xenophon** mentions the usefulness of sham deserters—presumably to gain information as well as to spread disinformation (Xen. *Hipparch.* 4.7). When the Greeks heard of the Persian preparations prior to the Second **Persian War** of 480-479, they sent out spies to discover the size of the Persian forces. Interestingly, they were caught but **Xerxes I** gave them a guided tour of his forces and sent them home to frighten their compatriots with the numbers marching against them (Hdt. 7.145-148).

Spies were used during the Classical period, but as campaigns usually involved a simple attack on an enemy's farmland and a **hoplite** engagement on the plain there was often no great need for information. Given the limited means of communication, spies must often have had difficulty providing timely intelligence. However, the need for information became more important from the **Peloponnesian War** of 431-404 on, as Greek warfare became increasingly sophisticated. **Philip II** of **Macedon**, for example, was reputed to have made considerable use of spies, but presumably to provide strategic rather than tactical information. **Sparta** had its own secret police, the *Crypteia*, but this was essentially an internal security organization, aimed at the helots. The Spartans nevertheless maintained fairly tight security over foreigners and military information—but in most other cities, information was generally readily available. In 330 **Demosthenes** (2) complained that he was at a great disadvantage in opposing Philip II, not least because the **Athenian** assembly publicly debated, and then published, their decisions concerning raising and deploying forces (*On the Crown*, 235-236).

During the Hellenistic period, the struggles between the various monarchies increased the need for intelligence services. At the same time, **Alexander the Great**'s conquest of the Persian empire had put the extensive and relatively sophisticated Persian intelligence services at Greek disposal. The limited evidence suggests that Alexander and his successors left these services in place and made considerable use of them.

IONIA. The central part of the west coast of Asia Minor, extending roughly from **Chios** in the north to **Miletus** in the south. Ionia contained many Greek colonies, dating in some cases from as early as the tenth century. In the sixth century Ionia was under considerable pressure from Lydia, and it lost its independence during the reign of Croesus (561-546). The region was conquered by Cyrus the Great in 546. Ionia was under varying degrees of **Persian** control from this point onward until **Alexander the Great** liberated it.

At the start of the fifth century, there was a widespread rebellion in the region, known as the **Ionian Revolt**. Although ultimately unsuccessful, this revolt caused the Persians serious concerns. Following the Second **Persian War** of 480-479 (in which Ionian ships fought on the Persian side), many Ionian cities and islands off the coast of Ionia joined the **Delian League** and gained their independence from Persia. Although Persia regained localized control in some cities from time to time, the **Athenian** domination generally lasted until the end of the **Peloponnesian War** of 431-404. In the final stages of this war, Ionia was the scene of an important campaign between Athens and **Sparta** that ultimately decided who would win the war. Under the terms of the 412 agreement between Persia and Sparta (Thuc. 8.18, 36-37, 57-58—the last reference probably gives the final version, the other references drafts), Ionia was abandoned to the Persians, whose claim to it was officially recognized.

In several of the **Common Peace** treaties of the fourth century, Persia succeeded in gaining Greek recognition for Persian ownership of Ionia. The Spartans campaigned against Persia in Ionia in the first decade of the fourth century under **Thibron**, **Dercylidas**, and **Agesilaus II**. Spartan naval supremacy was ended there in 394 at the battle of **Cnidus**. Alexander's victory at **Granicus River** in 334 opened the way for a campaign south into Ionia, which was wholly in his hands from the end of that year. Following this, the area was generally under the control of Seleucid **Syria** until its incorporation into the **Roman** empire.

Ionia was a wealthy and cultured part of the Greek world. The most effective military arm in the region during the Classical and Hellenistic periods was the naval one—the area generally produced indifferent **hoplites** and was always fairly weak on land.

IONIAN REVOLT. A revolt of the **Ionian** Greeks against **Persian** rule that lasted from 499 to 493 and led to the **Persian Wars**. The revolt began in 499 and was centered on **Miletus** and led by its ruler, Aristagoras (prompted by his father-in-law and ex-ruler of Miletus, Histiaeus). In winter 499/8 Aristagoras traveled to mainland Greece, seeking support. He was refused it at **Sparta**, where **Cleomenes I** sensibly decided it was too dangerous to risk a Spartan expeditionary force so far away from home. **Athens**, however, sent 20 ships to help and **Eretria** on **Euboea** sent five.

The 498 campaign started well for the Greeks, who marched inland and burned the Persian king's summer palace at Sardis. Although the garrison held out on the citadel, the symbolism of this act caused the revolt to spread to the Hellespont, **Cyprus**, and Caria. In 497, worried by the increasing scale of the revolt, the Persian king, **Darius I**, dispatched three armies to deal with it. The Cypriot cities that had revolted defeated an enemy fleet but lost a concurrent battle on land when part of their army deserted. The Persians besieged the rebellious cities and the last, Soli, surrendered in 496. The Carians had mixed success, losing two major battles (despite considerable Ionian Greek support in the second) but then destroying a Persian army in a night **ambush**. The region was pacified by mid-494. In summer 494 the Ionians decided to stake their success on a sea battle at **Lade** (1), but their disunity led to their complete defeat. By the summer of 493 the last Greek cities had been reduced. Most of the cities involved were sacked and punished by executions or massacres. Because of the role of the tyrants in the revolt, **Mardonius** (appointed to settle the area in 492) established democracies in many of the recaptured cities.

Although courageous, the revolt was an extremely risky venture because the resources of the Persian empire far outweighed those of the Ionian Greeks. If the Ionians had followed the advice of the historian Hecataeus and concentrated their efforts on **naval warfare**, they may have been more successful. However, it was their lack of unity that ultimately caused their defeat. The Athenian and Eretrian involvement led to Darius' expedition of 480 against them, beginning the Persian Wars. *See also* ARTAPHERNES.

IPHICRATES (IPHIKRATES; c. 415-353). An **Athenian** *strategos* (general) active in the first half of the fourth century; son of a cobbler. He seems to have had some association with **Conon** and first distinguished himself in the naval battle of **Cnidus** in 394 during the **Corinthian War**. During the course of his public life he elevated himself to a position of wealth, and one of his sons, Menestheus, married **Timotheus'** daughter in 362 and also served as a general at least twice and as a **trierarch** four times.

In 393 Iphicrates was given command of a force of **peltasts**, probably recruited from **Thrace**, and was sent to campaign in the Peloponnese. Iphicrates quickly established himself as a skilled commander of light troops, staging a successful **ambush** against the Phliasians, putting **Mantinean hoplites** to flight, and exerting a complete psychological domination over the **Arcadian** hoplites (Xen. *Hell*. 4.4.15-17). In 390 his peltasts played a major role in the destruction of a *mora* of Spartan hoplites at **Lechaeum**—one of Iphicrates' greatest triumphs. In 388, he commanded a force of 12 **triremes** and 1,200 peltasts (veterans of his Peloponnesian campaign) in the Thracian Chersonese, staging a very successful ambush of an army led by the **Spartan** Anaxibius near

Abydus (Xen. *Hell.* 4.8.34-39). Around 386, Iphicrates married the daughter or (more likely) the sister of the Thracian king, Cotys.

In 373 Iphicrates replaced Timotheus as commander of the expedition to **Corcyra**. Taking 70 triremes, he left the large sails behind, forcing the sailors to row and staging maneuvers and races on the way, toughening his sailors and giving them experience. **Xenophon** (*Hell.* 6.2.27-32) is particularly complimentary of Iphicrates' attention to security and his combining of **training** with swift movement to the campaign area. He rapidly secured Corcyra, capturing 10 triremes sent by **Dionysius I of Syracuse**.

Iphicrates' failure to prevent the **Theban** invasion of the Peloponnese in 369 was regarded as well below his usual performance (Xen. *Hell.* 6.9.49-52) and his failure to recapture **Amphipolis** (367-364) led to his temporary retirement to Thrace. In 356 he apparently failed to provide full support to **Chares** at **Embata** and was prosecuted on his return to Athens. Unlike his colleague **Timotheus**, Iphicrates was acquitted. He died between 354 and 352. An excellent general, Iphicrates is credited with reforms to peltast equipment (lengthening the spears and swords and introducing a new type of boot), which helped pave the way for **Philip II** of **Macedon's phalanx**. However, the evidence for this is dubious and the "reform" may simply represent a standardization of equipment already used by some Thracian peltasts (see J. G. P. Best, *Thracian Peltasts and Their Influence on Greek Warfare*, 102-110).

IPSUS (IPSOS). A battle fought in Phrygia in 301 during the war between **Antigonus I (Monophthalmus)** and a coalition comprising **Ptolemy I (Soter), Cassander, Seleucus I (Nicator)**, and **Lysimachus**. Ptolemy failed to supply troops but the other partners together provided 10,500 **cavalry**, 64,000 infantry, 400 **elephants**, and 120 scythed **chariots**. Antigonus and his son **Demetrius I (Poliorcetes)** commanded 10,000 cavalry, 70,000 infantry, and 75 elephants. Demetrius drove the enemy cavalry from the field but pursued too far and left the flank of the Antigonid **phalanx** exposed. Attacked by Seleucus' elephants and cavalry, part of the phalanx deserted to the enemy and the remainder was dispersed. Antigonus was killed and Demetrius was only able to rally 8,000 troops. The battle resulted in the division of Antigonus' territory between Seleucus and Lysimachus, creating the basis for future problems between Seleucus and Ptolemy. *See also* SYRIAN WARS.

ISSUS (ISSOS). A plain in eastern Cilicia (in modern-day Turkey), site of the second battle in **Alexander the Great's** conquest of **Persia**, fought at the Pinarus River (either the modern-day Deli Tschai or the Pajas) in

November 333. **Darius III** himself commanded the Persian army, whose size is grossly overestimated in the ancient sources but which may well have contained 30,000 Greek **mercenary** infantry. Alexander was certainly outnumbered and surprised by the appearance of the Persian force to his rear, cutting him off from his Cilician bases.

The Persians deployed with their **cavalry** on the right wing, flanked by the sea, the mercenary infantry in the center, and their native infantry on the left extending into the inland foothills and angled to outflank the **Macedonian** right wing. Alexander placed his **Thessalian** and allied Greek cavalry on his left, the seaward side, his **phalanx** in the center, and his Macedonian cavalry on the right. To guard his flanks and rear against the Persians posted in the hills, he threw back at an acute angle a line of **archers** and light infantry.

In the main battle Alexander attempted to disrupt the enemy by the speed of his attack. As soon as he closed within Persian missile range, he had the right wing charge across the river, causing the Persian troops to give way under the impetus. However, this opened up a gap in the Macedonian infantry phalanx in the middle, and disaster almost ensued when the Greek mercenary infantry began to penetrate the side of the phalanx. On the left the Persian cavalry steadily pushed back the outnumbered Thessalian and allied cavalry. Alexander's right wing, however, continued to advance toward Darius, and when he fled, the center and left of the Persian line began to crumble. Alexander pressed on diagonally toward the sea, relieving the pressure on his phalanx and left wing and causing a general Persian retreat.

The Persian army suffered great numbers of **casualties**, as their large numbers and the restricted area of the battlefield caused considerable crowding and confusion and slowed their withdrawal. Alexander lost 500 men. Although Alexander had handled his army with great skill, part of the reason for his victory was Darius' poor choice of battlefield. The narrow plain forced Darius to place most of his army in the rear, where it could do little and effectively nullified the advantage of his superior numbers. As a result of this battle Persian resistance in the west began to collapse and the Persian fleet essentially disintegrated, leading to the loss of **Ionia** and the coast of Asia Minor. *See also* GAUGAMELA; GRANICUS RIVER.

ITALY. Greek *poleis* (cities) established colonies in southern and western Italy (and in **Sicily**) from the eighth century onward. Despite some military and political successes—and exerting a major cultural influence on Italy and the Italian peoples—the Greeks were never able to fully penetrate the peninsula. They were blocked in the north by the Etruscans (the major power in the northwest of Italy prior to the rise of **Rome**) and in the west were restricted to the narrow coastal plain by the Oscan-speaking hill tribes. From around the middle of the seventh century the Greek cities were faced with Etruscan expansion, but they also came into early conflict with **Carthage**, particularly from around 600, with the foundation of Massilia (modern Marseilles). The Phocaean colonizers there defeated the Carthaginians in a major sea battle, but a subsequent attempt to move east into Corsica provoked a joint Carthaginian and Etruscan response. The Greeks were defeated in a naval battle off Alalia, their colony in Corsica, and the island was lost to the Etruscans. However, around 10 years later an Etruscan attempt to conquer Cumae (in Campania) not only failed but encouraged a Latin revolt against the Etruscans. Cumae assisted in this revolt and made a major contribution to the defeat of the Etruscans at Aricia circa 506.

The Greek colonies preserved the military traditions of the home-land, introducing **hoplite** warfare to the Italians. However, in general they had a greater emphasis on **cavalry** and the navy than most *poleis* in Greece—influenced by local tradition and the fact that most were located on the coast and had been founded by seaborne expeditions.

A major barrier to the success of the Greek states in Italy was their lack of unity. The original colonies had been founded by a variety of Greek cities, and they never united in any sort of common regional alliance and fought each other from time to time. Sybaris, for example, was totally destroyed by its neighbors in 510. This relative lack of unity hampered the cities' efforts to maintain independence. Several of the southern Italian Greek states fell under Syracusan control in the early fourth century following **Dionysius II**'s victory at the **Elleporus River** and the Oscans captured Capua in 423 and Cumae in 420. Despite **Syracuse**'s temporary domination in the area, by about 350 the Oscans had also extended their control over most of the Greek territory in southern Italy, with the notable exception of **Tarentum** (Taras).

In the fourth and third centuries, Tarentum, a very prosperous city, hired a series of Greek commanders to fight on its behalf. The first was **Archidamus III** of **Sparta**, who fought there from 343-338 (when he was killed at the battle of Mandionion), and the second was **Alexander I of Molossia**, the brother-in-law of **Alexander the Great**. Alexander of Molossia arrived in 334 and had some success against the Lucanians but was killed by them when the Tarentines, fearing his ambition, terminated their support to him. The last and most famous general to serve Tarentum was **Pyrrhus I** of Epirus, who agreed to help the city in its war against Rome. Pyrrhus won several battles against Rome (**Siris River** in 280; **Asculum** in 279; and **Beneventum** in 275), but the losses he sustained forced him to end the campaign. The garrison he left in Tarentum withdrew in 272 and the city was handed over to the Romans. Rome then controlled all the Greek areas of Italy, and the *poleis* were incorporated into the Roman alliance system. This effectively ended independent Greek action in Italy, although several of the Greek cities had a brief flurry of independence when they joined Hannibal after his defeat of the Romans at Cannae in 216.

ITHÔME. A mountain in **Messenia**. Circa 464 an earthquake caused considerable damage and casualties in **Sparta**, providing the opportunity for a helot revolt. The Spartans quickly reorganized and the rebellious helots took refuge on Mount Ithôme, joining Messenian rebels who had been there since 469. The Spartans were unable to take the helot stronghold and called upon their allies to assist in siege operations. The Athenians sent 4,000 **hoplites** under **Cimon**. During the siege the Spartans, aware of democratic reforms at **Athens** and apparently unsettled by the Athenians' radical ideas, dismissed them. When the Athenian force returned, outrage at Sparta's treatment of it led to Cimon's political demise and exile and to a change in Athenian policy. This directly contributed to the outbreak of the First **Peloponnesian War** in 459/8. Around 460/59, the helots on Mount Ithôme were allowed to evacuate under truce and the Athenians settled them at Naupactus, on the Corinthian Gulf, to extend their influence in the area. *See also* DIPAEA; MESSENIAN WARS (3).

- J -

JASON OF PHERAE (d. 370). Tyrant of Pherae in **Thessaly**, c. 385-370; *tagus* of Thessaly, 374-370. A physically strong and gifted leader, he led his troops by example and was noted for campaigning swiftly by both day and night. Checked on **Euboea** by **Sparta** in 380, Jason gradually united all Thessaly under his control, using a large **mercenary** army (6,000 strong circa 375) and a process of alliances to generate pressure on individual cities. The process was completed in 374, when Sparta rejected an appeal from Pharsalus, which was then forced to join Jason. He was elected *tagus* of the Thessalian League, controlling an impressive army of some 8,000 **cavalry**, at least 20,000 **hoplites**, and numerous **peltasts** (equipped with the **javelin**). With this army, comprising the combined Thessalian force and a large number of mercenaries, he became a major force in Greece for several years.

After the battle of **Leuctra** in 371 Jason arrived in **Boeotia** with an army, as requested, but refused to assist his Theban allies in following up their victory. Instead he negotiated a truce between **Thebes** and

Sparta and on his way home **ravaged Phocis** and destroyed the **fortifications** at Heraclea, apparently to allow his free access to the rest of Greece through its narrow pass (and also ending Spartan influence in central Greece). He mobilized his army during the Pythian games in 370 but the purpose of this remains obscure as he was assassinated before marching anywhere. His two brothers succeeded him, but both were killed shortly afterward, leaving Jason's nephew, **Alexander of Pherae** in control.

JAVELIN. The ancient Greek military javelin was lighter and generally shorter than the modern athletic javelin. It consisted of a simple wooden shaft and a leaf-shaped (or sometimes barbed) iron blade. Most, if not all, military javelins had a cord loop (*amentum* or *ankyle*) fixed at the point of balance. The thrower placed his first two fingers through the loop and used it to gain greater velocity and distance from the throw. Modern experiments suggest that the use of the *amentum* could increase the distance thrown by about 25 percent. Earlier experiments, conducted for Napoleon, found a 300-percent increase in range with the *amentum*. The maximum range of the military javelin was probably about 300 feet (92 meters), about three-quarters that of a **bow**.

The javelin was known from **Homeric** times, but in Classical and Hellenistic Greece was mainly used by troops from the wilder areas of Greece that were not dominated by the **hoplite** (e.g., **Aetolia**, Acarnania, **Arcadia**, **Thessaly**, and **Thrace**). From the fifth century onward javelin-equipped soldiers in Greek armies were often **mercenaries**. In addition to its use by *psiloi* (light armed troops), the javelin was also employed by Greek **cavalry**. Both *psiloi* and cavalry that were armed with the javelin carried two, and **Xenophon** (*On Horsemanship,* 12.12) recommends that cavalrymen use javelins made of cornel wood, which was strong enough to allow the second javelin to be used as a close-quarter thrusting weapon if necessary. In experienced hands the javelin could be an effective **weapon**. For example, a javelin pierced both of **Philopoemen**'s thighs at the battle of **Sellasia** and because of the *amentum* could not be withdrawn until the javelin was broken (Plut., *Philopoemen,* 6.4-6).

JEWS. *See* PALESTINE

- K -

KAMAX **(pl.** *KAMAKES).* Literally "pole," but also used for a spear (or spear shaft). The term is usefully employed (originally by N. V.

Sekunda) to differentiate between a **cavalry** thrusting spear—perhaps similar to the *sarissa* (but in some cases to be identified with the **hoplite** spear or *doru*)—and the **javelin**.

KING'S PEACE. *See* COMMON PEACE (1)

KOPIS. A sword of saber type, probably of **Persian** origin, of the same or similar type as the *machaira* (figure 5b). *See also* WEAPONS; *XIPHOS.*

- L -

LACEDAEMON (LAKEDAIMON). *See* SPARTA

LACONIA (LAKONIA). *See* SPARTA

LADE. A small island off **Miletus** in **Ionia**, the site of at least two major naval battles:

(1) 494. The final major battle of the **Ionian Revolt**. During the **Persian** combined land and sea operations against **Miletus** the **Ionian** rebels left the land defense to the Milesians but decided to challenge the 600-strong Persian fleet (composed of contingents from Phoenicia, **Cyprus**, Cilicia, and **Egypt**). The Ionian fleet consisted of 353 **triremes**, with the largest contingents drawn from **Chios** (100), Miletus (80), Lesbos (70), and Samos (60). Inspired by Dionysius of Phocaea, the Ionians began a rigorous **training** program for the battle. Unfortunately, it proved too rigorous and the Ionians abandoned it after only a week. The Samian contingent had been suborned by the Persians and in the ensuing battle, all but 11 of their ships deserted. This led to the disintegration of the Ionian fleet, although several contingents, including the Chian one, stayed and fought bravely. The loss of this battle removed all hope for Miletus, which fell soon afterward, effectively ending the revolt.

(2) 201/0. A battle between **Rhodes** and **Macedon** during **Philip V**'s attempt to establish naval supremacy in the Aegean. Following an indecisive engagement off **Chios**, Philip engaged the Rhodian fleet without its allies from Pergamum. When two **quinqueremes** were captured with their crews and another withdrew under sail after being holed, the rest of the Rhodian fleet fled, escaping through superior speed. Philip occupied the Rhodian camp on Lade and quickly secured territory such as Prinassus on the mainland.

LAMIA. The main city of Malis, in the south of **Thessaly**, Lamia became increasingly important in the fourth century because it controlled the main route from Thessaly to central Greece. It gave its name to the **Lamian War** of 323-321 because **Antipater** was besieged there in 323-322. Lamia was under the control of **Aetolia** in the third century and was sacked by Manlius Acilius Glabrio in 190 during his campaign against the Aetolians and **Antiochus III (the Great)**.

LAMIAN WAR. A war fought from 323 to 321 between a Greek coalition, led by **Athens**, and **Macedon**, led by **Antipater**. The Athenians referred to this war as "the Hellenic War." In 324 **Alexander the Great** had caused considerable unrest in Greece, particularly in Athens and **Aetolia**, with his decree restoring all exiles except those guilty of sacrilege. Worried at the prospect of losing Samos, where it had settled **cleruchs** on land confiscated from exiles in 365, Athens began preparations to resist. An Athenian **mercenary** general, **Leosthenes**, was charged with hiring troops among the large number of unemployed mercenaries at Taenarum. These men were recently returned from Asia Minor and themselves posed a considerable source of social unrest in Greece. An alliance with Aetolia boosted Athens' prospects and in 323, when news of Alexander's death reached Greece, Athens revolted against Macedon.

Although underprepared for war, Athens had considerable early success. In autumn 323, aided by **Demosthenes** (2), Athens rapidly created a coalition which included Aetolia (which provided 7,000 troops), much of **Thessaly**, **Argos**, **Sicyon**, **Elis**, Messene, **Phocis** and **Locris**, Carystus on **Euboea**, and **Rhodes**, along with other states. Leosthenes defeated a pro-Macedonian **Boeotian** army at **Plataea** (2). With his army expanded to about 30,000 men, he then defeated Antipater, the Macedonian regent in Greece, at **Thermopylae** (2). Antipater had marched south with as many troops as he could muster in light of the recent large Macedonian levies sent to Asia Minor, but was defeated when the **cavalry** of his Thessalian allies deserted to the coalition army. With his **phalanx** apparently intact and his fleet of 110 **triremes** operating off the coast, Antipater took refuge in the Thessalian city of **Lamia** (after which the war was named). As this city lay astride the route to Macedon, the Greeks could not advance northward until Antipater's army was neutralized.

A vigorous siege ensued, during which Leosthenes was killed and replaced by the Athenian **Antiphilus** and a Greek fleet blockaded Lamia from the sea. The siege was raised in spring 322 with the arrival of a Macedonian relief force under Leonnatus. Although Leonnatus was killed and his army fought to a standstill in battle at **Melitaea**, north of

Lamia, Antipater managed to break out and link up with it and then successfully withdrew to the north. With the arrival of additional forces under **Craterus**, Antipater was able to go on the offensive. He was aided in this by the success of his fleet, which apparently won engagements against the Greeks at **Abydus** in late spring and at at **Amorgos** in May/June and against the Athenian fleet in the Malian Gulf in late summer. Although the course of the naval campaign is unclear, it resulted in the end of Athens as a major naval power. The defeat at Amorgos allowed Macedonian reinforcements to be shipped from Asia Minor to mainland Greece.

On land, Antipater, now commanding a much larger army, advanced into Thessaly. An outnumbered Greek army, much depleted because various contingents had returned home, faced him at **Crannon**. The Greeks were defeated, effectively ending their resistance. Dealing individually with their opponents and capturing cities one by one, Antipater and Craterus broke up the Greek coalition. At Athens the democracy was replaced by an oligarchy, with leaders such as Hyperides and Demosthenes (2) being executed or committing suicide. A Macedonian garrison was installed at **Munychia** in Athens, a heavy indemnity imposed, and the border fort of Oropus handed over to Boeotia. The Aetolians continued to resist during the winter of 322/1 and, although close to surrender, were able to secure a favorable peace in 321 when Antipater and Craterus became involved in a war with **Perdiccas** (2).

LAODICEAN WAR. *See* SYRIAN WARS (4)

LAWS OF WAR. From at least the Archaic period, Greek warfare nominally operated under a system of rules or "laws," which covered activities or actions that were generally agreed to be contrary to natural custom (the "laws of the Hellenes") and were therefore unacceptable. However, given the exigencies of warfare, there was often a large discrepancy between theory and practice. In addition, there was no judicial or panhellenic mechanism to enforce these laws—only the weight of public opinion within a *polis* (city-state) or within Greece as a whole, or a fear that the gods would punish transgression. In addition, as the laws were unwritten, they were subject to different interpretations.

A good example of the opportunity for debate in this area occurred after the battle of **Delium** in **Boeotia** in 424. The **Athenians** had fortified the temple of Apollo at Delium, but their main army had been defeated while returning home. The Boeotians refused to return the bodies of the Athenian dead until the Athenians evacuated the temple. Both sides were therefore in breach of the normal rules of war—which required that the dead be returned under truce and that religious sites be excluded from military action. **Thucydides** records that the Athenians argued that withholding the bodies was worse than occupying the temple, which they had anyway only done for self-defense, and the site was now Athenian territory by right of conquest. They added that "it was entirely reasonable to think that even the god [i.e., Apollo] would make allowances for things which had been done under the pressure of war and danger" (Thuc. 4.98.6). The Thebans argued that occupying the temple contravened the "laws of the Hellenes" and that, if the site was now Athenian territory, the Athenians did not need a truce to recover the bodies—they should just come and get them (which was not possible without another battle). The issue was resolved by the Boeotian capture of the temple and subsequent return of the bodies.

The generally agreed laws of war prohibited the mistreatment of heralds, the mutilation of corpses, the defiling of religious sites (including the harming of soldiers/civilians who had sought sanctuary there), and the harming of priests and pilgrims. The laws also required returning of the dead or allowing access to the dead for burial, keeping the terms of truces or surrender agreements (especially concerning the safety of **prisoners**), and not executing prisoners in cold blood. While these undoubtedly provided some measure of protection and did help mitigate the brutalities of war, there are fairly frequent examples of every one of these rules being breached throughout Greek history. However, the Greeks themselves regarded early warfare rather more romantically and considered that standards had declined over time. In the second century, for example, **Polybius** (13.3) lamented the emphasis on trickery in his day and the change from the ancient "laws of war" which had prohibited hidden or projectile weapons and saw the only genuine victory as one arising from straightforward face-to-face combat. *See also* HALIARTUS; SOLYGIA.

LEAGUE OF CORINTH. *See* HELLENIC LEAGUE (2).

LECHAEUM (LECHAION/LEKHAION). The port of **Corinth**; site of a **Spartan** disaster in 390, during the **Corinthian War**. A contingent of about 600 Spartan **hoplites** from Amyclae returning home to celebrate a religious festival was **ambushed** by a mixed **Athenian** force of hoplites and **peltasts**. The peltasts under **Iphicrates** harried the Spartans, while the hoplites under Callias remained in formation, preventing the Spartan contingent from anything but a fairly slow pursuit in formation. The Spartans were unable to cope with this, continually losing men when they deployed *ekdromoi* to chase the peltasts and mishandling their cavalry support. Retiring to a hill, the survivors eventually broke when threatened by the Athenian hoplites. Around 250 Spartans died and the incident caused considerable damage to Spartan prestige and morale.

LELANTINE WAR. The first known major war in post-**Homeric** Greece. It was fought circa 700 by the Euboean cities of Chalcis and **Eretria** over possession of the fertile Lelantine plain which lay between them. Both cities had a wider importance as founders of major colonies in the Chalcidice and in **Italy**. Chalcis also controlled the Euripus Strait between **Euboea** and the mainland, an important shipping channel, while Eretria controlled the islands of Andros, Tenos, and Ceos. Because of this, many Greek states joined in the war, which lasted several years. Samos, **Thessaly**, and probably **Corinth** allied with Chalcis, and **Miletus** and probably **Megara** with Erctria. According to Strabo (10.1.12), the war was notable for an agreement prohibiting the use of missile weapons. It ended with the defeat of Eretria in a battle that was heavily influenced by the performance of the Thessalian **cavalry** assisting Chalcis. Eretria ceased to be a major power and Chalcis became dominant in the Chalcidice and increased her influence and commercial power in Italy and **Sicily**.

LEONIDAS I (d. 480). The **Spartan** king (reigned 490-480) who led the Greek land forces at **Thermopylae** (1). Although he managed to secure the withdrawal of the main body of his troops when the position was outflanked, he remained with his 300-strong Spartan contingent (possibly the *Hippeis*) and the Thespian and **Theban** contingents and died defending the pass to the last man. According to **Herodotus** he may have been influenced by the **Delphic** oracle, which predicted that "either Lacedaemon [Sparta] would be destroyed at the hands of the *barbaroi* [i.e., Persians] or their king would die" (7.220). *See also* PERSIAN WARS (2).

LEOSTHENES (d. 323/2). Athenian leader of the Greek forces against **Macedon** during the **Lamian War** (probably not to be identified with the Athenian *strategos* of the same name mentioned in an inscription of the 320s). Prior to the war the Athenians instructed him to hire the 8,000 discharged **mercenaries** located at Taenarum but to do so without officially involving **Athens**. On **Alexander the Great**'s death in 323 he marched north, reinforced by an additional 2,000 mercenaries and Athenian troops consisting of 5,000 **hoplites** and 500 **cavalry**. After defeating a **Boeotian** force at **Plataea** (2), he defeated **Antipater** at **Thermopylae** (2) and besieged him in **Lamia**. Unfortunately for the Greek cause, Leosthenes was struck in the head by a stone during the siege and died three days later. He was given a hero's burial and Athens voted him an official funeral speech by the orator Hyperides.

LEOTYCHIDES II (also LEOTYCHIDAS; c. 545-469). The **Spartan** king who commanded the Greeks at their decisive victory at **Mycale** in 479 during the second phase of the **Persian Wars**. Around 478 he led a joint Greek punitive expedition against the Aleuadae and other **Thessalians** who had brought the region over to the **Persians**. The expedition had some success at Pagae and perhaps Pherae, but he failed to take Larissa. On his return he was charged with accepting bribes from the Thessalians and was banished to Tegea, where he died. *See also* DEMARATUS; MEDIZE.

LEUCTRA (LEUKTRA). A small Boeotian town, southwest of **Thebes** and in the territory of Thespiae. It gave its name to a major battle fought nearby in 371, which ended **Spartan** military supremacy in Classical Greece. Following the Theban refusal to sign the restatement of the King's Peace (**Common Peace [1]**) unless allowed to do so on behalf of all the Boeotians, the Spartan king Cleombrotus I led a **Peloponnesian League** army into **Boeotia**. His campaign was initially very successful. Arriving from **Phocis** by an unexpected route, he seized the port of Creusis, along with 12 Theban **triremes**. He then marched inland and camped at Leuctra where he was faced by **Epaminondas**, commanding a Boeotian army unsupported by allies.

Unfortunately the details of the battle that followed are difficult to reconstruct because **Xenophon**'s account (*Hell.* 6.4.4-15), the only contemporary one, is more concerned with explaining why the Spartans lost than providing a complete description of the battle. **Diodorus Siculus**' account (15.51-57) is a fairly stock literary battle description and has several obvious mistakes (including stating that both Spartan kings were present). **Plutarch**'s description of Leuctra in his *Life of Pelopidas* (20-23) understandably concentrates on the role of **Pelopidas**, but does preserve some detail which may have been drawn from earlier Theban sources. Because of this range of incomplete information, various scholars have been able to reconstruct several different versions of what happened.

Using Xenophon's account, supplemented by Plutarch, as a basis, the most likely sequence of events appears to have been as follows. The Peloponnesian army, perhaps of 10,000 **hoplites** and 1,000 **cavalry** (probably outnumbering their Boeotian opponents, but perhaps not by very much), was camped on the plain. Inflamed by drinking wine and criticism that he had been pro-Theban in the past, Cleombrotus decided to attack. He placed his cavalry (which was of poor quality compared to the Boeotian horse) in front of the **phalanx**. This was unusual, but not unknown, and in this case seems designed to screen the outflanking maneuver Cleombrotus intended with his Spartan troops, occupying the right flank. He moved the right flank out to overlap the enemy and brought it back in against the enemy's left flank, in a repeat of the tactic that had proved successful at **Nemea**. However, the Boeotians had deployed with the Thebans 50 deep on the left flank, facing the Spartans, and with the **Sacred Band**, commanded by Pelopidas, deployed as a separate contingent. The Sacred Band was either behind the Theban phalanx or to its left, but apparently stationed with its front some distance to the rear of the front of the Theban phalanx.

Epaminondas advanced in an oblique line, with his left wing forward and his right wing further back. He launched his cavalry first and they quickly drove the Peloponnesian horse back onto the Spartan line while it was still carrying out its outflanking maneuver. The main Theban phalanx struck the part of the line where Cleombrotus was stationed, now disordered by its own cavalry, and after a hard struggle overwhelmed the Spartans, who were deployed 12 deep. Cleombrotus was killed in the process. At the same time, the Sacred Band took the outflanking Spartans in their right, unshielded flank as they were attempting to do the same to the main Theban phalanx. The battle was fought and won in the action between the Spartans and the Thebans; the Peloponnesian left wing and Boeotian right wing apparently did not engage in full combat as the battle had already been won by the time they reached each other. Peloponnesian casualties on the right wing were severe, probably because of the flank attack of the Sacred Band—1,000 Lacedaemonians and 400 Spartiates died, a heavy toll for Sparta with its small population of full citizens.

The substantial number of Peloponnesian survivors remained intact and awaited reinforcements but after a few days withdrew, influenced by the arrival of **Alexander of Pherae**. Although its effects were not immediately obvious to Greece, Leuctra not only demonstrated the effectiveness of the new Theban **formation, training**, and tactics but also led to the end of the Spartan hegemony and its replacement by Theban dominance in Greece.

LIGHT INFANTRY. *See PSILOI*

LOCHOS (pl. *LOCHOI*). An armed band of troops (generally infantry). The term was used for subunits of varying sizes in several Greek armies.

In the later tactical writers such as **Aelian, Asclepiodotus**, and **Arrian** a *lochos* seems to have been standardized at 16 men (or eight for light troops). In the **Spartan army** of the Classical period the *lochos* for a time may have been the largest subdivision, around 500 men strong (Thuc. 5.68.3), but probably more normally had around 150 men. At **Athens** a *lochos* may have been the *trittys* contingent of a tribal regiment. The army of the **Ten Thousand** also contained *lochoi* (each of 100 men), as did **Philip II**'s and **Alexander the Great**'s armies (each of 500 men). The commander of a *lochos* was called a *lochagos* (pl. *lochagoi*). *See also* FOOT COMPANIONS.

LOCRIS (LOKRIS). The name of two central Greek states (there was a third, in **Sicily**), located on either side of **Phocis**. Both were fairly poor regions and subject to domination by, and loss of territory to, larger neighbors. Although they sometimes acted together (for example, they both sent troops to fight against **Sparta** at **Nemea** and **Coronea** in 394), the two states were rather different and generally acted independently—for example, starting off on opposite sides during the **Peloponnesian War** of 431-404. Unfortunately, the evidence for the history of both the Locrian states is fairly patchy.

(1) Ozolian Locris, located on the northern coast of the Gulf of **Corinth**, was considered fairly backward (Thuc. 1.5.3). The main city in the region was Naupactus, which was seized by the Athenians shortly before 464 and used by them to resettle the **Messenians** from Mount **Ithôme** at the end of the Third **Messenian War** (Thuc. 1.103.1-3). The Ozolian Locrians traditionally operated as *psiloi* (light troops), predominantly armed with **javelins**—like their neighbors, the Aetolians (Thuc. 3.95.3, 97.2). During the **Peloponnesian War** of 431-404 they were allies of **Athens** and in 427/6 **Demosthenes** (1) used Locris as a base for his unsuccessful campaign against **Aetolia**, with the aim of invading **Boeotia** from the west, via **Phocis**. Shortly afterward a **Peloponnesian League** army forced the area into temporary alliance with **Sparta** (Thuc. 3.95; 100-101).

According to the *Hellenica Oxyrhynchia* (18), the Ozolian Locrians were heavily involved in the outbreak of the **Corinthian War (Xenophon** [*Hell.* 3.5.3] is probably wrong in identifying the Locrians involved in this as the Opuntian Locrians, although they also fought against Sparta and Phocis during the war). In 394 the Ozolian Locrians supplied 50 **cavalry** and an unspecified number of *psiloi* to support the anti-Spartan coalition at **Nemea**. They (along with the Opuntian Locrians) also sent troops to fight against Sparta at **Coronea** and performed quite well in harassing Spartan troops that invaded their territory after

Coronea (Xen. *Hell.* 4.2.17, 4.3.15, 4.3.21-23). Ozolian Locrians were allies of Boeotia by 370, when (along with the Opuntian Locrians) they sent troops to participate in **Epaminondas'** campaign in the Peloponnese (Xen. *Hell.* 6.5.23).

The strategic location of the pass though Amphissa (and the Locrians' traditional enmity with Phocis) meant that Locris was fully involved in the Third **Sacred War**. Subsequently it was dominated by **Macedon** and then became a member of the **Aetolian League**.

(2) Eastern Locris, divided into Opuntian and Epicnemidian Locris (although "Opuntian" is often used to include both), occupied the coast opposite **Euboea** and was a **hoplite** state. During the Second **Persian War** Locris supplied its entire army to help defend to **Thermopylae** in 480 but was forced to join the **Persians** after it fell, supplying hoplites to them for **Plataea** (1) (Hdt. 7.203, 9.31). Up to this point the only Locrian troops attested in the literature seem to be hoplites, but sometime prior to the **Peloponnesian War** of 431-404 the Locrians developed a **cavalry** force—this was the arm they supplied to their **Peloponnesian League** allies on the outbreak of the war (Thuc. 2.9.3). The troops they sent to assist at **Delium** in 424 were probably also mounted, as they arrived late but were still able to catch up with the fleeing **Athenian** hoplites (Thuc. 4.96.8).

The Locrians were traditional enemies of **Phocis** and fought against them during the Third **Sacred War**. Following this, they were in the **Macedonian** sphere of influence, sending cavalry to fight under **Alexander the Great** in his conquest of the Persian empire (Curt. Ruf. 4.13.29), and later allied with **Philip V** (Polyb. 11.5), apparently in the interim having been in the **Aetolian League**. Locris was freed from Macedonian domination by the **Romans** in 196 but restored to the Aetolian League (Polyb. 18.46-47).

LOGISTICS. In comparison to later periods of warfare, the logistic art was rather undeveloped in ancient Greece. The same basic limiting factors generally applied throughout the whole period: the daily nutritional requirements of soldiers and animals, the methods for preserving food, the pace of movement, and the available technology for repair of items. Although some improvements occurred over time in the means of sea transport available and, in some places, the quality of the roads, these were relatively minor. The major differences between logistic practices in **Mycenaean** and Dark Age warfare, the Archaic and Classical periods, and Hellenistic Greece were caused by organizational change and more professional practice. In general, Greek logistic practice became more sophisticated.

In **early Greek warfare** rations were probably initially provided by the soldier—each infantryman on the Mycenaean "Warrior Vase" (Athens, National Museum, 1426), which dates to circa 1200, carries a bag of rations tied to his spear. For more distant or longer expeditions, food could have been transported by cart, pack mule, or **ship**. Sea transport was by far the most economical, as land transport could carry only small quantities (the ancient harnessing system for horses meant they could not draw larger loads efficiently) and distance of course precluded transporting food with a limited life. **Thucydides** (1.111) states that the Greeks at Troy grew their own food during the siege, further reducing the numbers available for combat (and that the original numbers of the expedition had already been kept low through concern over supplies). However, short campaigns were probably the norm, reducing the logistic problems.

The resupply or repair of broken weapons and equipment was again probably also the responsibility of the individual, although the palace societies on Crete produced quantities of military equipment, including spare parts for **chariots**, on a scale that suggests they were issued out to soldiers. Troops could be transported on foot, in chariots (but this would have been only a few), or by sea.

These practices essentially continued through the Dark Age, in Archaic Greece, and into the Classical period. If anything, there was a reduction in the organization from the level attained in the palace society and an increasing emphasis on the responsibility of the individual. In the days of the true citizen-soldier and **hoplite** warfare, campaigns were short and sharp, with a very small logistics bill. When a city mobilized its army, the soldiers would be directed to turn up with rations for a specified number of days. After that the soldiers depended on **foraging** or buying food from markets. This began to change from the time of the Second **Persian War**, when the nature of the threat and the size of the forces meant that supplies had to be organized. The preliminaries to the battle of **Plataea** in 479 involved the **Persian** destruction of a Greek supply column (Hdt. 9.38-39).

About the middle of the fifth century, state **pay** for soldiers, presumably initially to purchase rations, was introduced at **Athens** and later became standard practice elsewhere. The treaty of alliance between Athens, **Argos**, **Elis**, and **Mantinea** provides a good example of logistic arrangements in 420. Under this agreement the state supplying troops was responsible for feeding them for 30 days after arrival at the city being helped and on the return journey. In the event of a longer campaign, the state that requested help would be responsible for the supplies past 30 days "at the daily rate of three Aeginetan obols for a hoplite, **archer**, and

psilos (light infantryman) and an Aeginetan drachma for a **cavalryman**"
(Thuc. 5.47.6)—who also had to feed his horse.

The logistic requirements for **naval warfare** were particularly large
and Athens maintained state dockyards and supply pools. This was a
precursor of things to come. In the fourth and following centuries, as
warfare became more complicated, with longer campaigns across wider
distances and extended sieges, the logistics burden on states and monarchs increased. The **Spartans** had always taken craftsmen and a supply
train in the field (Xen. *Lac. Pol.* 11.2), being better organized than the
average Greek army in this regard, but this was brought to a fine art in
the **Macedonian** army under **Philip II** and **Alexander the Great**.

Philip had trimmed his logistic train by forbidding his troops the
use of wagons and limiting servants to one for each group of 10 men
(Frontinus, *Strategemata*, 4.1.6). Unlike other Greek armies, the Macedonian soldiers had to carry their own arms, **armor**, and equipment. Alexander maintained these practices, and his success in Persia arguably resulted in large part from his detailed logistic planning and a relatively
sophisticated logistics organization. He used mules, horses, and camels
to transport supplies and seems to have arranged provisions in advance
when accepting the surrender of various Persian provinces (this is covered in detail in D. W. Engels' *Alexander the Great and the Logistics of
the Macedonian Army*).

Practice under the Hellenistic successor kingdoms was pretty similar, as might be expected given their descent from Macedon. However,
there was a tendency to increase the baggage train and the number of
camp followers, losing some of the mobility enjoyed by Philip and Alexander. The increasing variety of troop types and the inclusion of chariots and **elephants** in some armies also increased the logistic burden.
However, unlike the average Greek *polis* (city-state), the successor kingdoms were able to mobilize larger resources and build on Eastern organizational practices to supply their troops. Given the swelling of the camp
followers and baggage train, naval resupply became increasingly important. By the time **Rome** conquered the Greek world, the Greeks were
outclassed by the Roman logistics organization as well as by their fighting prowess.

LYDIADAS (d. 227). Son of Eudamas. He led Megalopolis against
Sparta in 251 and became tyrant of the city around 243. He abdicated
the tyranny in 235 and brought Megalopolis into the **Achaean League**.
Lydiadas became a rival of the prominent **Achaean** commander **Aratus**
and was elected general of the Achaean League in 234, 232, and 230.
Enraged at Aratus' inaction at Ladocaea in 227, during the **Cleomenic**

War, Lydiadas led the **cavalry** in an unauthorized charge across rough country against the Spartans under **Cleomenes III** and was killed.

LYSANDER (LYSANDROS; d. 395). A Spartan **officer** active in the naval campaign that ended the **Peloponnesian War** of 431-404 and in **Sparta**'s expansionary policy that led to the **Corinthian War** (395-386). Appointed **navarch** (admiral) in 406, Lysander quickly established good relations with **Cyrus the Younger**, persuading him to increase the **pay** for the Peloponnesian fleet. Lysander defeated the **Athenian** navy in the opportunity naval battle of **Notium** but sensibly avoided a subsequent engagement against **Alcibiades'** much larger fleet.

Following the death of his successor, **Callicratidas**, in 405 Lysander was reappointed to naval command in Asia Minor at the request of Sparta's allies there. Technically Abacus was navarch and Lysander *epistoleus* (deputy admiral), but Lysander actually exercised command (Xen. *Hell.* 1.17). Lysander rapidly reestablished his good relations with Cyrus, secured funds, and refitted the fleet. He then sailed to the Hellespont to interfere with Athens' vital grain imports and annihilated the Athenian fleet at **Aegospotami**. This victory destroyed Athens' naval capacity and meant that the city could no longer import the food it needed. It also led to the defection of the greater part of the Athenian empire. Lysander secured the Hellespont by garrisoning **Byzantium** and **Chalcedon** and giving any Athenians in the area safe conduct back to Athens. In doing so he simultaneously cut off Athenian food supplies and increased the population of Athens. Sailing to Athens, detaching states from its empire on the way, he blockaded the Piraeus. This, in conjunction with the land blockade of **Pausanias** (2), quickly starved Athens into surrender.

Lysander played an important role in establishing an oligarchy at Athens (as he had in the states he had detached from the Athenian empire). However, a quarrel with King Pausanias had reduced Lysander's influence by about 402. He was apparently involved in an unsuccessful move to make the Spartan kingship elective and alienated his protege, King **Agesilaus II**, by his behavior in Asia Minor. Despite these problems Lysander seems to have played a central role in Sparta's generally expansionary policy in northern and central Greece, Asia Minor, **Sicily**, and **Egypt**. In 395 this policy was a major contributor to the outbreak of the Corinthian War. Sent to **Phocis**, Lysander raised an army from Sparta's allies in the area and invaded **Boeotia** from the north. Before Pausanias' army could link up with him from the south, Lysander was killed in battle outside **Haliartus** in 395. He was an able admiral and general but his behavior alienated may of his contemporaries and he suffers from a generally bad press in the ancient sources.

LYSIMACHUS (LYSIMACHOS; c. 360-281). One of the *diadochoi*. His father, Agathocles, was probably of Thessalian origin but held estates in Pella. Lysimachus was one of **Alexander the Great**'s bodyguards and after Alexander's death received the territory of **Thrace**. His power reached its height in 284 when he ruled over Thrace, the western and southern coasts of the Black Sea, **Macedon**, **Thessaly**, and Anatolia.

The early years of his reign were spent securing Thrace, building his capital Lysimacheia and extending his control to the Black Sea. In 302 Lysimachus invaded Asia Minor as the spearhead of an attack on **Antigonus I (Monophthalmus)** by a coalition which also included **Cassander**, **Seleucus I (Nicator)**, and **Ptolemy I (Soter)**. Lysimachus, leading his own army and a contingent sent by Cassander, fought a clever campaign, avoiding a pitched battle against Antigonus until joined by Seleucus' army the following year, and then helped defeat Antigonus at **Ipsus**. As a reward Lysimachus gained most of Anatolia. In 299 he allied with Ptolemy and in 288 he captured **Amphipolis** and eastern Macedon from **Demetrius I (Poliorcetes)**. In 285 Lysimachus expelled **Pyrrhus I** of Epirus from the western half of Macedon to secure control over the whole country.

The last years of his rule were marked by struggles between court factions, which resulted in the execution of his generally popular eldest son, Agathocles, in 283 or 282 and a significant drop in Lysimachus' support. Agathocles' widow, Lysandra, fled to Seleucus with her children and other defections followed. Seleucus seized the opportunity to invade Asia Minor and Lysimachus was defeated and killed by him at the battle of **Corupedium** in 281.

- M -

MACEDON. A country in the north of Greece, separating Greece proper from **Thrace**. It is quite fertile and in antiquity supported a large population. The nature of this population is hotly debated, but on balance it seems likely that the Macedonians spoke a Greek dialect. In the mid-seventh century it was ruled by the Temenid dynasty, which was probably of Greek origin (it certainly claimed descent from **Argos**). Macedon was generally regarded as rather backward by the rest of Greece and had only a relatively small impact on the military history of Greece until the fourth century.

Late in the sixth century (circa 512 or in 492) Macedon was occupied by the **Persians**, and under **Alexander I of Macedon** (circa 495-452) it was forced to contribute troops to fight against the Greeks in the Second **Persian War**. Following the Persian defeat at **Plataea** (1), Macedon regained its independence, although internal dissension, pressure from Thracian and Illyrian tribes, and the activities of Greek colonies in the Chalcidice and Thermaic Gulf helped keep it relatively weak. **Athenian** ambitions in the north were also a major threat and kept Alexander busy for some time after 479. The situation noticeably worsened with the foundation of the Athenian colony of **Amphipolis** in 437/6. Macedon was not militarily strong enough to end this threat by force and, under **Perdiccas II** (circa 450-413), for most of the rest of the century attempted to weaken its enemies by a mixture of diplomacy (involving several changes of alliance during the **Peloponnesian War** of 431-404) and occasional military action. In 429 Perdiccas was successful in using his **cavalry** in a mobile defense role to absorb a major invasion by the Thracians under Sitalces, who was forced to withdraw by the cold and lack of supplies.

Macedon was to some extent modernized, especially in terms of road building and **fortifications**, under King Archelaus (circa 413-399), who also reorganized the cavalry and introduced **hoplite** equipment and tactics. Despite this and Macedon's possession of major manpower resources and the existence of rich mineral deposits on its borders, it was not until the accession of **Philip II** that it began to exert a real influence in Greek affairs. Faced by several major challenges, both internal and external, Philip restructured the **Macedonian army**, turning it into a highly professional force and creating the famous *sarissa*-armed **phalanx**. He combined heavy infantry, *psiloi* (light infantry), and high-quality cavalry (always a Macedonian strength) and supplemented these with good **artillery** and a siege train. The resultant flexible and highly efficient force ultimately replaced the traditional hoplite army as the model for Greek warfare. Philip used his army, along with bribery and diplomacy, to first secure Macedon's safety and then expand its control over Thrace, **Thessaly**, and eventually most of the rest of Greece. Greek opposition was crushed at **Chaeronea** in 338 and although Philip's son, **Alexander the Great**, had to use some force to gain recognition of his inheritance, the dominance of Macedon in Greek affairs lasted for nearly 150 years. Alexander's conquest of the Persian empire also ensured Macedonian influence on Asia Minor for many years.

Following Alexander's death the Greek failure in the **Lamian War** left most of Greece firmly under Macedonian control. However, the third century saw Macedon, the ultimate prize for Alexander's successors, pass through several hands in a series of power struggles that ultimately weakened the country. Despite this, all of the successor kingdoms were ruled by Macedonians and secured by Macedonian troops. Enduring Macedonian dynasties were founded by **Seleucus I (Nicator)** in **Syria** and by **Ptolemy I (Soter)** in **Egypt** and retained a distinctly Macedonian character. In Greece, the rise of confederacies such as the **Aetolian League** and the **Achaean League**, which could compete on more equal terms than could the individual *poleis* (city-states), challenged Macedon, but under the Antigonid dynasty (founded by **Antigonus I [Monophthalmus]**) it continued as the major power. This remained so despite frequent challenges from various parts of Greece and the occasional interference by Seleucid or Ptolemaic monarchs.

Macedon was eventually finished as a major power by **Rome**. The two states first came into conflict just prior to the Second Punic War and **Philip V** of Macedon's alliance with Hannibal during that war sparked a series of wars with Rome, all of which Macedon lost. Its power was gradually reduced until the Fourth **Macedonian War**, after which Macedon was incorporated into the Roman empire in 148.

MACEDONIAN ARMY. At the time it first appears in the historical record, the best section of the Macedonian army was the **cavalry**. The infantry levy was generally of indifferent quality, and during the sixth and fifth centuries the Macedonians often chose to defend their country by retiring to strongholds and using the cavalry in a mobile defense role to harry and harass the enemy. The army was organized on territorial lines, with contingents recruited regionally. At the end of the fifth century, King Archelaus (circa 413-399) reorganized the cavalry, built roads and a network of forts, and considerably improved the infantry by the introduction of **hoplite** equipment and tactics. The infantry, however, was still generally inferior to the hoplites of the Greek *poleis* (city-states) to the south.

The reforms under **Philip II** (359-336), which created a heavy infantry to match the quality of the cavalry, made the Macedonian army the best in Greece. Philip introduced the *sarissa*, the Macedonian **phalanx formation**, and, very importantly, a rigorous program of **training**. In 358 he was able to field 10,000 men against **Bardylis I** of Illyria. The core of Philip's new army was eventually the six brigades (*taxeis*) of the **Foot Companions** (*pezhetairoi*), totaling 9,000 men. This, along with the 3,000-strong division of **hypaspists**, became a professional force, supplemented when necessary by mobilization of the wider levy, drawn from all able-bodied Macedonians. Using the phalanx to fix the enemy foot, Philip would then destroy them using his cavalry. The core of this cavalry force was the elite Companion cavalry, which included a Royal Squadron drawn from Philip's personal companions and the Macedonian aristocracy. This squadron fought beside the king, who frequently commanded it. Philip also developed a very effective siege train, introducing **artillery** in considerable quantity into the Macedonian army.

Alexander the Great further refined and developed this force. The army with which he conquered the **Persian** empire consisted of the six *taxeis* of Foot Companions, the six companies (*lochoi*) of hypaspists, and eight squadrons of Companion cavalry (each squadron consisting of 200 men, except the Royal Squadron, which was 400 strong). This Macedonian core was supplemented by troops provided by areas under Macedonian control, such as **Thessaly** and **Thrace**, and by allies and **mercenaries**; these included 1,800 Thessalian cavalry, 800 Thracian *prodromoi* (in four squadrons), light Thracian cavalry from Paeonia and Odrysia (initially totaling around 400 men), 600 Greek cavalry, at least 800 **archers**, and 7,000 other light troops (*psiloi*).

When reinforcements arrived after **Gaugamela** in 331, Alexander reorganized his forces, adding a seventh *taxis* to the Foot Companions and grouping the *lochoi* into larger groupings of 1,000 men called a chiliarchy. Each cavalry squadron (*ile*) was further subdivided into two *lochoi* for easier control. He also reduced the regional character of the various Macedonian contingents by appointing commanders on the principle of merit (and political reliability). Prior to his death Alexander had been training Persians in Macedonian phalanx tactics and had intended to mix archers and **javelin** troops in the phalanx, to provide it with extra flexibility and hitting power.

In the armies of the *diadochoi* the Macedonian troops retained their political importance (as they provided the basis of authority for the leadership). However, while keeping many of the organizational features of Alexander the Great's army, over time the successor kingdoms' armies, starved of Macedonian recruits and influenced by local conditions, ceased to be Macedonian in nature. For example, at **Paraetacene** in 317, only around 14,000 of the 65,000 troops involved were Macedonian. However, under the Antigonids, who came to control Macedon, the Macedonian army proper generally preserved the organization and traditions of Philip II and Alexander. For some time, though, the army was understrength because of losses under Alexander (including some 23,000 who stayed in Asia Minor after his death).

The last fling of the Macedonian army was against **Rome** under **Philip V** and his successors. By this time the famed phalanx had become rather more rigid than in Philip II and Alexander's time, when it could, and did, traverse rough terrain. In a series of defeats at **Cynos-**

cephalae (2) and **Pydna** it proved no match for the more flexible formations of the Roman army. At this time the field army, probably numbering around 20,000 infantry and 2,000 cavalry, was still organized regionally and backed up by a considerable number of troops in garrisons. At Cynoscephalae in 197, Philip was able to field 18,000 heavy infantry, 2,000 cavalry, and 5,500 light troops (which included 1,500 mercenaries). However, he could only achieve this by a general levy which included boys from 16 years of age and older men.

At the start of the Third **Macedonian War** of 171-167, **Perseus** was able to field a force of 43,000 men, the Macedonian core of which included 3,000 cavalry, 21,000 phalangites, and 5,000 elite infantry. It was a well-equipped force but represented the entire forces of the kingdom; there were no reserves to speak of. At this stage the cavalry was organized with a Royal Squadron, elite ("sacred") squadrons, and ordinary squadrons. These were trained to fight with *hamippoi*. The elite infantry, numbering some 5,000, were called *peltastai* (**peltasts**), and were the equivalent of Alexander the Great's hypaspists. The phalangite core of the army was still divided into brigades (*taxeis*), perhaps still of 1,500 men. This army was destroyed at Pydna in 168, and the subsequent Roman annexation of Macedon ended the existence of an independent Macedonian army.

MACEDONIAN WARS. A series of wars between **Rome** and **Macedon** in the late third and early second centuries that resulted in Macedon's annexation as a Roman province.

(1) First Macedonian War (215-205). Fought between **Rome** and **Philip V** of **Macedon**. Philip, already on bad terms with Rome and believing the Romans were finished after their defeat at Cannae in the Second Punic War, promised military assistance to the Carthaginian Hannibal. However, the Romans discovered the alliance and used their navy to prevent Philip from sending troops across the Adriatic. Rome allied with the **Aetolian League** and, apart from a few naval **raids** on Epirus, carried on the war by proxy through its new allies. The Aetolians and Philip made peace in 206 and in 205 the Romans followed suit. Although the war had essentially died of inaction, it left the Romans angry with the Aetolians for making peace without consulting them. More particularly, the result failed to satisfy Roman anger with Philip for joining their archenemy Hannibal during Rome's greatest crisis.

(2) Second Macedonian War (200-197). Fought between **Rome** and **Philip V** of **Macedon**. Philip's ruthless naval operations in the Aegean in 202-201 alienated **Rhodes** and **Attalus I (Soter)** of Pergamum. After suffering heavy naval casualties in 201/0, Attalus (supported by Rhodes

and later **Athens**) appealed to Rome for assistance. Although the Roman people initially rejected the idea of yet another war, the consul for 200, Publius Sulpicius Galba, managed to persuade them that Philip was a threat. War was declared when Philip, unsurprisingly, rejected a Roman ultimatum requiring him to pay an indemnity to Rhodes and Pergamum and to refrain from future acts of war.

The Romans invaded Illyria in 200, establishing a base at Apollonia, which caused Philip to return to Greece from Asia Minor. In 199 a Roman strike against Macedon was forced to turn back because of lack of supplies. The following year Philip held up the Roman commander, Titus Quinctius Flamininus, for several weeks until outflanked and forced to withdraw. Flamininus occupied **Thessaly** and, after abortive peace negotiations, decisively defeated Philip at the battle of **Cynoscephalae** (2) in 197. Philip had to pay an indemnity, surrender his fleet, and withdraw his garrisons from Greece, which Rome declared free. **Aetolian** discontent with the limited reward gained for their assistance to Rome in this war contributed to the subsequent outbreak of war between Rome and **Antiochus III (the Great)** and Roman involvement in Greek Asia Minor. *See also* CHIOS; LADE (2).

(3) Third Macedonian War (171-167). Fought between **Rome** and **Perseus** of **Macedon** (son of **Philip V**). Perseus had angered the Romans by his involvement in the death of his brother Demetrius, who was popular in Rome. Suspicion increased over his alliances with **Thracian** and Illyrian kings, the expansion of his army, his marriage to Laodice, daughter of Seleucus IV (Philopator) of **Syria**, the marriage of his half-sister to Prusias II of **Bithynia**, and his rapid improvement in state finances. **Eumenes II (Soter) of Pergamum** also kept up a barrage of anti-Perseus propaganda at Rome. In 171 Rome declared war when negotiations over Perseus' activities broke down. The war was fought by a relatively small Roman force, and Perseus generally remained on the defensive. Perseus was initially able to prevent the Romans from penetrating his borders but in 168 the Roman consul, Aemilius Paullus, landed at **Delphi**, marched north, and defeated Perseus in a pitched battle at **Pydna** (1). Perseus was deposed and Macedon split into four republics, financially crippled by the closure of the mines at Mount Pangaeum and restrictions on trade in salt and shipbuilding timber. Although this was designed to prevent a Macedonian resurgence, the republics were left too weak to provide for their own self-defense. **Rhodes**, which had attempted to intervene diplomatically on Perseus' behalf, was punished by the loss of Lycia and Caria and by the establishment of Delos as a free port in direct competition to its own commercial activities.

(4) Fourth Macedonian War (149-148). Fought between **Rome** and **Andriscus**, a pretender to the throne of **Macedon**. Andriscus took control of the four Macedonian republics, which Rome had left too weak to properly defend themselves, and overran **Thessaly** in 149. The Macedonians were defeated at the battle of **Pydna** (2), and Andriscus was captured and later executed in Rome. Following this war, the Romans finally ended Macedonian independence, incorporating it into the Roman empire as a province in 148.

MACHAIRA (pl. *MACHAIRAI*). A slashing sword, probably with curved blade, not unlike a larger version of the Gurkha kukri (see figure 5b). Although a common **cavalry weapon**, it was also used by infantry. *See also KOPIS; XIPHOS.*

MAGNESIA. More fully, Magnesia ad Sipylum (to distinguish it from other Magnesias); a Lydian city situated on slopes of Mount Sipylus. It was the site of a major battle fought in 190 between **Antiochus III (the Great)** and **Rome**. The only real account of the battle is in Livy (37.37-44), not always the most accurate of military historians. Magnesia was the final battle in the war between Rome and Antiochus and followed his defeat in 191 at **Thermopylae** (3) and withdrawal from Greece.

The Roman army was led by Publius Scipio and consisted of 22,000 legionaries, around 3,000 **Achaean** light infantry and auxiliaries (the latter supplied by **Eumenes II [Soter] of Pergamum**), 2,000 **Macedonian** and **Thracian** volunteers, about 500 Trallians and Cretan **archers**, more than 3,000 **cavalry** (800 Pergamene, the rest Roman), and 16 **elephants**. Antiochus led a force of 60,000 infantry, more than 12,000 cavalry, and 54 elephants. Antiochus had established a fortified camp, protected by a palisade, then a ditch, and finally a wall with towers. Although outnumbered, the Romans decided to attack in order to end the war before the onset of winter. The battle proper was preceded by minor skirmishes caused by contingents of the king's troops attacking the Romans while they were constructing their camp—on each occasion without success.

Antiochus drew up his army in response to a Roman advance, with 10,000 men in a *sarissa*-equipped **phalanx** in the center, deployed in 10 contingents, 32 ranks deep, with two elephants in the interval between each. From the phalanx to the right flank (personally commanded by Antiochus) were arrayed 1,500 Galatian foot, 3,000 **armored** cavalry, 1,000 mixed cavalry, 16 elephants, the royal guard (*argyraspides*), 1,200 mounted archers, 3,000 light infantry, 2,500 archers, and finally a mixed group of 4,000 **slingers** and archers. From the phalanx to the left flank were arrayed another 1,500 Galatian foot, 2,000 Cappadocian infantry, 2,700 mixed auxiliaries, 3,000 armored cavalry, and 1,000 mixed cavalry, with scythed **chariots** and archers mounted on camels deployed in front of them. The extreme right was held by a group similar to that on the left and an additional 16 elephants were apparently held slightly back in reserve. The Romans were deployed with the legions in the center, and to the right the Achaean light troops and Pergamene auxiliaries, 3,000 cavalry, and the 500 Trallians and Cretans on the extreme flank. On the left flank Scipio deployed only four squadrons of horse (of unknown number), relying on riverbanks for protection. The Thracians and Macedonians guarded the camp under Roman command.

Antiochus' larger force was hampered by the poor visibility caused by a mist, which also reduced the efficiency of bowstrings, slingshots, and the **javelin** throwing loops—a bigger problem for his men than for the Romans, who were predominantly heavy infantry. Eumenes began the battle with an attack on the scythed chariots by a very loose formation of slingers and archers, successfully causing the horses to panic and the force to disperse from its assigned position—which led to the flight of the light infantry nearby. Once the battle proper began, Antiochus' left flank, deprived of its light protection, was fairly easily driven off. These fleeing troops disordered the phalanx in the center and opened it up to the experienced Roman legionaries. Meanwhile, on the right Antiochus had taken advantage of his superior numbers and mobility to outmaneuver those facing him and had broken through to the Roman camp, which was only saved by some hard fighting by its defenders. The Roman camp commander stabilized the fleeing troops, and when Eumenes' brother arrived with 200 cavalry, Antiochus was forced to give way.

Livy reports Antiochus' **casualties** as 50,000 infantry and 3,000 cavalry dead and a total of 1,400 taken captive, along with 15 elephants. The Romans had numerous **wounded** but only 300 infantry and 24 cavalry killed, while Eumenes lost 25 cavalrymen dead. Whether these figures are accurate or not, it was a decisive defeat and shortly led to the Treaty of **Apamea** which formally ended the war and seriously and permanently reduced Seleucid power. Magnesia yet again demonstrated the superiority of the Roman legions over Hellenistic armies.

MANTINEA (MANTINEIA). An **Arcadian** city, located in the middle of a plain in the central Peloponnese. Mantinea became part of the **Peloponnesian League** in the sixth century and fought on the Greek side during the Second **Persian War**—500 Mantinean **hoplites** served at **Thermopylae** (1), but their contingent arrived just too late for **Plataea**

(1) (Hdt. 7.202, 9.77). The Mantineans supported the Spartans in the helot revolt of 464 (Third **Messenian War**) and initially fought alongside the Spartans during the **Peloponnesian War** of 431-404, but from around 450 their democratic constitution sat uneasily with the Spartan alliance. In 421, after a quarrel with Tegea (which involved a major battle, fought in 423) and afraid that the Spartans would make them return territory they had recently seized in Arcadia, the Mantineans joined with **Argos, Athens**, and **Elis** against **Sparta**. The Mantineans were heavily involved in the fighting against Sparta, which was ended by the coalition defeat at the battle of Mantinea in 418 (see [1] below). Mantinea made peace with Sparta, relinquishing its newly acquired territory (Thuc. 4. 134; 5.29, 3, 47, 65, 81).

The Mantineans fought on the Spartan side at **Nemea** in 394 (Xen. *Hell.* 4.2.13), but otherwise were fairly reluctant allies. **Xenophon** (*Hell.* 4.4.15) records Spartan contempt for the performance of Mantinean hoplites against **peltasts** prior to their own defeat at **Lechaeum** in 390. In 385-384, following the King's Peace (**Common Peace** [1]) the Spartans took the opportunity to reduce the power of Mantinea. The Mantineans were defeated in battle outside their city and forced to surrender when the Spartans dammed the exit of the river that ran through the city, causing flooding and erosion of the city's mudbrick walls. Mantinea was then broken up into the separate villages from which the city had originally been formed (Xen. *Hell.* 5.2.1-7; Polyb. 4.27). The citizens, though, reconstituted the city (with stone walls and diverting the river around the walls) after the Spartan defeat at **Leuctra** in 371 (Xen. *Hell.* 6.5.3-5).

Mantinea was prominent in the re-creation of the **Arcadian League** and the foundation of Megalopolis (achieved with **Boeotian** assistance), but after a quarrel with Tegea, split the league in 363 by rejoining the Spartan alliance (Xen. *Hell.* 7.4.33-5.3). Despite their involvement in the defeat at Mantinea in 362 (see [2] below), the Mantineans remained a major force in Arcadia. It joined the **Achaean League** on its formation but later became a member of the **Aetolian League** and in 229 was handed over to Sparta, helping cause the **Cleomenic War** between the Achaean League and Sparta (Polyb. 2.43). The city fell to a surprise attack by **Aratus** in 226 and, in the interim having ejected the Achaean garrison, around 223 was captured by **Antigonus III (Doson)**, who sacked the city, sold the inhabitants into **slavery**, and modestly renamed it Antigonea (Polyb. 2.54, 57-58). By 207, though, under **Philopoemen**, it was again a part of the Achaean League and became the site of another battle (see [3] below), which removed the Spartan domination of the league.

Mantinea fell under **Roman** control, along with the rest of the Peloponnese, in the second century and was the only Arcadian city to fight on the side of Octavian (Augustus) at **Actium**. Under Hadrian, the name Mantinea was restored in place of Antigonea. Mantinea gives its name to at least five major battles fought in its territory, including two for which we have no details—both Spartan defeats, against **Demetrius I (Poliorcetes)** in 295 and against the Achaean League, led by Aratus, in 242—and the following:

(1) A battle fought in 418 between a **Peloponnesian League** army led by **Agis II** of **Sparta** and an anti-Spartan coalition force drawn from **Argos, Athens,** Mantinea, and several other **Arcadian** cities. The Spartans were attempting to reassert their authority in the Peloponnese during the Peace of **Nicias** and more particularly to provide assistance to Tegea, which was on the point of going over to the anti-Spartan coalition. **Thucydides** gives a detailed account of the battle (5.64-74), although unable to give precise numbers, while **Diodorus Siculus** (12.79) provides a brief description focussing on Agis and the Argive *epilektoi*.

Agis led his army into Mantinean territory, successfully drawing the coalition army away from Tegea. The coalition army drew up in a very strong position on high ground and Agis was dissuaded from attacking by the remark of an old soldier who shouted out "that he was trying to cure one evil with another"—making up for his earlier ignominious withdrawal from Argos by a foolhardy attack. Agis withdrew and after some hesitation (for which their own troops blamed them) the Argive generals followed. The next day both armies drew up in battle order on the plain. The Spartans deployed with the *Sciritae* on the left, then the troops who had served under **Brasidas** in the Chalcidice, along with other freed helots equipped as **hoplites**. The Spartan contingent was in the center-right position, with their allies to their left. The Tegeans held the right wing, with some Spartiates as stiffening on the extreme right. The **cavalry** was posted on both flanks of the force. The coalition army deployed with the Mantineans on the right wing, then the Arcadian troops, the 1,000-strong Argive *epilektoi*, the main body of Argives, and to their left, other allies. The 1,000 Athenian hoplites occupied the left wing, with 300 of their own cavalry (apparently the only coalition horse present) providing flank protection.

During the advance Agis decided to correct the natural rightward drift of his hoplites by moving the *Sciritae* and Brasidas' old troops further to the left and filling in the gap with two regiments drawn from the right wing. This maneuver failed when the two **officers** ordered to move from the right refused because of the short notice and proximity of the enemy (both were subsequently banished from Sparta for this). The

Mantinean **phalanx** broke the now isolated *Sciritae* and accompanying troops and, along with the Argive *epilektoi* and other troops, penetrated the gap in the Spartan line causing considerable **casualties** and driving the enemy back to their wagons. In the center, though, Agis and his men quickly routed their opponents. On the other wing the Spartans and Tegeans, who were extended out past the Athenians, swung inward and took them in the flank and rear. The Athenians escaped great loss only because of the presence of their cavalry and the fact that Agis recalled the rest of his force to assist on the left wing. The intact Spartan center and right inflicted significant casualties on those coalition contingents that had broken through and were now returning to their original positions.

The allies lost around 1,100 men, the Spartans around 300. The Spartan success here checked Athenian anti-Spartan maneuverings (made under **Alcibiades'** influence), restored Spartan prestige in the Peloponnese, and led Argos to make peace. The outflanking tactic that occurred at Mantinea, apparently accidentally, was later used deliberately by the Spartans—successfully at **Nemea** and unsuccessfully at **Leuctra**.

(2) A battle fought in 362 between an anti-**Boeotian** army, which included contingents from **Sparta**, **Athens**, Mantinea, **Elis**, and **Achaea**, and a Boeotian army, supported by **Euboeans**, **Thessalians**, and a considerable number of Peloponnesian states, including **Argos**, Messene, **Sicyon**, Tegea, and several of the smaller **Arcadian** cities. This was the major battle of the Boeotian campaign to restore its position in the Peloponnese, prompted by the split in the **Arcadian League** and Mantinea's defection to Sparta.

Xenophon provides the only contemporaneous account of the campaign, in which his sons served in the Athenian **cavalry** (*Hell.* 7.5—the battle is described at 7.5.18-27). **Diodorus Siculus** (15.84-87) provides another account which is highly dramatic, and almost certainly wrong in several respects (including the positioning of the cavalry on the flanks), but which does preserve some detail omitted by Xenophon. **Polybius** (9.8) also describes the general course of the campaign, but not the battle that concluded it.

From its base at Tegea, the Boeotian army, led by **Epaminondas**, staged a lightning strike against Sparta while the bulk of the Spartan army was en route to join its allies at Mantinea. He did not take the city but this attack caused King **Agesilaus II** to return to Sparta with the bulk of his force. Epaminondas then marched north again and engaged the enemy at Mantinea. Diodorus Siculus (15.84.4) gives the Boeotian force as numbering around 30,000 infantry and 3,000 cavalry and the anti-Boeotian coalition force as about 20,000 foot and 2,000 horse.

Epaminondas deployed his army behind a screen of cavalry. This not only hid his movement but, judging his cavalry to be far superior, he believed they would also be very useful in boosting the morale of his own infantry and damaging the enemy morale by driving the coalition cavalry from the field at the start of the action. To assist this, Epaminondas posted *psiloi* (light troops) among the horsemen. He then led his army to the left as if to make camp for the night, but once the enemy had relaxed, thinking there would be no engagement that day, launched his attack. The **Thebans** occupied the left wing, which operated in a very deep **phalanx**, and Epaminondas advanced in an oblique line with this wing forward and the other back. He posted a mixed force of **hoplites** and cavalry on some high ground on his right flank to prevent the enemy left wing (consisting of the Athenians) from interfering with his attack. The Boeotian cavalry, as planned, drove the coalition cavalry from the field and the subsequent oblique hoplite attack was as successful as it had been at **Leuctra** in 371. However, Epaminondas was killed during the breakthrough of the Mantinean and Lacedaemonian hoplites on the coalition right wing and the disheartened Boeotians did not make a determined pursuit of the defeated enemy.

Although the battle confirmed the result of Leuctra, it was remarkably indecisive. The loss of Epaminondas crippled the Boeotian cause and led to another **Common Peace** (4), signed in 362/1, which again excluded the Spartans, whose continued refusal to recognize **Messenia**'s independence left them isolated (Polyb. 4.33).

(3) A battle fought in 207 between **Achaean League** forces under **Philopoemen** and the Spartans under the tyrant Machanidas. It was the first test of Philopoemen's reforms of the league army, now used to restore the **Achaean** position in relation to **Sparta**. **Polybius** (11.11-18) gives an account of the battle, and the main outline of events is clear, although some of the details are open to debate. When Machanidas learned of the opposition of the Achaeans and that their army was at Mantinea, he advanced from Tegea, which he had just captured. Machanidas' army was probably around 15,000 strong, consisting of a **Macedonian**-style **phalanx**, light-armed **mercenaries**, **cavalry**, and **artillery**. The Achaean army was probably around 15,000-20,000 strong and consisted of a newly trained citizen phalanx equipped with the *sarissa*, cavalry, and both light and heavy mercenary infantry.

Machanidas deployed his army from column into line, with his artillery at intervals along the front, facing Philopoemen's army, which was on high ground in front of Mantinea. Philopoemen launched his cavalry forward very early to prevent Machanidas' artillery from disrupting the Achaean infantry formations. The cavalry engagement was apparently hard fought and both sides gradually fed mercenary light infantry into the mêlée. Machanidas' mercenaries, however, gained the upper hand and drove the Achaean light troops and more heavily armed support back toward the city, causing the collapse of the Achaean left wing. Machanidas, sensing victory, pursued with all of his mercenaries, allowing Philopoemen to deploy his intact phalanx where his left wing had been, occupying the high ground behind Machanidas' pursuing troops but with a ditch between his troops and the Spartan phalanx. Encouraged by the success of the mercenaries and not realizing the extent of the obstacle provided by the ditch, the Spartan phalanx attacked across it. The Achaean phalanx fell on them while they were emerging uphill and disrupted by crossing the ditch and inflicted considerable **casualties**, causing the phalanx and the rest of the army to break. Philopoemen sealed his success by personally killing Machanidas as he tried to cut his way to safety back through the Achaeans.

The Spartans lost 4,000 men and all their baggage. The next day the Achaeans recaptured Tegea and invaded Spartan territory—taking the offensive against their old enemy for the first time in some years.

MANTIS (pl. *MANTEIS*). *See* SACRIFICE

MAPS. The first Greek maps, traditionally dating from Anaximander of **Miletus** (c. 611-546), were apparently of the entire known world. However, practical maps were soon developed from them. For example, Aristagoras of Miletus used one in his briefing of **Cleomenes I** of **Sparta** in 499/8 (Hdt. 5.49) and maps were known in late fifth-century **Athens** (Aristophanes, *Clouds*, 200ff.; Plut., *Alcibiades*, 17). These maps were normally painted on wood or, less usually, incised on bronze and were without a scale. Given this, they were of limited use in military operations, and the Greeks seem to have generally relied on local knowledge, guides, or written descriptions of routes and **terrain** in the planning and conduct of campaigns. **Alexander the Great**, for example, had the distances between stopping places on his march into Asia meticulously recorded by two specialist road-measurers (Pliny, *Natural History*, 6.61.4). The inclusion of detailed distances in **Arrian**'s account of each stage of **Nearchus**' voyage from the Indus to the Tigris (*Indica*, 20-43) suggests that Nearchus was also very careful to record distances for future use. *See also* RECONNAISSANCE.

MARATHON. A plain in northeastern Attica, site of a battle in 480 during the first round of the **Persian Wars** between **Athens**, supported by **Plataea**, and a **Persian** army. The Athenians, with around 10,000 **hop-**

lites (including some from Plataea), marched north to meet a Persian force, probably about 25,000 strong, commanded by **Datis**. This force had just captured **Eretria** on **Euboea** and then, as advised by **Hippias** (the ex-tyrant of Athens), landed at nearby Marathon, which was suitable for their **cavalry**. Although the **polemarch Callimachus** was apparently the supreme commander, **Miltiades**, one of the 10 generals, or *strategoi*, seems to have been the driving force behind the Athenian strategy. The Athenians took up a position in the foothills, safe from the Persian cavalry, and on Miltiades' day for command (each *strategos* held command for one day in rotation) the Athenians charged out of the hills at the Persian force, which was probably starting to move south along the main route to Athens. The Persians were initially stunned, "thinking the Athenians insane . . . charging with such a small force, without either cavalry or archers" (Hdt. 6.112). The Persians had some success in the center, where the Athenian **phalanx** had been weakened by stretching the wings to cover the larger Persian frontage. However, the Greek wings were eventually able to encircle a large part of the Persian force, trapping and annihilating it. The remaining Persians fled and were evacuated by sea. One of the mysteries of the battle is the location of the Persian cavalry, which played no part and were apparently away—perhaps off scouting or already loaded onto the ships. The 192 Athenians who died were (unusually) buried on the battlefield as a mark of respect. The Persians withdrew, leaving 6,400 dead. After an abortive attempt to sail around to Athens before the Greek army could march home, they abandoned the expedition. *See also* IONIAN REVOLT; PHEIDIPPIDES.

MARDONIUS (d. 479). Nephew and son-in-law of **Xerxes I** of **Persia**. He restored Persian control in the **Ionian** Greek cities immediately after the suppression of the **Ionian Revolt** through political reforms and then brought **Thrace** under Persian domination. **Herodotus** portrays him as an aggressive member of Xerxes' court and the driving force behind the invasion of Greece in 480. After **Salamis**, Xerxes placed him in command of the Persian army left behind in Greece. He was defeated and killed by the Greeks at the battle of **Plataea** in 479. *See also* PERSIAN WARS (1).

MARINES. *See EPIBATES*; NAVAL WARFARE

MEDICINE. *See* WOUNDS

MEDIZE. A term used for Greeks joining the **Persian** side during the Second **Persian War**. Prior to the war, those Greeks who had decided to resist voted to punish medizers after the war. However, many states, especially those in the north of Greece, were forced to medize in 480 because of the Greek strategy of defense at **Thermopylae** and then the Isthmus of Corinth. In **Thessaly** the ruling clan (the Aleuadae) was held responsible for the region's medizing but in the case of more southerly states, especially **Thebes**, the taint of medizing applied to the whole city. Even in the fourth century the Thebans' desertion of the Greek cause was remembered and used as an insult against them. The generally negative role of the **Delphic** oracle in the Greek war effort in 480-479 also led to suspicions that it was pro-Persian, which probably affected Greek perceptions of the oracle for some time after this. Immediately after the war the **Hellenic League** sent expeditions north to punish medizers. Although successful at Thebes, in Thessaly the league forces were unable to cope with the mobile **cavalry** defense adopted by the Thessalians and were forced to retire without performing their mission.

MEGARA. A city strategically located on the Isthmus of Corinth. The north-south routes from central Greece to the Peloponnese ran through its territory (the Megarid), which bordered **Boeotia** and Attica in the north and **Corinth** in the south. Megara had good access to the Saronic Gulf and reasonable access to the Gulf of Corinth. It was an early maritime and colonizing power, particularly in **Sicily** and in the northern Aegean, but a period of internal fighting in the early part of the sixth century saw the loss of territory to Corinth and the island of **Salamis** to **Athens**. Toward the end of the sixth century Megara joined the **Peloponnesian League**.

The city played an active part in the Second **Persian War**—contributing 20 **triremes** at **Artemisium** and Salamis and 3,000 **hoplites** at **Plataea** (1) (Hdt. 8.1, 45, 9.28). In 459 a serious border dispute with Corinth caused Megara to leave the Peloponnesian League and ally with Athens. This was an important alliance, which allowed Athens to protect its southern border during the First **Peloponnesian War**. This strategy collapsed when Megara revolted against Athens in 446, one of the major factors in Athens' decision to end the war. Athenian anger over this undoubtedly helped influence a decree in 432 to exclude the Megarians from the Athenian *agora* and from all ports in the Athenian empire.

Megara's complaints (along with those from other states) to **Sparta** about this embargo helped spark the Peloponnesian War (2) of 431-404. Megara bore the brunt of Athenian activity in the early years of this war, as Athens marched into the Megarid in full strength and **ravaged** it in retaliation after each Peloponnesian League attack on Attica. Athens also blockaded Nisaea, Megara's main port on the Saronic Gulf, and in 427 **Nicias** occupied the island of Minoa just off Nisaea, making the blockade total. In 424, Athens, helped by the democratic party in Megara, occupied the Megarian long walls and Nisaea. The city itself was saved only by the swift action of **Brasidas**, but the Megarians recaptured their long walls shortly afterward (Thuc. 4.66-74, 109). Athens retained Nisaea under the terms of the Peace of Nicias in 421. Megara fought on the Spartan side at **Mantinea** in 418, and with the resumption of the war against Athens seems to have played a fairly active part in the naval campaigns in Hellespont and **Ionia**.

The evidence for Megara after the Peloponnesian Wars is thin. It was a democracy after the King's Peace (**Common Peace** [1]), and this could indicate Megara had joined the anti-Spartan coalition during the **Corinthian War**. Conversely, several references in **Xenophon** indicate that the Spartans were apparently able to use Megara as a base for military activity during this period. Megara joined the Athenian-led alliance against **Philip II** of **Macedon** and surrendered to him after **Chaeronea** (1). After **Alexander the Great** it was occupied at various times by different contenders for power, including **Cassander** and **Demetrius I (Poliorcetes)**—who proclaimed the city free in 307. In 243, Megara was brought into the **Achaean League** by **Aratus**. It surrendered without a fight to the **Romans** in 146 and was subsequently incorporated into the empire.

MELITAEA (MELITEIA). A city in Phthiotic Achaea, just south of **Thessaly**. In spring 322, during the **Lamian War**, it was the site of a major battle between a **Macedonian** force on its way to relieve **Antipater** and the Greek besiegers of **Lamia**. The Macedonian army of 20,000 infantry and 1,500 **cavalry**, led by Leonnatus, faced a Greek army of 22,000 foot and 3,500 cavalry, led by **Antiphilus**. Although the Greek cavalry was victorious and Leonnatus was killed, his **phalanx** remained largely intact and managed to withdraw to high ground, safe from further cavalry action. When Antipater managed to link up with this force, he was able to withdraw from Lamia and subsequently to successfully renew the offensive against the Greeks.

In 217, **Philip V** launched a surprise attack on Melitaea during a campaign in Upper Macedon and Thessaly. This attack failed because his scaling ladders were too short to reach the top of the unusually high walls—an error of which **Polybius** (5.98) is particularly critical. *See also* CRANNON.

MEMNON (d. 333). A native of **Rhodes** who served as a high-ranking commander with the **Persians** in the mid-fourth century. His elder brother, Mentor, was also a general in Persian service and the brother-in-law of the satrap Artabazus. Both Memnon and Mentor went into exile with Artabazus in 353 after the failure of his revolt. Mentor returned to Persian service first, fighting in **Egypt** in 343, and he secured the return of both Memnon and Artabazus (Diod. Sic. 16.52.1-3). After Mentor's death (c. 340?) Memnon married his widow, Barsine, and succeeded him as general in Asia Minor. Although both Memnon and his brother were **mercenaries**, they were clearly accepted into the upper echelons of Persian society.

Memnon had a good reputation as a commander and secured some successes against **Philip II** of **Macedon**'s troops in Asia Minor in 336 (Diod. Sic. 17.7.2, 8-10; 17.19.2). He proposed dealing with **Alexander the Great**'s invasion in 334 by a scorched-earth policy—which may well have worked but it was rejected by the satraps in favor of a pitched battle at the **Granicus River** (Arr. *Anab.* 1.12.9-10; Diod. Sic. 17.18.2-4). Memnon served with the **cavalry** at Granicus (Arr. *Anab.* 1.15.2) and according to **Diodorus Siculus** (17.19.4) was one of the commanders of the left wing.

Following Granicus, **Darius III** appointed Memnon commander of lower Asia Minor and of the Persian fleet (Arr. *Anab.* 1.20.3; Diod. Sic. 17.23.5-6). Memnon defended **Halicarnassus** vigorously, if ultimately unsuccessfully (Arr. *Anab.* 1.23.1-3; Diod. Sic. 17.24.5-27.6), and after its fall put his efforts into prosecuting the naval war against Alexander. **Chios** was surrendered to him, and he secured all of Lesbos except Mytilene and used bribery to foment unrest against Alexander in mainland Greece (Arr. *Anab.* 2.1.1-3; Diod. Sic. 17.29). It was at this point that Memnon suddenly fell ill and died. **Arrian** comments of his death that "nothing else during this time did greater harm to the king's cause", (*Anab.* 2.1.3) and Diodorus Siculus makes a similar comment (17.29.4).

MERCENARY. Mercenaries, non-citizen-soldiers serving for **pay**, were a feature of the Greek world from early times, although they were seldom prominent until the fourth century. **Herodotus** (2.152ff.) records **Ionian** Greek soldiers serving **Egyptian** kings circa 650. At about the same period, the poet **Archilochus** claims to have earned his living through his spear. Greek tyrants in the seventh and sixth centuries were early employers of mercenaries, using them as bodyguards and as an internal security force to keep possible opponents in check. However, the numbers of mercenaries involved here were probably not large because of the cost of maintaining them.

While citizen-soldiers were traditionally expected to bear their own expenses except on protracted campaigns, mercenaries received pay and rations or ration money. This expense restricted the use of mercenaries throughout much of Greek history. Another factor limiting their use down to the **Peloponnesian War** of 431-404 was the general view that it was the duty of the citizen to protect the state. Employing mercenaries therefore tended to be seen as a sign of failure by the citizen body. This is why, when a state did have to resort to mercenaries, they were often referred to euphemistically as *epikouroi* (helpers).

Following the Peloponnesian War (2), there was a general rise in the Greek use of mercenaries. The semiconstant state of warfare in much of Greece and Asia Minor provided a market and the Peloponnesian War had created a pool of experienced soldiers to supply it. Many of these men were poor and/or exiles or fugitives, and some had developed a taste for adventure. These were the main motives for mercenary service, with poverty probably the most important (and a taste for adventure the least).

At the end of the fifth century, a large contingent of Greek mercenaries, the **Ten Thousand**, served the **Persian** prince **Cyrus the Younger**. Following Cyrus' death, 6,000 of these men were employed by the **Spartans** in Asia Minor, setting the pattern for an increasing use of non-citizen-soldiers. Mercenaries were widely used in **Sicily** throughout the fourth century as the mainstay of various tyrants' armies. **Phocis** employed large numbers of mercenaries in the Third **Sacred War** (355-46), **Philip II** of **Macedon** also used mercenaries. **Alexander the Great**, had Greek mercenary troops with him in Asia and the best infantry facing him were Greek mercenary **hoplites** in Persian service.

Mercenaries continued in regular use under the Hellenistic monarchies, both in the **phalanx** and as light troops. For example, at **Raphia** in 217, **Antiochus III (the Great)** had at least 5,000 mercenaries while **Ptolemy IV (Philopator)** had around 18,000. As garrison troops, **cleruchs**, and settlers, such men also played an important part in the hellenization of the eastern frontiers of the Greek world.

The Peloponnese (especially **Arcadia**) was a good recruiting ground for mercenary hoplites in the fifth and fourth centuries. Other areas (including those on the borders of the Greek world) tended to supply specialist light troops such as **slingers** (**Rhodes**, Crete), **archers** (Crete, **Scythia**), and **peltasts** (**Thrace**, Chalcidice, **Thessaly**). Thessaly and Scythia also supplied cavalry and mounted archers, respectively. **Syracusan** dynasts also hired mercenaries from the **Italian** peoples and Spain. In later periods, the tendency to specialization declined, as did the number of Peloponnesian mercenaries. In the third century the evidence suggests mercenaries from northern and central Greece, as well as non-Greeks such as **Gauls** and Thracians, provided a large proportion of those hired. Gauls were first employed in large numbers by **Antigonus II (Gonatas)**. He was criticized by other Greeks for using what were considered dangerous barbarians, but Gauls were soon in common use in Hellenistic armies. *See also* CORCYRA; DION; DIONYSIUS (1); EGYPT; GRANICUS RIVER; *HIPPOTOXOTAI*; LAMIAN WAR; MEMNON; NABIS; TARENTINES; THIBRON; TIMOLEON.

MESSENIA. An area in the southwest Peloponnese. The main city was Messene, founded in 369 on Mount **Ithôme**. The history of Messenia is inextricably bound up with its neighbor **Sparta**, which conquered Messenia circa 720 in the First **Messenian War** and controlled it down to 370/69. During this period Messenia was a subject state and most of its inhabitants had the same status as helots, although some had the same status as *perioeci*. Despite several revolts (the Second and Third Messenian Wars), Messenia was unable to regain its independence until Sparta's military power was broken at **Leuctra** in 371. The subsequent **Boeotian** invasion of Laconia freed Messenia, whose independence was specified in the Peace of Pelopidas of 366/5 (**Common Peace** [3]) and reaffirmed in 361 after the battle of **Mantinea**—although Sparta consistently refused to sign treaties with this clause.

Messenia retained its independence from Sparta by means of its own military force and by alliances with other powers. In the middle of the fourth century (c. 346) it apparently received assistance from **Philip II** of **Macedon** against Sparta; it was allied to Philip in 338 but did not take part at **Chaeronea** (1) (although Messenia was subsequently awarded some Spartan territory). Messene may have had a pro-Macedonian tyrant under **Alexander the Great**. Messenia apparently joined the Greek alliance during the **Lamian War** but may well not have contributed troops because of the threat from Sparta (this certainly prevented them from sending troops to assist against the **Gauls** in 279; Paus. 4.28.3).

Messenia managed to remain independent from **Demetrius I (Poliorcetes)** and around 240 was linked with **Aetolia** through a treaty with Phigalea. However, only 20 years later Messenia joined the **Achaean League** because of Aetolian raids in the lead up to the Second **Social War**. During the next few years Messenia was subject to attack from a variety of powers, including Demetrius of Pharos, Sparta, and Macedon.

In 183 Messenia seceded from the Achaean League. Although **Philopoemen** was captured and killed in the early stages of this revolt, Messene was forced to surrender in 182. Apart from those involved in

Philopoemen's execution, the Messenians were treated leniently. It presumably regained its independence when the **Romans** dissolved the Achaean League in 146 but shortly afterward was absorbed into the Roman empire, along with the rest of the Peloponnese.

MESSENIAN WARS. A series of wars (or more accurately a war and subsequent rebellions) by which **Sparta** gained and then maintained control over its western neighbor, **Messenia**. The possession of Messenia, which remained in Spartan hands until 370/69, provided the basis of Sparta's power and allowed it to be a dominant state in Greece for centuries. However, details of the first two conflicts are obscure—the main account is in **Pausanias** (3) who was writing a very long time after the events and when a variety of alternative and at times conflicting traditions were extant.

(1) First Messenian War (c. 740-720). **Pausanias** (2) (4.4.1-14.5) firmly dates this war to 743-724 and preserves two accounts of its causes, one from the Spartan side, one from the Messenian side. The Messenian version was that **Sparta** coveted the rich Messenian territory and was looking for an excuse for war. This seems quite plausible, given Sparta's actions at the end of the war in annexing **Messenia** and reducing the inhabitants to serf-like status as helots—as well as her general attitude to border territory (especially with **Argos**). The war lasted for 20 years, during which the Messenians established themselves on Mount **Ithôme** and made good use of light troops to harass the Spartans. The Spartan victory almost doubled its territory and the land was divided into allotments (*klaroi*) for Spartiates, with the Messenians working these for them. The Spartan poet **Tyrtaeus** records the fate of the Messenians who fell under the "painful necessity to bring their masters | full half the fruit their ploughed land produced" (Paus. 4.14.5; trans. Peter Levi).

(2) Second Messenian War (685-668). The dates for the war are from the long account of it in **Pausanias** (2) (4.14.6-23.4) and may well be inaccurate. The war was actually a revolt by the **Messenians**, who found their subjection after the First Messenian War intolerable. Led by Aristomenes and aided by **Arcadia** and **Argos**, the Messenians fought for 14 years. The first action was an inconclusive battle against **Sparta** at Deira, after which Aristomenes was appointed king. He led them to victory in the next battle at Boar's Grave (somewhere in Upper Messenia) but was decisively defeated at the battle of the Great Foss. The loss was caused by the desertion of Aristomenes' Arcadian allies, whose king had been bribed by the Spartans—according to Pausanias, the first occasion when bribery was used in war. The Messenians established themselves

on Mount Eira, where they held out for 11 years before the fortress was betrayed by a runaway **slave**. Some scholars have doubted whether this war actually occurred, as there is no specific support for it in the extant poems of **Tyrtaeus** and Pausanias' account seems influenced by postliberation Messenian versions of the event. However, the tradition of the war was clearly well established in antiquity and it seems likely that some sort of Messenian revolt occurred in the seventh century.

(3) Third Messenian War (464-460). In 464 a major earthquake struck **Sparta**, causing considerable loss of life and the **Messenian** helots seized the opportunity to revolt. Although they were fairly quickly defeated in battle, they occupied a naturally strong defensive position on Mount **Ithôme**, which had had some helot rebels on it since 469. The Spartans, not noted for **siege warfare**, were unable to dislodge them and called in help from their allies in the **Peloponnesian League** and from **Athens**. However, the Athenian **hoplite** force, led by **Cimon**, was dismissed early, apparently because the Spartans found their radical democratic ideas disturbing. This slight hastened democratic reforms at Athens, led to Cimon's political demise and exile, and helped create the circumstances for the First **Peloponnesian War**. Still unable to capture Mount Ithôme, around 460 the Spartans let the rebels evacuate the stronghold under a truce. The Athenians settled them in Naupactus, on the northern coast of the Corinthian Gulf, to boost Athenian influence there. They played an active part on the Athenian side during the Peloponnesian War (2) of 431-404 but were expelled from Naupactus after the Athenian loss at **Aegospotami** in 405 (see Thuc. 1.101-103).

MILETUS (MILETOS). A wealthy maritime **Ionian** Greek colony in Caria which founded quite a few colonies of its own in the Euxine (Black Sea) and its approaches. In the sixth century it resisted several attacks by Lydia, eventually reaching a voluntary agreement recognizing the authority of the Lydian king and paying tribute but exercising considerable autonomy—and, alone among the Ionian Greeks, managing to secure the same arrangements when the **Persians** conquered Lydia around 546 (Hdt. 1.15-22, 141). Although the city apparently enjoyed considerable prosperity under the Persians, it was the focal point for the outbreak of the **Ionian Revolt** in 499 and the Persian military threat to the city led to the decisive battle of the war at **Lade** in 494. Following the Ionian defeat at Lade, Miletus was sacked and many of its inhabitants massacred. The survivors were moved to a distant part of the Persian empire, and the city became predominantly Carian (Hdt. 5.35-38, 6.18-20).

Following the Persian defeat in the Second **Persian War**, Miletus joined the **Delian League**, but revolted in 412 along with many other places in Asia Minor. Although the city was never again as powerful as it was prior to the Ionian Revolt, it did manage to establish its independence for a while. The **Athenian** force sent to recover it failed to do so (see below) and the Milesians subsequently prevented **Tissaphernes** from establishing a fort in Milesian territory (Thuc. 8.17, 25-27, 84.4-5).

Miletus again fell under Persian control in 386, under the terms of the King's Peace (**Common Peace** [1]), and in 334 resisted **Alexander the Great**. The city fell to a **Macedonian** assault but Alexander treated the inhabitants leniently (Arr. *Anab.* 1.18). Under the *diadochoi*, and despite fairly consistent efforts by Ptolemaic **Egypt** and Seleucid **Syria** to acquire it, the city managed to maintain a state of semi-independence. However, throughout the third century it frequently moved between Seleucid and Ptolemaic control or influence. It supported **Rome** against **Antiochus III (the Great)** in 190 and **Perseus** of Macedon in 170 (Livy 37.16, 43.6), becoming part of the Roman province of Asia in 129.

In addition to the sieges of 494 and 334, Miletus was the site of one major battle (described in Thuc. 8.25-27) fought in 411 between an Athenian force under Phrynichus and the Milesians, supported by Peloponnesians and Tissaphernes. Phrynichus, with a fleet of 48 **triremes**, landed a force of 3,500 **hoplites** (1,000 Athenian, 1,000 allies, and 1,500 from **Argos**—500 of whom were *psiloi* [light troops] who had been supplied with hoplite equipment). The Milesians marched out with 800 hoplites and an unknown number of allied troops—Peloponnesians under Chalcideus and **mercenaries** and **cavalry** commanded by Tissaphernes. The Argives, contemptuous of the Milesians opposite them as mere Ionians, charged forward precipitately, lost **formation**, and were defeated with the loss of about 300 men. The Athenians, though, defeated the Peloponnesians opposite them and then drove off Tissaphernes' troops, which caused the victorious Milesians to retire. The Athenians erected a **trophy** and began to construct a blockading wall, but prudently withdrew when they heard of the impending arrival of a Peloponnesian relief fleet.

MILITARY LAW. *See* DISCIPLINE; LAWS OF WAR

MILITARY MANUALS. Down to the fourth century there was no recognized body of military theory and no military manuals. With the relatively slow pace of **hoplite** warfare, the general view was that any educated citizen with a normal military experience was capable of exercising command. In fact those who exhibited a keen interest in perfecting their military skills were often looked down upon as overly concerned with war—this was certainly the reputation **Philopoemen** acquired (Plut., *Philopoemen*, 3.1-4, 4.5-6).

As warfare became increasingly sophisticated from the fourth century onward, there was an increasing need for manuals and military theory. However, there was no organized means of disseminating these and **officers** essentially continued to learn on the job (although the **Spartans** obviously had a more formalized system of passing on experience through their education system, the *agoge*, and their messing system).

Xenophon was one of the first writers of military manuals (although at least one, the *Cyropaedia*, was in the form of an historical novel). His works on leadership (the *Cyropaedia*) and on the **cavalry** (*On Horsemanship*; *Cavalry Commander*) presumably had a reasonable circulation in **Athens**, but the extent of their circulation elsewhere was probably initially fairly limited. Later there were numerous military manuals written by authors such as **Aelian**, **Aeneas Tacticus**, **Asclepiodotus**, and **Onasander**. **Alexander the Great**, however, is reputed to have carried **Homer's** *Iliad* with him as a military "manual" (Plut., *Alexander*, 8.2). In addition to the specialist military manuals, though, histories by men such as **Thucydides** and Xenophon were also designed to be of practical use to politically (and militarily) active people and had a major focus on warfare. *See also* TRAINING.

MILTIADES (c. 550-489). An Athenian aristocrat who served in the first phase of the **Persian Wars**. Around 524 he became absolute ruler in the **Thracian** Chersonese (Gallipoli Peninsula), encouraging Athenian settlement there. He was driven out by the **Persians** in 493 after the failure of the **Ionian Revolt**. Surviving a political trial on his return to **Athens**, he was elected *strategos* (general) for 490 and was regarded as the expert on Persian military matters. His plan was adopted by the Athenian-**Plataean** force at **Marathon** and the Greeks successfully attacked on his day in command. He led a naval expedition to Paros in 489 but failed to capture it and was tried and heavily fined on his return. He died shortly afterward from **wounds** received on Paros.

MINDARUS (MINDAROS; d. 410). A Spartan **officer** sent out to command the Peloponnesian fleet in Asia Minor in 411/10. Frustrated by the behavior of his supposed ally, the **Persian** satrap **Tissaphernes**—who was in fact deliberately trying to keep both **Athens** and the Peloponnesians weak—Mindarus transferred his fleet from **Ionia** to the Hellespont. Here he received much better support from Tissaphernes' rival, **Pharnabazus**. Although defeated by Athens in two naval battles off **Abydus** (1 and 2), he managed to capture **Cyzicus**. However, while **training** there, his 60 ships were surprised by **Alcibiades** and Mindarus was killed on shore unsuccessfully trying to prevent their capture. The Athenians captured the postbattle dispatch to **Sparta** from Mindarus' deputy, which read, "Ships lost, Mindarus dead; men starving; don't know what to do" (Xen. *Hell.* 1.1.23).

MISTHOS. See PAY

MITHRIDATES (also MITHRADATES). Name borne by members of the Hellenistic rulers of Pontus. The family was probably **Persian** in origin, but its rule took on a Greek character, especially under Mithridates II (c. 250-220) and Mithridates V (152-120). The kingdom of Pontus was founded under Mithridates I (302-266), developed important alliances with Seleucid **Syria** under Mithridates II and V, and was dismembered by **Rome** in 63-62. The most militarily important rulers of this name were:

(1) Mithridates V (Euergetes). King of Pontus, 152/1-120. He considerably extended Pontus and gave the court an increased Greek character (including, it is thought, marrying a member of the Seleucid royal family). He provided military support to **Rome** during the Third Punic War (149-146) and against the **slave** revolt in Pergamum led by Aristonicus (133-129). In return, the Romans awarded him Phrygia, considerably increasing the territory and power of Pontus. During his reign he also acquired Galatia and gained indirect control of Cappadocia after his daughter's marriage to the young king. Mithridates was assassinated in 120.

(2) Mithridates VI (Eupator). King of Pontus, 120-63. Although details of the early part of his reign are uncertain, it appears that by around 112 Mithridates had managed to throw off his mother's control, eliminate his brother, and become sole king. He embarked on a policy of expansion, first into the Crimean area in the north of the Black Sea—probably responding to an appeal from the Greek cities in the area for assistance against **Scythian** domination. Using the revenues and manpower resources from this area, he then annexed Lesser Armenia, Colchis, and eastern Pontus. These areas formed the core of Mithridates' empire.

Sometime in the period between 108 and 103 Mithridates and Nicomedes of **Bithynia** partitioned Paphlagonia and by a series of diplomatic maneuvers deflected a Roman demand to restore its independence. Down to around 99 Mithridates was moving to control Cappadocia, competing against Nicomedes, and finally securing it by personally murdering its young king Ariarathes VIII (Philometor) during peace negotiations. About 96-95 **Rome** proclaimed Cappadocia free from both Pontus and Bithynia. Apparently in an attempt to circumvent this without openly defying Rome, Mithridates allied with Tigranes of Armenia and had him intervene in Cappadocia for him. Lucius Cornelius Sulla was sent out and restored Cappadocian independence, but a year or so later Tigranes again intervened on Mithridates' behalf. At the same time Mithridates acted directly against Bithynia, taking advantage of the accession of a new king, Nicomedes IV (Eupator). The Roman commission sent out to settle affairs set up the conditions for the outbreak of the first of three **Mithridatic Wars**. Despite temporary success from time to time, Mithridates lost all three wars and in 63, cornered in the Crimea and abandoned by his subjects and his son, he committed suicide. Unable to kill himself with poison, because he had built up immunity over a long period of taking small doses, he had one of his own guards kill him. His coinage shows he modeled his personal appearance on **Alexander the Great**. Despite some successes against Rome, he ultimately proved incapable of resisting its advance—which had largely been provoked in the first instance by his ambitious expansionism.

MITHRIDATIC WARS. A series of three wars fought between **Rome** and **Mithridates VI (Eupator)** of Pontus in the first century.
    (1) First Mithridatic War (88-85). Following a vigorously expansionist policy, curbed only by a need not to directly confront **Rome**, Mithridates intervened in **Bithynia** in 91 or 90, expelling Nicomedes IV (Eupator). Under Roman pressure, Mithridates withdrew but was provoked to war by the Roman commissioner, Manlius Aquilius, who pressured Nicomedes to repay his debt to Rome by **raiding** Pontus and imposing a toll on shipping using the Dardanelles. In 88, Mithridates expelled the small Roman force from Bithynia and Cappadocia and rapidly overran the Roman province of Asia. There he staged a massacre (known as the "Asiatic Vespers") of all Romans and **Italians** there—perhaps as many as 80,000. However, he was unable to capture **Rhodes**.
    Mithridates then invaded Europe, rapidly gaining the allegiance of **Athens**, which he used as his base of operations. His general, **Archelaus**, secured most of central Greece and the Peloponnese but was starved out of Athens by Lucius Cornelius Sulla, who landed in Greece in 87. In 86-85, Sulla defeated Archelaus in battles at **Chaeronea** (2) and **Orchomenus (Boeotia)** and advanced as far as the Dardanelles. At the same time, another Roman force (from the anti-Sullan faction) had

opened the offensive against Mithridates in Asia. In 85 Mithridates' forces were defeated at Rhyndacus, but Mithridates escaped when the Sullan fleet under Lucius Licinius Lucullus failed to cooperate with the non-Sullan land forces. The same year, Sulla, who was under considerable pressure from his enemies in Italy, made peace with Mithridates on fairly favorable terms to the king. Under this treaty of Dardanus, Mithridates agreed to pay an indemnity, surrender his Aegean fleet, and evacuate all the territory he had taken in Asia. In return, he was recognized as king of Pontus.

(2) Second Mithridatic War (83-82). Following the first war, Murena, the **Roman** officer left in charge in Asia by Sulla, invaded Cappadocia and Pontus. Mithridates successfully drove him back but refused to be drawn into a general war. This stand was justified when Sulla disowned Murena's actions. This war ended with no change to the previous terms.

(3) Third Mithridatic War (74-63). Suspicious of **Rome**, after the senate's refusal ratify the treaty of Dardanus in 78, Mithridates had his armies **trained** along Roman lines. To aid his pirate allies and the Sertorian forces in Spain and to preempt Rome from occupying **Bithynia** (recently bequeathed to it by Nicomedes IV [Eupator]), Mithridates invaded the province of Asia in 74. He quickly overran Bithynia and took **Chalcedon**, but was held up at **Cyzicus**. This allowed Lucius Licinius Lucullus, the senior surviving Roman commander in the area, time to levy forces and march against Mithridates. The bulk of Mithridates' army was destroyed at Cyzicus or during the withdrawal from there. Lucullus followed up with a naval victory off Lemnos in 73 and Mithridates' fleet was decimated by a storm shortly afterward. Late in 73 Lucullus struck into Pontus itself and, although his army was nearly lost because of the harsh conditions and guerrilla warfare waged against it, Lucullus, with the assistance of a Galatian prince, Deiotarus, retrieved the situation and won a major battle at Cabira. By 70 Lucullus had secured Pontus itself, although Mithridates had escaped to Armenia, where he was supported by its king, Tigranes. In 69 (without authorization from the senate) Lucullus invaded Armenia and although heavily outnumbered won a decisive victory outside the capital, Tigranocerta. During a difficult pursuit into the interior of Armenia, Lucullus' weary troops mutinied, ending his offensive. Continually starved of reinforcement, and subject to political attack at home, Lucullus was crippled and Mithridates recovered a lot of his own territory in 68-67. However, when Pompey (Gaius Pompeius Magnus) took over Lucullus' command in 66 and was provided with ample resources, Mithridates was soon defeated. Abandoned by Tigranes, Mithridates was decisively beaten at Nicopolis

but escaped to the Crimea. His demands on his subjects there caused a revolt against him, led by his son Pharnaces, and he committed suicide.

MONUMENTS. In addition to **trophies** set up to commemorate victories, monuments to commemorate war dead are also recorded in ancient Greece. Several state **casualty** lists survive from fifth-century **Athens** (e.g., *IG* I^2 929, II^2 5221), some surmounted by impressive reliefs depicting combat scenes (e.g., Athens, National Museum, 2744). Private monuments commemorating individuals and depicting them in military dress also survive (e.g., Athens, National Museum, 835, 884, 1674, 2586, 2744, 3620a; Kerameikos Museum, P1130). The Athenian **cavalry** also set up a monument to its members who died at **Corinth** and **Coronea** in 394/3 (*IG* II^2 5222), even though they appeared on the state monument for that campaign (*IG* II^2 5221). **Alexander the Great** had bronze statues made as memorials to the 25 Companion cavalrymen killed in action at **Granicus River** (Arr. *Anab.* 1.16.4).

MORA (**pl.** *MORAI*). A unit of either infantry or **cavalry** in the **Spartan army**. Although the term is used loosely in many sources, a cavalry *mora* was probably 120 strong, while an infantry *mora* at full strength (which was apparently unusual) may have had as many as 1,280 men (Lazenby, *The Spartan Army*, 10). However, most of the *morai* attested in the ancient literature were between 600 and 900 strong. In the fifth and fourth centuries the army had six infantry *morai*.

MOUNTED ARCHERS. See *HIPPOTOXOTAI*

MOUNTED INFANTRY. Considerable artistic evidence, along with anachronisms in **Homer**'s *Iliad* (identified in Greenhalgh, *Early Greek Warfare*), demonstrates that prior to the **hoplite** reform Greek states made considerable use of mounted infantry. Sixth-century **Athenian** vase paintings, for example, predominantly depict riders as mounted infantry rather than true **cavalry**. While this can in part be explained by heroic survivals in the art, it seems likely that prior to the introduction of the hoplite **phalanx**, and perhaps for a little time after that, men equipped as heavy infantry rode horses to battle, dismounted, and fought on foot. The advantages of mounted infantry were increased mobility, allowing for rapid deployment and redeployment, and that the soldiers arrived on the battlefield fresher. The disadvantage, of course, was cost—horses were expensive and only the wealthy could therefore afford to join the mounted infantry. Once the hoplite phalanx was introduced, the need for mounted infantry disappeared; there was little point in having part of a phalanx arrive before the rest, as they all needed to operate together. Mounted infantry basically disappeared from the Greek order of battle until the fourth century, when **Alexander the Great** occasionally used mounted infantry to take part in pursuits during his conquest of the **Persian** empire. These, however, were not men who **trained** or even normally operated as mounted infantry, but footsoldiers temporarily mounted for a specific purpose when mobility was vital.

MUNYCHIA (MOUNYCHIA/MOUNYKHIA). A suburb of **Athens**; site of a battle in 404 between democratic exiles based in **Phyle** and an oligarchic force under the **Thirty Tyrants**. The democrats under **Thrasybulus** numbered around 1,000 and had no **cavalry**. Their opponents included the **Spartan** garrison and Athenian **hoplites** and cavalry. Although the oligarchic numbers are unknown, they deployed their hoplites 50 men deep, while the democrats were only 10 men deep. The democrats, however, had numerous **peltasts**, **javelin** men, and **slingers** deployed behind the **phalanx**. Their *mantis* (seer), aware of a prophecy that the democrats would be victorious after one of them had died, leapt forward alone and was killed. The democrats won the ensuing fight, pursuing the oligarchs down to the level ground, where the pursuit was abandoned, probably due to the cavalry threat. This victory led the Thirty Tyrants to withdraw to Eleusis and hastened the restoration of democratic government in Athens.

MYCALE. A promontory in **Ionia**; site of a battle between the **Persians** and the Greeks during the **Persian Wars**. Traditionally, it was fought on the same day as **Plataea** (1) in September 479. The Persians decided they were not strong enough to fight the Greek fleet heading for Samos and beached their ships at Mycale, where Tigranes was camped with a 60,000-strong army. They built a defensive wall around their ships but the Greeks, under **Leotychides II**, landed, destroyed the Persian army, and burned the fleet. During the battle the Ionian Greeks serving with the Persians defected and turned on them. This battle ended the Persian naval threat in the Aegean, ensuring Greek naval supremacy, and caused the revolt of the Aegean Islands from Persia and their enrollment into the **Hellenic League** (1).

MYCENAE. A city built on a rugged hill commanding the northeast of the plain of **Argos**. Its strategic location made it an important city in early Greece, earning it the epithets "of the wide ways" and "of the strong-founded citadel" (**Homer**, *Iliad*, 2.569, 4.52) and resulting in its name being given to the Mycenaean period (c. 1600-1100). Its early importance is also indicated by the position of its ruler, **Agamemnon**, who was commander in chief of the Greek expeditionary force in the **Trojan War**.

The city consisted of a citadel and surrounding settlement. Mycenae was protected by massive **fortifications** and a well-organized military. Its army employed **chariots** and heavy infantry, equipped with the spear, sword, **shield**, and body **armor**. **Weapon** finds there include daggers and long swords, richly decorated with gold inlay. In the mid-thirteenth century the citadel and parts of the settlement were damaged by fire, but there is no other evidence of attack and the damage was subsequently repaired. Around 1200 the citadel again suffered extensive damage and although occupation continued, within about 100 years Mycenae was apparently little more than a village.

In the Dark Age Mycenae's previously dominant position in the region was taken by Argos. Mycenae, however, remained independent—while Argos remained neutral during the Second **Persian War**, Mycenae sent 80 hoplites to **Thermopylae** and 400 (combined with Tiryns) to **Plataea** (1) (Hdt. 7.202, 9.28). In 468, Argos ended this rather anomalous situation with an extended siege of Mycenae. Still protected by their massive walls, the Mycenaeans were eventually starved into submission. Half fled to **Macedon**, the remainder to Cleonae and Cerynia, and the Argives ensured the fortress was left deserted (Diod. Sic. 11.65; Paus. 2.16.4, 5.23.3, 7.25.3). *See also* EARLY GREEK WARFARE.

- N -

NABIS (d. 192). Regent for Pelops of **Sparta** after the death of Machanidas in 207. Crowned king after Pelop's death, with the aid of **mercenaries** he forced a return to the system of **Cleomenes III**. From 204-200 he had mixed success against the **Achaean League**, suffering a defeat against **Philopoemen** in 200. During the Second **Macedonian War** he seized **Argos** with the help of **Philip V** but then defected to the **Romans**. In 195 he was charged with tyranny and the Romans forced him to surrender Argos and the Laconian ports. He attempted to regain the ports in 193, but was defeated by Philopoemen and Flamininus. He was assassinated in 192. This essentially marked the end of an independent Sparta, which was forced to join the Achaean League and subsequently formed part of the Roman province of Achaea.

NAUCRARIES (*NAUKRARIAI*). Pre-Solonic, local administrative areas of Attica, numbering 48. Each naucrary provided one **ship** and two **cav-**

alrymen for the **Athenian** armed forces. They probably ceased to exist as administrative units after Cleisthenes' reforms around 506, and their naval function was probably ended by **Themistocles'** reforms around 483.

NAVAL WARFARE. The expansion of trade, the colonization movement, and the development of less purely agricultural economies led to an increasing involvement by Greek states in naval activities. Navies were needed to suppress piracy, to create and protect foreign possessions, and to protect one's own coastline. However, a continual limiting factor on naval warfare was its expense. **Xenophon** expresses this succinctly in his comment on **Timotheus** in **Corcyra** in 373 that "he continually sent to Athens for money—he needed a lot because he had a lot of ships" (*Hell.* 5.4.66). Because of this, and despite a strong seafaring tradition in the coastal areas of Greece, few *poleis* (cities) became major naval powers. Major exceptions were **Corinth** and **Athens**, along with **Sicilian** Greek cities such as **Syracuse**, although the Hellenistic kingdoms, with the financial resources of Asia at their disposal, often later supported large navies.

The works of **Homer** and **Thucydides** (1.10.4) suggest that the earliest naval warfare was actually amphibious warfare, using undecked ships, with the rowers doubling as soldiers. This provided good tactical and strategic mobility (although, like later ships such as the **trireme**, these boats were drawn up on the beach at night), but actual combat at sea must have been fairly rare. Such engagements would only have involved boarding with part of the crew, and the prime targets would presumably have been merchant ships, not other **warships**.

The first evidence for the development of the ram is on vases dated between 850 and 700 (Casson, *Ships and Seamanship in the Ancient World*, 49). This new **weapon** had considerable implications for ship design and the **training** of crews, both of which added to the expense of naval warfare. Rowers became increasingly important in providing maneuver for use of the ram, rather than mere motive power to deliver soldiers to land. This meant that they had to be better trained and better protected, and ships also had to be heavier to withstand the shock of ramming. This probably led to the development of the bireme and trireme—both of which provided more rowers (and therefore more power) with no increase in ship length. A deck helped protect the rowers but also allowed the carriage of some *epibatai* (marines). The numbers on a trireme were not great, with the standard Athenian complement in the fifth century being 10 *epibatai*, equipped as **hoplites**, and four **archers**. This gave the ships some combat power in addition to the ram, as well as providing soldiers to be used to protect the crews on land or to operate as amphibious troops.

The heyday of maneuver warfare at sea was in the fifth and early fourth centuries, where the trireme and skilled crews reigned supreme. Maneuvers such as the *diekplous*, which involved sailing through an enemy line and wheeling sharply to catch its ships in the flank or stern, were perfected and used by experienced navies, especially Athens. However, even after the introduction of the ram, sea battles could still be fought primarily by the marines carried on board. This could arise from unsophisticated maneuver, as at **Sybota** in 433, or because of cramped space, as in the battle in the harbor of Syracuse in 413 during the Athenian **Sicilian Expedition** (Thuc. 1.49.1-3, 7.70.2-7).

After the mid-fourth century, naval warfare tended to revert more to battles between the marines. This was a definite feature of Hellenistic warfare, which saw the use of larger ships—the four, five, six, even ten. These provided considerable stability and allowed the carriage of **artillery** and greater numbers of troops (allowing large-scale amphibious landings). The skill of the rowers became less important (removing an important training/experience liability) and the bigger ships were more suited to the dignity of the monarchs who owned them. The ram was still used, but increasingly alongside artillery, archers, and boarding parties—as the account of **Chios** in 201/0 (Polyb. 16.2-7) demonstrates. In the battle of **Actium** in 31, the last fought by a Hellenistic Greek-style navy, Marcus Antonius initially relied on the size (and height) of his ships, their towers, and the marines operating from them (Cassius Dio, 50.23.1-3). *See also* PENTECONTOR.

NAVARCH (*NAUARCHOS*, pl. *NAUARCHOI*). Commander of a fleet, or admiral, especially at **Sparta**. Unlike the Spartan land forces, the fleet was not normally commanded by a king, although this did occasionally happen when the fleet included allied contingents. The Spartan navarch was assisted by a vice admiral, the *epistoleus*. The navarchy (office of navarch) may have been an elected one, as an individual held it for only one year and apparently could only serve once. **Lysander**, for example, finished a very successful term as navarch in 406, and in order to retain his expertise, he was appointed *epistoleus* in 405 with de facto command. *See also* EURYBIADES; LEOTYCHIDES II; MINDARUS.

NAXOS. A fertile island in the Cyclades, southeast of Attica, famous for its wine and marble; the major settlement on the island is also called Naxos. Under the tyrant Lygdamis (c. 545-524) Naxos dominated the Cyclades. In 499 the island withstood a four-month siege by a **Persian**

force whose failure helped start the **Ionian Revolt**. In 490 the island was ravaged by the Persians, who burned the city of Naxos and sold as **slaves** those of its inhabitants who had not fled into the hills. Naxos was one of the original members of the **Delian League** and the first member to revolt—losing its independence (c. 468/7). An Athenian **cleruchy** was established there around 450. Naxos was a member of the Second Athenian Confederacy and again revolted unsuccessfully. It played a minor role in Greek military affairs from this point on.

Naxos was the site of at least one major naval battle, fought in 376 during the hostilities which followed the abortive Spartan **raid** on the Piraeus, led by Sphodrias. This raid had caused an Athenian-**Boeotian** alliance against Sparta and the founding of the Second Athenian Confederacy. An Athenian fleet of 43 **triremes** under **Chabrias** (although there were several other *strategoi* present) attacked Naxos, apparently to draw off the Spartan fleet blockading the Piraeus. The Spartan fleet of 65 triremes under Pollis overwhelmed the left wing of the Athenian fleet but **Phocion**, perhaps a *strategos* and probably commanding part of the second line of the Athenian fleet, launched an aggressive counterattack and turned the tide. The Athenians sank 24 Spartan ships and captured another eight, complete with crews, for the loss of 18 of their own ships. However, apparently mindful of the fate of the generals after **Arginusae**, Chabrias did not pursue the enemy. This victory was the first truly Athenian naval success since the **Peloponnesian War** of 431-404 (**Cnidus** was won largely with Persian ships). It ended the blockade of Athens and gave considerable impetus to the newly formed Second Athenian Confederacy, which gained 17 new members after the battle.

NEARCHUS (NEARCHOS/NEARKHOS; d. 312?) A Cretan (later granted **Macedonian** citizenship) who was a close and loyal friend of **Alexander the Great**. Nearchus was one of four of Alexander's companions exiled by **Philip II** in 337 during the family turmoil caused by Philip's marriage to Cleopatra. On Philip's death these men were recalled and rewarded by Alexander; Nearchus was made governor of Lycia (Arr. *Anab.* 3.6.5; Plut., *Alexander*, 10).

Nearchus appears at the start of Alexander's Indian expedition bringing a contingent of Greek **mercenaries** to join Alexander, and after the capture of the rock of Aornos, he was sent out with light troops on a **reconnaissance** (Arr. *Anab.* 4.7.2, 30.5-6). In November 326, at the end of the Indian campaign, Nearchus was appointed admiral and commanded the fleet which made the epic voyage from the Indus River to the Tigris in 325—a feat for which he was decorated by Alexander (Arr. *Anab.* 6.2.3, 7.5.6; *Indica*, 20-43). Sometime before 312 Nearchus completed a description of India, which included an account of his voyage. This was apparently well regarded for its accuracy and honesty and was used both by Strabo the geographer and in **Arrian**'s history of Alexander's Indian campaign. After Alexander's death Nearchus served under **Antigonus I (Monophthalmus)** and **Demetrius I (Poliorcetes)** and was probably killed at the battle of **Gaza** in 312.

NEMEA. A river in the northeast Peloponnese, south of **Sicyon**; site in 394 of a full-scale land battle during the **Corinthian War**. **Xenophon** preserves the only detailed account (*Hell.* 4.2.14-23) but some supplementary information can be drawn from **Diodorus Siculus** (14.83.1-2). The battle was fought between a **Spartan**-led **Peloponnesian League** army, which had marched north to reestablish Spartan control, and an allied force attempting to win a victory before **Agesilaus II** and his army could return from Asia Minor. The anti-Spartan coalition force was drawn from **Argos, Athens, Boeotia, Corinth,** and **Euboea,** with smaller contingents from other states. It consisted of 24,000 **hoplites,** 1,550 **cavalry,** and an unknown number of *psiloi* (light troops). The Peloponnesian League force, commanded by the regent Aristodemus, consisted of at least 16,000 hoplites (probably closer to 23,000), 600 cavalry, 300 Cretan **archers,** and 400 **slingers**.

The two armies camped opposite each other, separated by the dried-up bed of the Nemea. There was some argument on the coalition side concerning the depth of each state's **phalanx,** probably arising from suspicion of the **Theban** practice of deploying in a very deep phalanx, with a narrow frontage. Each state seems to have held command for a day and when the Boeotian turn arrived and they took up their position on the right of the line, they deployed their deep phalanx, contrary to the agreement, and initiated the battle.

The Boeotians led off at an oblique angle, with the right wing leading and the left wing trailing. The Spartans were initially unaware of the advance because of the vegetation, but speedily deployed when they heard the enemy *paean*. With the Spartans deployed on the right wing, Aristodemus also moved his army to the right, attaining a considerable overlap against the Athenian contingent facing him on the enemy left wing. This seems to have been a deliberate maneuver and once achieved, Aristodemus swung his right wing in against the rear of the Athenians. The Spartans caused considerable **casualties** among the Athenians as they now traversed the battlefield from (their) right to left at a right angle to their initial orientation. On the other wing the coalition force had driven Sparta's allies back, no doubt aided by the deep Boeotian phalanx, but as they returned from the pursuit, several contingents, notably the Corinthians and Argives, but also the Thebans, were also caught in their flank by Aristodemus. Xenophon does not give casualty figures, but the 1,100 Peloponnesian and 2,800 coalition casualties cited by Diodorus Siculus (14.83.2) is plausible. Although the battle was not as decisive as the Spartans had hoped, it did check coalition plans and gained valuable time for the arrival of Agesilaus from Asia. Nemea is interesting for the Spartan tactic of a flank attack, perhaps deriving from their success at **Mantinea** (1) in 418, and the Boeotian tactic of an oblique attack. Both foreshadowed the battle of **Leuctra** in 371, where the Boeotian tactic proved superior.

NICIAS (NIKIAS; c. 470-413). An **Athenian** general and politician, prominent during the **Peloponnesian War** of 431-404; son of Niceratus. Politically moderate, he was continually elected *strategos* after the death of **Pericles** and was the main opponent of **Cleon**. In 427 he led a successful operation to occupy the island of Minoa, just off **Megara**, to use as an observation post and base for the ships blockading Nisaea, Megara's port. The following year he **ravaged** Melos, failing to bring it to terms, and landed his troops at Oropus, from where they marched inland into **Boeotia**, linked up with other Athenians and won a minor battle against a force from **Tanagra**, supported by some Thebans. In 425 he was one of the three Athenian commanders at the battle of **Solygia**, near **Corinth**, and opposed Cleon over the action required at **Pylos**. In 424 he was one of the Athenian commanders who captured the island of Cythera to use as a base against **Sparta**, his contacts in the main city apparently accelerating the surrender.

Nicias was one of the generals who negotiated the armistice with Sparta in 423 after **Brasidas'** capture of **Amphipolis**. After Cleon's death he was the prime instigator in 421 of the peace with Sparta, commonly called the Peace of Nicias. Nicias was a cautious commander and **Thucydides** (5.16) suggests one of his main reasons for advocating peace in 421 was that he had been successful in the war to date and wanted to avoid the risk of a future defeat and loss of reputation if it continued.

In the period of peace, Nicias opposed **Alcibiades'** policy of forming an anti-Spartan alliance in the Peloponnese. He also argued strongly against Athens' decision to send an expedition to **Sicily** in 415. Ironically, his arguments only persuaded the Athenians to increase the size of the force, making the magnitude of the final disaster even greater. Elected one of the three generals to command the **Sicilian Expedition**, his overcautious policy after Alcibiades' recall was a major factor in the Athenian loss. His decision to relocate the Athenian camp to Plemmyrium caused considerable problems for the army and his decision to delay the evacuation because of an eclipse proved fatal. Nicias was captured when the expeditionary force was destroyed at the **Assinarus River** and then executed. Thucydides comments that Nicias was "the least worthy of all the Greeks of my time to come to such an unfortunate end because he had spent his whole life studying and practicing virtue" (7.86.5).

NICIAS, PEACE OF. *See* NICIAS; PELOPONNESIAN WAR (2)

NICOMEDES. *See* BITHYNIA

NIGHT ATTACKS. Greek armies generally tended to avoid night operations and attacks because of the difficulties involved. For example, the **Athenians** are known to have launched only two night attacks during the entire **Peloponnesian War** (2) of 431-404, and both of these were planned by the same *strategos* (general), **Demosthenes** (1).

The difficulties are illustrated by several failed efforts, including the Athenian one on Epipolae at **Syracuse** in 413 during the **Sicilian Expedition** and **Pyrrhus I** of Epirus' defeats at **Beneventum** in 275 and **Argos** in 272. At Epipolae the Athenian attack was initially successful because of the element of surprise, but the army lost cohesion as it pressed on to its subsequent objectives. A major problem was recognizing friend from foe, compounded for the Athenians by the presence in their army of allied contingents who spoke the same Dorian dialect as the enemy. In the resulting chaos the Syracusans were able to regroup and inflicted major losses on the Athenians who were unfamiliar with the territory (Thuc. 7.43-44).

At Beneventum the approach march was longer than Pyrrhus expected, the torches burned out and his troops became lost (Plut., *Pyrrhus*, 25). Pyrrhus was killed in 272 in the aftermath of a night attack on Argos. Although his army was initially successful in forcing entry to the city, it lost valuable time and alerted the enemy, while trying to fix a problem with the **elephants** in the dark. His main force then became lost and disorganized in the unfamiliar streets, allowing the Argives to mount a determined resistance at daybreak (Plut., *Pyrrhus*, 32-33). A similar failure occurred in **Sicily** in 312 when forces hostile to **Agathocles** failed in an attempt to storm Centoripa (Diod. Sic. 19.103.2-3).

These examples illustrate that although surprise could be gained under cover of darkness, navigation, especially on unfamiliar terrain, was difficult—as was distinguishing friendly troops from the enemy. These difficulties may well have been the real reason for **Alexander the Great**'s rejection of **Parmenio**'s suggestion of a night attack at **Gaugamela** (see Arr. *Anab.* 3.10.2-4 on this and for an excellent analysis of the risks involved in a night action). The popular explanation was that he considered a night attack dishonorable (Arr. *Anab.* 3.10.2; Curt. Ruf. 4.13.4-10), but he certainly did not hold this view on other occasions (see below).

Not all night attacks failed, however, particularly when carefully planned. **Herodotus** (8.27) records a very successful **Phocian** one against the **Thessalians** some time prior to the Second **Persian War**. The Phocians selected 600 men (500, according to Paus. 10.1.5) and covered them with whitewash so they could easily recognize friend from foe. The enemy's panic at the unexpected attack was greatly increased by their fear that the whitened men were ghosts, and 4,500 were killed. Alexander the Great also destroyed an army of Dardanians and Taulantians in a successful night attack during his campaign against **Cleitus** (3) in 335 (Arr. *Anab.* 1.6.9-11). Demosthenes (1) destroyed an Ambraciot army in 426 when he surprised them just before dawn (Thuc. 3.112.2-6). The tactic of attacking at dawn after moving into position in the dark retained the chance of achieving a similar degree of surprise as in a night attack, while reducing some of the dangers by timing the final assault for daybreak (Pritchett, *Greek State at War*, vol. 2, 161, table 3, has a useful list of dawn attacks).

NOTIUM (NOTION). A place in **Ionia**, north of Samos and west of **Ephesus**; site of a naval battle fought between **Athens** and the **Peloponnesian League** in 406, during the **Peloponnesian War** (2) of 431-404. **Alcibiades**, the Athenian *strategos* in the area, was using Notium as a naval base and the battle occurred while he was absent on a liaison visit to **Thrasybulus**, commander in the Hellespont. Alcibiades had left Antiochus in command with instructions not to precipitate a naval battle against the Peloponnesian fleet at Ephesus. However, Antiochus sailed insultingly past the prows of the Peloponnesian fleet with two **triremes**. **Lysander**, the Spartan admiral launched several ships in pursuit and, when the Athenians responded in kind, he followed up with his whole fleet in good battle order. Unprepared for this, the Athenians launched the remainder of their ships as best they could and sailed individually into action. The result was an embarrassing defeat for the Athenians, who lost 15 triremes, although most of the crews escaped. The Athenians blamed Alcibiades for the defeat and deprived him of his command; he went into exile in the Chersonese.

- O -

ODYSSEUS. A prince of Ithaca; one of the main characters in the epic accounts of Troy, and the main character in **Homer**'s *Odyssey*, an account of Odysseus' return home after the **Trojan War**. Odysseus was a powerful warrior, but his main characteristics were his wise counsel and his cunning. He feigned madness to avoid joining the expedition to Troy but failed and appears in Homer's *Iliad* as a wise adviser in the council of the chiefs. In Classical Greece he was less popular, apparently because his ruses and subterfuges were felt to be dishonest—even in warfare. *See also* LAWS OF WAR.

OENOPHYTA (OINOPHYTA). A decisive battle fought in **Boeotia** in autumn 457 between an **Athenian** army under Myronides and a **Boeotian League** army under **Thebes**. The Boeotians were decisively defeated and their whole territory, except for Thebes, fell under Athenian domination. This battle reversed the result of the recent battle of **Tanagra** and also caused **Phocis** to ally with Athens.

OFFICERS. Greek armies operated a variety of military hierarchies which differed from state to state and over time. During the earlier periods there was no real need for subordinate officers and the generals were drawn from the royal family or aristocratic class. In Classical **Sparta**, because of their common **training**, all Spartiates in theory could serve as officers. In Classical **Athens** the officers were elected officials, holding office for one year, and there was a tendency for the higher ranks to be drawn from the upper classes—at least in the fifth century. However, given that their armed forces were made up of citizens, many voters were also soldiers or sailors and had a vested interest in the election of competent men to military rank. Senior officers in the **Boeotian** and **Achaean Leagues** were also elected, and it was not unheard of for a man to be a general one year and serving in the ranks the next. **Epaminondas** led Boeotian invasions of the Peloponnese in 369 and 366 and of **Thessaly** in 367, but served as an ordinary **hoplite** in the expedition to Thessaly in 368. In mercenary armies, such as the **Ten Thousand**, officers were sometimes also elected. In Hellenistic monarchies the senior officers and those in key posts were normally appointed by the king. As might be expected, officers received higher **pay** than their men. *See also* DISCIPLINE; HIPPARCH; *HIPPARMOSTES*; NAVARCH; PHYLARCH; POLEMARCH; *STRATEGOS*; TAXIARCH.

Plate 3. Walls, Eleutheria (Attica, fourth century B.C.)

Plate 4. Leuctra Monument (Boeotia, 371 B.C., restored)

OLPAE (OLPAI). A town in Amphilocia; site of a battle in 426 during the **Peloponnesian War** of 431-404. A Peloponnesian-Ambraciot army, led by the Spartan Eurylochus fought a mixed force predominantly of Acarnanians and Amphilocians, with some **Athenians** and **Messenians,** led by **Demosthenes** (1). Outnumbered, Demosthenes placed a force of around 400 **hoplites** and *psiloi* (light troops) in an **ambush** position. Eurylochus, commanding the left wing, managed to outflank Demosthenes, but most of Eurylochus' army broke and ran when the hidden contingent attacked them from the rear. Under a secret agreement made with Demosthenes the night after the battle, Eurylochus withdrew with his Peloponnesian troops, abandoning his Ambraciot allies. Another force of Ambraciots en route to join Eurylochus and unaware of his recent defeat was surprised and destroyed by Demosthenes while they slept. These two actions considerably reduced the power of the Ambraciots and the reputation of **Sparta** in the area.

OLYMPIA. There was no distinct settlement of Olympia; the site consisted of the temple and sacred grove of Zeus in **Elis** in the Peloponnese, along with other temples and public buildings associated with the Olympic games. Greek participation in the games was one of the reasons **Herodotus** (7.206) gives for the small size of the contingent sent to defend **Thermopylae** against the **Persians** in 480. Quarrels over the presidency of the games caused several wars in the region, either within Elis or between Elis and **Arcadia.** There were two battles at Olympia during the Classical period:

(1) 401. Fought between **Elis** and **Sparta.** The details of the battle are unknown but **Pausanias** (5.27.11) records the existence of a bronze **trophy** set up by Elis to commemorate its victory. The action apparently took place during the invasions of Elis led by **Agis II** (Paus. 5.4.8), although **Xenophon**'s account of this war (*Hell.* 3.2.22ff.) does not record an engagement at Olympia. Pausanias (5.20.4-5) relates that years later, during restoration work on the roof of the temple of Hera, the Eleans discovered the well-preserved corpse of a soldier who had died in this battle; apparently quite a few Eleans had occupied the rooftops (as the **Arcadians** were to do in 364) and this man had been wounded, slipped into a crevice, and died.

(2) 364. Fought between **Elis** and allies and the **Arcadians** and their allies. **Xenophon** (*Hell.* 7.4.28ff.) describes the action, which actually interrupted the conduct of the pentathlon. Numbers on both sides are unknown, although the Arcadians had the support of 2,000 **hoplites** from **Argos** and 400 **Athenian cavalry.** The Eleans, not noted as brave soldiers prior to this, routed the Arcadians, stood firm against and then

pushed back the Argives, and were stopped only when they became bogged down in the built-up area and were pelted with stones and missiles from the rooftops. They withdrew, leaving their enemy shaken. The next day the Eleans formed up for battle again, but the Arcadians refused to leave the shelter of the precinct, which they had fortified overnight using the pavilions set up for the games. The Eleans withdrew and Xenophon's remarks that their courage was clearly divinely inspired probably reflects general Greek opinion at the time. Their strong showing here had an important impact on the Arcadian decision to return the presidency of the games to them (although internal dissension within the **Arcadian League** over the illegal use of the treasures from Olympia to fund the war was also a major factor).

OLYNTHUS (OLYNTHOS). A city in the Chalcidice. Originally a Bottiaean settlement, **Herodotus** remarks (7.122) that it was Greek when **Xerxes I**'s invasion fleet passed by in 481. However, he also states (8.127) that it was occupied by Bottiaeans when it revolted and was captured in the winter of 480/79 and was then taken over by the Chalcidian Greeks. Part of the **Athenian** empire, it revolted in 432 when **Perdiccas II** of **Macedon** persuaded the inhabitants of smaller towns in the region to move to Olynthus. The city gained its independence, although continuing to pay tribute, under the Peace of **Nicias** following **Brasidas'** campaign in the area (Thuc. 1.58.2, 5.18.5). From 392-379 Olynthus was the major city in the **Chalcidian Confederacy** (although under attack from Sparta from 382 until its surrender in 379). The city had joined the Second Athenian Confederacy by 375 but lost **Pydna**, Methone, and **Potidaea** to Athens between 368 and 363. For a while it vacillated between Athens and **Philip II** of Macedon before allying with Athens in 349. Olynthus was eventually destroyed by Philip in 348. The city and surrounding area, known for its high-quality **cavalry**, was the site of several battles:

(1) 381. A **Peloponnesian League** army of 10,000 men, led by the **Spartan** Teleutias and reinforced locally by Elimia and **Macedon**, deployed outside the city, with the Peloponnesians on the left and the allied contingents on the right. Teleutias personally commanded the left wing and kept the 400-strong Elimian **cavalry** under their ruler, **Derdas**, with him; the Laconian, **Theban**, and Macedonian cavalry were deployed on the right wing. The **Chalcidian Confederacy** forces elected to fight and started the battle with a successful cavalry charge against the enemy cavalry on the right. The Spartan cavalry commander, Polycharmus, was killed and his cavalry routed. The infantry on the right was also starting to break when Derdas prevented a Spartan defeat by launch-

ing a cavalry charge at the gates of Olynthus. This threatened to cut off the Olynthian horse and caused them to retire. The Olynthian infantry also retired—apparently in good order, as they suffered few losses (Xen. *Hell.* 4.2.37-43).

(2) 348. Although the details are unknown, during **Philip II** of **Macedon**'s attack on Olynthus, the two **hipparchs**, Lasthenes and Euthycrates, betrayed 500 **cavalrymen** into Philip's hands. This greatly contributed to the capture of the city, which was destroyed and the surviving inhabitants enslaved (Hyperides, fr. 76, [*Against Demades*], 1; Demosthenes, 8 [*On the Chersonese*], 40, 9 [*Third Philippic*], 66).

ONASANDER (ONASANDROS). A Platonic philosopher and author of a book (the *Strategikos*) on generalship. Written in the middle of the first century A.D., probably shortly before A.D. 59, the book is a compilation of principles of leadership and the duties of a general, with particular emphasis on ethics. Although lacking in much originality overall, the principles do provide some useful information on the general employment of various arms, including **cavalry** and *psiloi* (light troops), and on security and tactics. The *Strategikos* was widely read and quite influential in later periods, including the Renaissance. *See also* AELIAN; AENEAS TACTICUS; ASCLEPIODOTUS.

OPSONION. See PAY

ORCHOMENUS (ORCHOMENOS/ORKHOMENOS).
(1) A city in north **Boeotia**, the second most important city in the region after **Thebes**. Along with most Boeotian cities it **medized** during the Second **Persian War**. Orchomenus generally followed a pro-Thebes policy, but in 395 separated from the **Boeotian League** and Thebes and fought on the **Spartan** side in the **Corinthian War**. Orchomenus supplied troops to assist **Lysander** at **Haliartus** and in 394, alone of all the Boeotians, fought for **Agesilaus II** at **Coronea** (2) (Xen. *Hell.* 3.5.6-7, 3.5.17, 4.2.17, 4.3.15). The city became formally independent in 386 under the terms of the King's Peace (**Common Peace** [1]). Following the battle of **Leuctra** in 371, Orchomenus narrowly avoided destruction by the Thebans, who instead brought it back into the Boeotian League (Diod. Sic. 15.57). However, in 364 the Thebans accused Orchomenus of plotting with Theban exiles to overthrow the democratic government at Thebes and destroyed the city. All the adult males were executed and the women and children sold into **slavery** (Diod. Sic. 15.79; Paus. 9.15.2).

Orchomenus was apparently rebuilt by the **Phocians** during the Third **Sacred War** and used as a base to **ravage** the rest of Boeotia. The Phocian general Phayllus was defeated near the city in 352. In 346 **Philip II** of **Macedon** handed the city over to Thebes, which destroyed it for a second time, again selling the population into slavery. Following the battle of **Chaeronea** in 338 the city was rebuilt once again, either by Philip II or **Alexander the Great** (Paus. 4.27.10, 9.37.3; Arr. *Anab.* 1.9).

From this point on, Orchomenus played little part in the military history of Greece, apart from being the site in 86/5 of a major battle during the First **Mithridatic War**. This battle was fought on the open plain between approximately 90,000 troops of **Mithridates VI (Eupator)** of Pontus, led by **Archelaus**, and a **Roman** army of around 20,000 men led by Lucius Cornelius Sulla. Outnumbered in **cavalry**, Sulla dug trenches to protect his flanks from this threat and to reduce the enemy's room to maneuver. In a series of engagements Sulla routed the enemy and blockaded their camp, which he soon captured. The enemy army was destroyed, ending Mithridates' campaign in Greece (Plut., *Sulla*, 20-21).

(2) A city in **Arcadia**, to the north of **Mantinea**. In remote antiquity it was a powerful city, but it played a minor role in the historical period. Orchomenus contributed 120 **hoplites** to the Greek force at **Thermopylae** in 480, and 600 at **Plataea** in 379, during the Second **Persian War** (Hdt. 7.202, 9.28). The city was used by **Sparta** to secure Arcadian hostages during the political and diplomatic maneuvering in the Peloponnese after the Peace of **Nicias**. However, in 418, because the city's **fortifications** were in a poor state of repair, the Orchomenians were forced to surrender the hostages to a force from **Athens, Argos, Elis**, and Mantinea and to join the anti-Spartan alliance (Thuc. 5.61). It is unclear whether they participated at the battle of Mantinea (1) in the same year and, if so, on which side they fought.

Orchomenus was the site of at least one major battle in the Classical period, fought in 370 and described by **Xenophon** (*Hell.* 6.5.13-14; cf. Onasander, 27). At that time Orchomenus, through enmity with Mantinea, had refused to join the **Arcadian League** and was pro-Sparta. The city had been reinforced by a **mercenary peltast** force under the command of a Spartan, Polytropus. At the same time that **Agesilaus II** invaded Arcadia, the Mantineans attacked Orchomenus to neutralize it and the mercenary force. The Mantineans were beaten back from the walls and retired in some disorder under considerable pressure from Polytropus and his men, who harassed them with **javelins**. The Mantineans were able to re-form their **phalanx** and charged out at the enemy, killing Polytropus and driving back his troops. The **casualties** would have been considerable if the Phliasian **cavalry** with Polytropus had not prevented a sustained pursuit by the Mantinean hoplites.

Orchomenus apparently also refused to join in the founding of Megalopolis after the battle of **Leuctra** and lost territory to the new city (Paus. 8.27.4). Because of its strategic location, Orchomenus was considered a prize in the struggles to dominate the Peloponnese following the death of **Alexander the Great** and changed hands several times. In 313 it was controlled by **Cassander**, who allowed the oligarchic party to slaughter the partisans of Polyperchon (Diod. Sic. 19.63.5). Orchomenus fought on the Greek side in the **Chremonidean War** and joined the **Achaean League** in 234, shortly afterward leaving and joining the **Aetolian League**. It was captured by **Cleomenes III** of Sparta in 229, with the connivance of the Aetolian League (Polyb. 2.46), and **Aratus** failed to recapture it—although in 226 he did defeat a Spartan force near the city (Plut., *Aratus*, 38.1). Orchomenus was taken by **Antigonus III (Doson)** in 223 and garrisoned by him (Polyb. 2.54, 4.6). **Philip V** of **Macedon** returned the city to the Achaean League in 199 to free up the garrison for use against the **Romans** in the Second **Macedonian War**.

- P -

PAEAN (PAIAN). A battle hymn, sung before engaging the enemy to avert evil and ensure success. Later, it came to play an important role in keeping **hoplites** in step and **formation** as they marched toward the enemy (and presumably also helped maintain morale and a sense of unity). This use is particularly associated with the Dorian Greeks (who had their own distinctive *paean*) and the **Spartans** supplemented their singing of the *paean* with rhythmic music to help preserve order and discipline in the advance. The *paean* was also used as a signal to advance or on the departure of a military or naval force (see **Thucydides'** description [6.32] of the departure of the **Sicilian Expedition** in 415). *See also* NEMEA.

PALESTINE. An ethnically mixed area strategically located astride the land route to **Egypt** from Asia Minor. Palestine was sometimes regarded as part of Coele-Syria in antiquity. It was removed from **Persian** rule by **Alexander the Great** in 332. In the third century its possession was consistently disputed (along with Coele-Syria) between the **Ptolemies** of Egypt and the Seleucids of **Syria**. The Ptolemies generally controlled Palestine until the Fifth **Syrian War** (202-195), when it fell under Seleucid control.

Conflict between the Jews (who were a minority in Palestine but dominated the area around Jerusalem) and a hellenizing party among the priestly ruling class led to major civil disturbances during the reign of **Antiochus IV (Epiphanes)**, culminating in a Jewish rebellion. Antiochus ruthlessly crushed this in 167, taking Jerusalem with considerable bloodshed. This did not quell the uprising, though, and Antiochus installed a garrison in Jerusalem and enacted harsh anti-Jewish measures—including dedicating the Temple in Jerusalem to Zeus.

This led to the revolt of Mattathias and his sons, including Judas Maccabaeus, who became the leader on his father's death in 166. Under Maccabaeus the Jews waged successful guerrilla warfare and also won several pitched battles against the Seleucids. In 164, motivated in part by internal Seleucid politics, the rebels were offered an amnesty and the lifting of the anti-Jewish laws. That same year Maccabaeus purified the Temple in Jerusalem and on the death of Antiochus went on the offensive. However, he was not as successful as before and only survived because Lysias, the Seleucid general, was in danger of being replaced and needed a quick settlement. In 161 Maccabaeus, at war with Jewish rivals backed by the Seleucids, secured an alliance with **Rome**, which failed to save him from the new king, Demetrius I (Soter), who defeated and killed him in 160.

Taking advantage of the power struggle between Demetrius and Alexander Balas, Maccabaeus' brother Jonathan secured control in Jerusalem and in 141 the last surviving brother, Simon, expelled the Seleucid garrison from Acra, the fortress in Jerusalem. With this the Jews had finally achieved self-rule—although they were temporarily reconquered by **Antiochus VII (Sidetes)** in 135-134 (Jerusalem holding out until 131/0). On Antiochus' death in 129, Palestine regained its autonomy again under Hyrcanus. However, the Jews lost control of several conquered Greek cities in 63-62, with Pompey's settlement of the area. In A.D. 6 Palestine was incorporated into the Roman empire. Palestine's strategic location made it the site of several major clashes during the Hellenistic period, the most important being **Gaza** in 312 and **Raphia** in 217.

PANDEMEI. A term (literally "with all the population") used of Classical Greek armies, denoting that the entire state's forces had been mobilized.

PANION. A battle between **Antiochus III (the Great)** and the **Egyptian** army, led by the **mercenary** Scopas, on the northern border of **Palestine** in 200, during the Fifth **Syrian War**. Scopas lost and, taking refuge in Sidon with the 10,000 survivors of his army, was besieged there. The battle led to Antiochus' occupation of the whole of Coele-**Syria** as far south as **Gaza**.

PARAETACENE (PARAITAKENE). A battle fought northwest of Persepolis in autumn 317 between **Eumenes of Cardia** and **Antigonus I (Monophthalmus)**. Eumenes had around 35,000 foot (including 3,000 Macedonians)—but only half of them were heavy infantry—6,000 **cavalry**, and 114 **elephants**. Antigonus led 28,000 well-equipped foot (including 8,000 Macedonians), 8,500 cavalry, and 65 elephants. Eumenes had rather the better of things and Antigonus withdrew to Media.

PARMENIO (PARMENION; c. 400-330). A **Macedonian** noble and important general under **Philip II**. Little is known of his service under Philip, although in the summer of 356 he defeated the Illyrians while Philip was campaigning in the Chalcidice (Plut., *Alexander*, 3). He captured Halus in **Thessaly** in 346 and in 342/1 fought in **Euboea** (Demosthenes, 19 [*On the False Embassy*], 2, 9ff.). He apparently also campaigned in **Thrace** around 340. In spring 336 Parmenio was sent to Asia Minor with Philip's advance guard and served as **Alexander the Great's** second-in-command during his Persian campaign.

The sources (perhaps bordering on a literary device) portray Parmenio as a voice of caution during this campaign, advocating a pause at **Granicus River** (Arr. *Anab.* 1.13.2-7) and a night attack at **Gaugamela** (Arr. *Anab.* 3.10; Curt. Ruf. 4.13.4-10). He commanded the left wing at Granicus, **Issus**, and Gaugamela (Arr. *Anab.* 1.14 2.8 3.12; Curt. Ruf. 3.9.7, 4.15.6). When Alexander continued his eastward march in 330, Parmenio was left in Media. That same year Parmenio's son, Philotas, was executed for plotting against Alexander. Parmenio, although apparently not implicated in this plot, was a member of Philip II's generation and, like **Cleitus** (1), probably out of sympathy with some of Alexander's aims and methods. Given this, and his position in the army, Alexander had Parmenio executed as a precautionary measure (Arr. *Anab.* 3.26-27; Curt. Ruf. 7.2.1-33). Parmenio had a reputation as an excellent soldier and tactician (Curt. Ruf. 7.2.33-34) and Philip is recorded as saying that, in contrast to the **Athenians**, who were able to find 10 generals to elect every year, "he himself over many years had found only one general—Parmenio" (Plut., *Mor.* 177 c).

PASIMACHUS (PASIMACHOS/PASIMAKHOS; d. 392). A Spartan *hipparmostes* killed in action at **Corinth** during the struggles over the unification of Corinth and **Argos**. He dismounted his **cavalry** and, requisitioning shields from fleeing **Sicyonian hoplites**, led his small force against the advancing Argives (Xen. *Hell.* 4.4.10).

PAUSANIAS.

(1) (d. c. 470/69) Nephew of **Leonidas I** of **Sparta**, appointed regent after his uncle's death at **Thermopylae** (1) during the Second **Persian War**. He was victorious at **Plataea** in 479 and in 478 led a Greek expeditionary force to the Hellespont, where he captured **Byzantium**. His arrogant behavior caused a mutiny among the allied Greek contingents there, allowing **Athens** to supplant Sparta as leader of the continued hostilities against **Persia**. Recalled to Sparta and twice acquitted on charges of treasonable negotiations with the Persians, he was starved to death after taking refuge in a temple when accused of planning a helot revolt. *See also* ARISTIDES; CIMON; DELIAN LEAGUE.

(2) (c. 450-post 380) Agiad king of **Sparta**, 409-395. Although nominally king as a minor during the temporary deposition of his father, Pleistoanax, in 445-426, Pausanias was active in his own right in the final stages of the **Peloponnesian War** of 431-404 and down to 395. He besieged **Athens** from land in conjunction with **Agis II**'s forces at **Decelea** and **Lysander**'s naval blockade, starving it into submission in 404. He subsequently disagreed with Lysander over arrangements for Athens, reversing Lysander's policy of supporting the **Thirty Tyrants**.

On the outbreak of the **Corinthian War** in 395, Pausanias was supposed to invade **Boeotia** from the south with the main **Peloponnesian League** army at the same time Lysander invaded from the north. However, Pausanias delayed his invasion because of bad omens and Lysander was killed outside **Haliartus**. When Pausanias did arrive, he decided to negotiate for the return of the bodies of Lysander and those killed with him. His decision was based on the fact that, even if he had won a battle, the bodies were very close to the walls of Haliartus and therefore covered by enemy fire. In addition, he was short of **cavalry**, the Corinthian contingent had refused to cross into Boeotia, and the other allied contingents were only there reluctantly. In return for the bodies, Pausanias withdrew his army from Boeotia. This lack of action shortly led to the defection of **Corinth** and other allies. On his return to Sparta Pausanias was sentenced to death but went into exile in Tegea. He died in exile sometime after 380. *See also* SACRIFICE.

(3) (fl. c. A.D. 150) A Greek author, perhaps from Lydia. He traveled extensively and wrote *A Description of Greece*. This work gives a historical background, topographical description, and account of buildings, statues, and ruins for the regions of mainland Greece. He preserved a lot of interesting information about Greek military events, and the ac-

curacy of his description of **monuments** and buildings is confirmed by surviving remains. Pausanias is the main source of information on the **Messenian Wars**.

PAY. Pay for soldiers and sailors consisted of two elements: pay proper and ration money. From the end of the fifth century, a clear distinction in the terminology emerges between pay (*misthos*) and ration money (*siteresion* or *sitos*). In the Hellenistic and Roman periods, the term *opsonion* replaced *misthos*. Although the citizen-**hoplite** traditionally served his state at his own expense, the changing nature of warfare after the **Persian Wars** meant that this was increasingly difficult to maintain. At **Athens**, for example, the overseas expeditions and protracted sieges required by the acquisition and maintenance of empire after 479 led to the introduction of state pay for military service. The exact chronology is not known, but in the absence of a state supply service, it seems likely that ration money was provided first. However, by 431, on the outbreak of the **Peloponnesian War** (2), Athenian soldiers were also receiving pay. This had apparently always been the case for the Athenian navy, whose sailors were drawn from the poorer sections of society and could not be expected to take time away from their civilian occupations without recompense.

The general introduction of military pay in Greek cities from the time of the Peloponnesian War of 431-404 caused considerable strains on state finances. Athenian *strategoi* (generals) of the fourth century were regularly sent on campaign with insufficient money to pay their troops and had to raise the difference by extortion from allies, **raids**, or even hiring out men as laborers. It was not uncommon for the soldiers' pay to be dependent on **booty** taken from the enemy.

Pay rates varied according to market forces and specific conditions. During the Peloponnesian War the basic rate was three obols per day. More would be paid, or offered, if campaigning was in areas where food was likely to be scarce (e.g., Athenians serving at **Potidaea** and **Syracuse** received double ration money) or when trying to attract sailors away from the enemy into one's own fleet. **Officers** received higher pay (in the **Ten Thousand**, a soldier's daily rate was five obols, while *lochagoi* [company commanders] received 10 obols and *strategoi* 20). Elite troops, such as **cavalrymen** and the **hypaspists** under **Alexander the Great** and his successors, also received higher rates of pay than the standard infantryman. Part of a soldier's or sailor's pay was often withheld until the end of the campaign, partly to reduce the amount of money taken on an expedition but also to reduce the risk of desertion. Pay could be forfeited for **disciplinary** infractions.

Mercenaries in theory normally received their *siteresion* at the start
of the month and *misthos* at the end of the month. However, it was not
uncommon for the pay component to be in arrears. **Xenophon** (*Anab.*
1.2.11) records that at one stage **Cyrus the Younger** was three months
behind with his soldiers' pay, although he had been providing them
regularly with rations. There is evidence in the Hellenistic period that
mercenaries who constituted a standing force were paid only for 9-10
months per year, on the basis that campaigning did not occur in the
depths of winter.

PELOPIDAS (c. 405-364). A Theban general and political leader. Born
into a wealthy and prominent family, Pelopidas distinguished himself in
battle at **Mantinea** in 385. He went into exile in 382 when **Phoebidas**
occupied the Cadmea and Ismenias, the leader of the anti-**Spartan** group
in **Thebes**, was arrested. Pelopidas played a major role in the coup
against the pro-Spartan group in Thebes in 379, during which the Cadmea was retaken and Thebes liberated from Spartan control. Elected as a
boeotarch for 378 in partnership with **Epaminondas**, Pelopidas exerted
a major influence on Thebes' rise to hegemony in Greece over the following years.

Pelopidas led the **Sacred Band** with distinction at **Tegyra** in 375
and **Leuctra** in 371, and **Plutarch** (*Pelopidas*, 19) credits him with being the first to ensure its members fought together as a single unit rather
than forming the front rank of the Theban **phalanx**. He was also joint
commander with Epaminondas in the first invasion of the Peloponnese
in 370/69. After this, he was apparently largely involved in Theban activity to the north of **Boeotia**.

In 369, after an appeal from various Thessalian cities, Pelopidas was
sent against **Alexander of Pherae.** Pelopidas captured Larissa, significantly weakening Alexander's influence in **Thessaly**, and then intervened in **Macedon**, settling affairs there and taking the king's brother,
the future **Philip II**, back to Thebes as a hostage. In 368, though, he
was treacherously seized and imprisoned by Alexander. Freed by an expedition led by Epaminondas, in 367 Pelopidas scored a major diplomatic triumph by persuading the **Persians** to transfer support from
Sparta to Thebes. The resulting treaty, the Peace of Pelopidas (**Common
Peace** [3]), affirmed the independence of **Messenia** from Sparta. Although Sparta and most other states refused to ratify the treaty, it provided considerable moral authority to Thebes and may have been partially accepted in 366/5. Although his army was victorious, Pelopidas
was killed at the battle of **Cynoscephalae** (1) in 364, when he pressed

forward to personally attack Alexander of Pherae. *See also* BOEOTIAN LEAGUE.

PELOPIDAS, PEACE OF. *See* COMMON PEACE (3)

PELOPONNESIAN LEAGUE. A military alliance, originally of Peloponnesian states, with **Sparta** as the dominant power. The league, known in antiquity as "the Lacedaemonians and their allies," was a Spartan initiative of the mid-sixth century designed, at least in part, to provide insurance against revolt by its helots and **Messenia**. In addition to the internal security benefits, Sparta was able to use the combined power of the league members to generate considerable military power and political muscle. The league was a mutual defense pact, but each member of the league was in a treaty arrangement with Sparta rather than with each other and there are cases of league members fighting each other.

In 506 **Cleomenes I** summoned the allies and led them against **Athens**, without revealing the mission until the army reached Eleusis. There, the Corinthians withdrew and the expedition collapsed (Hdt. 5.74-76). Following this the league operated more as an alliance, with the allies meeting to vote on proposals and the Spartan assembly doing the same. If both bodies agreed, the proposal was adopted; if they differed, it was not. By the time of the Second **Persian War** (480-479) the Peloponnesian League included Aegina, **Thebes**, and **Megara**—all outside the Peloponnese—and Spartan prestige was responsible for the Greeks accepting its command on both land and sea during this war.

The league fought against Athens, and her alliance system, during the **Peloponnesian Wars**, although suffering some dissension during these conflicts. For example, Megara left the league in 459 over a quarrel with **Corinth** (rejoining in 446). Corinth, Thebes, Elis, and Megara refused to follow the Spartan lead and sign the Peace of **Nicias** in 421 and **Mantinea** left the league around the same time. Although both Corinth and Thebes joined with Sparta again when hostilities resumed in 413, Thebes disagreed with the decision in 404 not to destroy Athens and left the league. From this point on, the league essentially contained states from within the Peloponnese.

The league was effectively disbanded in 366 when Sparta, crippled militarily by its loss of the battle of **Leuctra** and the subsequent removal of Messenia from its control, released its allies from their obligations (Xen. *Hell.* 7.4.9).

PELOPONNESIAN WARS. Two wars fought in the fifth century between **Athens** and its alliance system/empire and the **Peloponnesian League, Sparta**'s alliance system.

(1) First Peloponnesian War (459-445). A war fought between **Sparta** and her allies and **Athens** and her allies, lasting from 459-445, with a five-year truce from 451 to 446. From 464, relations between the two rival alliance systems steadily deteriorated over Athens' increasing power and Sparta's discomfort with this. In 459, **Megara** quarreled with neighboring **Corinth** and deserted the **Peloponnesian League** for Athens. Athens willingly accepted Megara's alliance, as this enabled her to control the best land routes along the Isthmus of Corinth, protecting her southern border from invasion. The war saw Athens, assisted by Megara, fighting principally against Corinth and **Thebes**, two of Sparta's main allies, and also against Aegina, a long-time enemy of Athens. Success for Athens would have given it a secure southern border, with Megara as a buffer state, and substantially reduced Sparta's power. As the war developed, Athens also attempted to neutralize Sparta's northern ally, Thebes, by conquering **Boeotia**. This would have given Athens a secure northern border and a land empire on mainland Greece to match her naval alliance system, the **Delian League**.

The war was fought in two phases, the first ending in a five-year truce in 451. Despite an early minor reverse at Halieis in 459 and a more serious defeat at **Tanagra**, by 457 Athens had forced Aegina to join the Delian League, gained control of Boeotia, and allied with **Phocis**. Athens was clearly winning at this point, but in 454 an Athenian-Delian League expedition to **Egypt** was destroyed by the **Persians**. Both sides were now weakened and in 451 signed a truce for five years.

The second phase began in 449 with a war between Sparta and Athens' ally, Phocis (the **Sacred War** [2]), followed soon after by Athens' defeat at **Coronea** (1), which left Boeotia independent. In 446, the Spartans **ravaged** Attica and both **Euboea** and Megara revolted against Athens. Although **Pericles** recovered Euboea, Megara remained free, opening up the isthmus route to easy Peloponnesian access again. Now threatened by land from both north and south, Athens could not sustain hostilities. The war ended with the **Thirty Years' Peace** of 446/5 by which Athens lost Achaea, Troezen, and the Megarid. Aegina remained a member of the Delian League but Athens guaranteed her autonomy. *See also* ITHÔME; OENOPHYTA.

(2) The Peloponnesian War (more rarely known as the "Second Peloponnesian War") fought between **Sparta** and her allies and **Athens** and her empire/allies from 431 to 404, with a period of quasi-peace from 421 to 413. The underlying reason for the war was the jealousy and fear

aroused in Sparta by Athens' rise to rival it as the preeminent military power in Greece in the period after the **Persian Wars**. In the years leading up to the outbreak of the war, Athens, led by **Pericles**, had placed increasing pressure on Sparta's allies, to the point that the **Peloponnesian League** was in danger of collapse if Sparta did not take action. Technically, Sparta was to blame for the outbreak of the war because it refused arbitration when it was offered by Athens and in doing so contravened the terms of the **Thirty Years' Peace**. However, Sparta had little choice but to go to war or abdicate its position as head of the Peloponnesian League.

The immediate events that led to the outbreak of war were **Corcyra**'s quarrel with **Corinth** and subsequent appeal for help to Athens. Athens' interference alienated Corinth, an ally of Sparta, and led to Athenian pressure on **Potidaea**, a colony of Corinth that was also a member of the **Delian League**. Potidaea revolted and requested Corinthian help. As relations between the Peloponnesian and Delian Leagues rapidly deteriorated, **Thebes** launched a surprise attack on **Plataea**, Athens' ally in **Boeotia**. This led to the outbreak of war, which was fought in two parts, the Archidamian and the Decelean Wars, separated by a period of uneasy peace.

A. The Archidamian War. The first part of the Peloponnesian War is known as the Archidamian War, after the Spartan king **Archidamus II**, who led the first invasion of Attica in 431. This involved the traditional method of waging war, with the Peloponnesian League army marching into Attica and **ravaging** the countryside. The Athenians, however, refused to come out to fight. Instead, Pericles persuaded them to adopt what is commonly referred to as the "Island Policy." This involved remaining inside the walls, replacing the lost produce with imports, and restricting military action to seaborne **raids** on the coast of the Peloponnese and land invasions of **Megara** when the Peloponnesian army had returned home. The Athenian **cavalry** was also deployed to provide some measure of protection to Attica.

Despite this, the Peloponnesians persisted in the annual invasions, clearly hoping that the Athenians would be forced to come out and fight. The Athenians refused to do so and for the most part held the initiative. From 431 to 429 the Athenians followed Pericles' island policy. From 429, when Pericles died from the plague, until 427, the Athenians continued to follow Pericles' policy but did send a 20-ship expedition to **Sicily**, heralding further expansionist activity there. During this period an Athenian army was defeated at **Spartolus** in 429 and Mytilene, the main city on Lesbos, revolted but was recovered.

The period 427-424 saw Athens following a policy of expansion on mainland Greece. This policy, associated with **Demosthenes** (1) and **Cleon**, aimed at neutralizing Boeotia, Sparta's northern ally, thereby allowing Athens to focus on the threat from the south. To bring the war home to Sparta, in 425 Athens fortified **Pylos**, a headland on the Peloponnese, and used it as a base for raids on Spartan territory. The securing of Pylos resulted in the capture of 292 Spartans on the nearby island of Sphacteria. The Athenians prevented further attacks on Attica by threatening to execute these **prisoners** if the Peloponnesians ever invaded again. However, this phase of the war ended in 424 with Athenian defeat at the battle of **Delium** (which ended her attempt to win the war on land) and the Spartan capture of **Amphipolis**, the heart of the Athenian possessions in the Chalcidice. The loss of Amphipolis sparked off several revolts by states subject to Athens in this region. Ironically, Sparta had only sent a force north under **Brasidas** because Attica was now off limits and the Chalcidice was the only part of the empire that could be reached from land.

Because of the loss of Amphipolis, Athens spent 424-421 on the defensive, protecting her empire. Cleon led an expedition to the Chalcidice but despite some local successes he was defeated and killed outside Amphipolis in 422. Brasidas, too, died in this battle and Athens and Sparta negotiated a truce, signing the Peace of **Nicias** in 421.

B. "Peace." The Peace of Nicias held from 421 to 413, but it was an uneasy one. Sparta's key allies, Thebes and Corinth, refused to sign the treaty and instead negotiated with Athens a seven-day truce, which they continually renewed. There were also disputes over the fulfillment of the terms, notably in the Chalcidice, where several cities that had revolted after the peace was signed but before the news of it had reached the north refused to be handed back to Athens. With the rise of **Alcibiades'** influence, the Athenians embarked on a program of securing alliances to strengthen their position against the Peloponnesian League. An alliance was made with **Argos** in 420 and in 418 Athenian troops assisted Argos and its allies against Sparta at the battle of **Mantinea** (1). Although defeated at Mantinea, Athens soon directed its attention toward Sicily, sending an expeditionary force of 94 **triremes** and nearly 6,500 **hoplites** and *psiloi* (light troops) against **Syracuse** in 415. This **Sicilian Expedition** ended in disaster in 413 with the loss of almost the entire expeditionary force, which had been substantially reinforced.

C. The Decelean War. The second phase of the Peloponnesian War, which lasted from 413-404, is known as the Decelean War after the Peloponnesian fortification of **Decelea** in Attica. Following the Athenian disaster in Sicily and provoked by Athenian involvement in the Peloponnese at Mantinea in 418 and Athenian raids on the Peloponnesian coastline, Sparta declared war on Athens in 413.

Athens was much more on the defensive during this phase of the war and Sparta fairly quickly gained the initiative by supporting the revolts of members of the Athenian empire. From 413 to 411, under a council of ten men, the *probouloi*, which as an emergency measure temporarily replaced the democratic government, Athens concentrated on the protection of its empire. Athens put down revolts in Mytilene (Lesbos), Lampsacus, and elsewhere and defeated the Peloponnesian fleet in battles at **Abydus** (1 and 2). Over the winter of 412/11, Sparta agreed to recognize Persian sovereignty in Asia Minor (effectively abandoning the **Ionian** Greek cities to Persian control) in return for **Persia** helping to expand and maintain the Peloponnesian fleet. From this point onward, Persia became the paymaster for the Spartan fleet, tipping the financial balance firmly toward the Peloponnesians.

This was the beginning of the end for Athens, whose overall strategy of avoiding land battles and maintaining Athens by importing food relied upon naval superiority and possession of an empire. Both of these were increasingly challenged by the Peloponnesians and by revolts in the empire. Athens was also beset by internal problems. The democratic government, temporarily replaced by the *probouloi*, was overthrown in an oligarchic coup in 411. However, the army and fleet stationed in Samos remained loyal to the democracy and political infighting, complicated by Alcibiades' interference, distracted Athens from the war effort.

Despite this, Athens was a very resilient city, and a major Athenian naval victory over the Peloponnesians at **Cyzicus** in Ionia in 410 led to the restoration of full democracy. The mood of confidence was such that the Athenians even refused Spartan offers of peace—in hindsight a mistake. Athens no longer possessed the same depth of resources as before, and any major losses to the fleet were likely to be unsustainable. The Spartans, on the other hand, were in a much better financial position than before because of the Persian subsidies to their fleet.

In 406 the Athenians managed to survive a minor defeat at **Notium**, which led to the final exile of Alcibiades, and then won a major victory at **Arginusae**. In 405, though, the Peloponnesian naval buildup inevitably paid off. Under the leadership of **Lysander**, the Spartans decimated the Athenian fleet at **Aegospotami**, capturing 160 of their 180 triremes. Lysander followed this up by interdicting the Athenian grain supplies from the Black Sea and clearing Athenian **cleruchs**, settlers, and garrisons from their locations in the Aegean. These were not imprisoned or killed, but rather expelled and driven toward Athens—swelling the population there and compounding the shortage of food.

Athens was besieged from land and sea and starved into surrender in April 404. The population was spared (despite the wishes of Thebes and Corinth), but Athens had to pull down large sections of its Long Walls, dismantle the Piraeus, surrender all except 12 triremes, restore all exiles, and relinquish all her overseas possessions.

This marked the end of Athens as a power to rival Sparta, although it did attempt to resurrect its position in the fourth century with the creation of the Second Athenian Confederacy. Sparta was left in a dominant position for some years until challenged by its ex-ally, Thebes, which had been increasingly out of step with Sparta in the final years of the war.

PELTAST (*PELTASTES*, pl. *PELTASTAI*). Light infantry of a type originating in **Thrace**. Peltasts were equipped with a **shield** (*pelte*), usually of wicker and crescent shaped, although shields of other shapes and covered with animal skin are also attested; their offensive **weapons** included a sword, **javelin**, or thrusting spear (plate 2b). Those equipped with a thrusting spear were capable of fighting in **formation** and may have influenced the development of the **Macedonian phalanx**. Attested on vase paintings from the mid-sixth century onward, the peltast first appears in literary sources in **Thucydides'** account of the **Peloponnesian War** of 431-404. Widely used from the fifth century onward, the peltast really came into its own in the more fluid warfare of the fourth century. *See also* CHABRIAS; IPHICRATES; MACEDONIAN ARMY.

PENTECONTOR (*PENTEKONTOROS* also **PENTECONTER/** *PENTEKONTEROS*). The pentecontor, or "fifty-oared" ship, had a long history in Greece, dating from **Homeric** times until the **trireme** replaced it as the main **warship** type in the second half of the sixth century. There were two types of pentecontor, the single-level and the two-level (or bireme). To compensate for its length, the single-level pentecontor probably had a broader beam and deeper hull. This meant it could carry a large cargo or good numbers of people, although it was presumably not particularly maneuverable. This difficulty could be overcome (at the sacrifice of carrying capacity) by adding a second level of oars and reducing the length of the ship. This probably explains the development of the two-level pentecontor, which, with a ram added on the bow, was a very maneuverable and useful warship. Attested from the eighth century, the two-level pentecontor was the dominant warcraft until superseded by the trireme. *See also* NAVAL WARFARE; SHIPS.

PERDICCAS (PERDIKKAS). Name used in the **Macedonian** aristocracy and royal family. Two kings were called Perdiccas, including the first king of Macedon (about whom very little is known). The most militarily important men with this name are:

(1) Perdiccas II. King of **Macedon**, c. 450-413. Much of the information about his rule comes from **Thucydides** and focuses on the **Peloponnesian War** of 431-404. Perdiccas was king of a fairly weak Macedon, subject to rebellion and to external pressure from **Thrace**, **Athens**, and the Greek colonies on its borders. Under Perdiccas, Macedon survived these pressures through a sound mixture of military and diplomatic skill. A feature of his reign is the frequent switching of alliances to meet the needs of the moment.

Perdiccas was allied with Athens until 437/6, when he (rightly) judged its foundation of **Amphipolis** to be a threat to Macedon. His breach with the Athenians caused them to support his exiled brother Philip and Derdas of Elimia (in Upper Macedon) against him. In turn, Perdiccas helped foment the rebellion of **Potidaea** and other Chalcidian cities in 432. According to Thucydides, it was Perdiccas who persuaded the Chalcidians to abandon their cities and settle at **Olynthus**. The success of his intrigues bore immediate fruit with the diversion of an Athenian force of 30 **triremes** and 1,000 **hoplites** from its intended target, Macedon, to deal with the trouble in the Chalcidice (Thuc. 1.56-59).

With the help of Sitalces, the Odrysian ruler of Thrace, Perdiccas switched back into alliance with Athens in 431, gaining territory and probably also the allegiance of Derdas. However, in 429 Sitalces invaded Macedon, apparently because Perdiccas had not fulfilled his side of the bargain agreed in 431. With a force claimed to have been an incredible 150,000 strong (including about 50,000 **cavalry**), Sitalces rapidly overran Macedon (and subsequently the Chalcidice) but Perdiccas adopted the strategy of mobile defense. The population retired to various strongholds and the Thracians were constantly harassed by the high-quality Macedonian cavalry. Running short of food, suffering from the cold, and unable to bring the Macedonians to a decisive engagement, Sitalces retired. Perdiccas is also supposed to have bribed Sitalces' son, Seuthes, to persuade his father to withdraw by promising his daughter to him as a bride (Thuc. 2.95-101).

According to Thucydides, between 429 and his death around 413, Perdiccas switched sides at least four more times. He was allied with Sparta in 425/4, helping persuade the Spartans to send the army led by

Brasidas north, but joined Athens again in 422, on the failure of his joint expedition with Brasidas into Lyncus. To counter Athenian influence in the north, Perdiccas allied with Sparta and **Argos** in 417—leading to Athenian-supported **raids** on his territory by Macedonian exiles. He was again allied with Athens in 414/3, when he supported an Athenian attempt to recapture Amphipolis. Ultimately, this frequent switching of allegiance did not gain much territory for Macedon, but it did preserve the kingdom and blunt external pressure at a time when it was not strong enough to resist on a purely military level.

(2) Perdiccas (d. 321), son of Orontes. A Macedonian aristocrat and military commander under **Alexander the Great** who was initially the most powerful of the *diadochoi*. He began his career under Alexander as a battalion commander in the **phalanx**, progressing from there to a **cavalry** contingent commander, one of Alexander's bodyguards, and command of all the cavalry (Arr. *Anab.* 1.6.8, 5.12.2, 6.28.4). He had a distinguished career, acting as commander at **Tyre** for a while, being **wounded** at **Gaugamela**, and being entrusted with independent command in India (Curt. Ruf. 4.3.1, 4.16.32, 9.1.9; Arr. *Anab.* 6.5.4-6, 6.9.1, 6.15.1). On Alexander's death he was proposed as king, provoking a serious mutiny by the infantry and leading to a compromise solution. Perdiccas was made *epimeletes* (guardian or regent) of the monarchy (i.e., of Philip Arrhidaeus and Alexander's as-yet-unborn heir by Roxane) and given command of the core of the army in Asia (Diod. Sic. 18.2-3; Curt. Ruf. 10.6-7).

Perdiccas seems to have played the role of loyal maintainer of Alexander's tradition and supporter of the Macedonian monarchy, while attempting to establish himself and his family as the most powerful of the successors. In 322 he successfully brought Cappadocia and Pisidia under Macedonian control but then accused **Antigonus I (Monophthalmus)** and **Ptolemy I (Soter)** of treason. Ptolemy stood trial and was acquitted, while Antigonus, also facing a financial and administrative audit, fled to Greece, where he persuaded **Antipater** and **Craterus** (perhaps correctly) that Perdiccas' ambition was to control the whole of Alexander's empire and that they ought to attack him immediately (Diod. Sic. 18.16.1-3, 18.23.3-4, 18.25.3-6).

Perdiccas sent **Eumenes of Cardia** to the Hellespont to oppose any crossing and decided to deal with Ptolemy personally by invading **Egypt**. After a disastrous opposed crossing of the Nile in which 2,000 soldiers were killed—many drowned or taken by crocodiles—his troops mutinied and killed him. In this he was rather unlucky. After an initial failure to prevent Craterus from crossing into Asia, Eumenes redeemed himself by defeating and killing Craterus in their first engagement. The news of Craterus' defeat, which would have considerably strengthened Perdiccas' grip on power (and would probably have prevented the mutiny) arrived two day's after Perdiccas' death (Diod. Sic. 18.29-37, although with two quite different dates for the arrival of the news). Perdiccas was an able commander (although not apparently particularly popular with his troops) who had the chance to become the dominant successor to Alexander, losing it through a combination of poor personal skills, some political mistakes, and bad luck.

PERICLES (PERIKLES; c. 495-429). An **Athenian** statesman and *strategos* (general) who was extremely influential during the third quarter of the fifth century. Born into a wealthy family, Pericles chose the democratic side of Athenian politics. In the late 460s he was involved in the prosecution of **Cimon** and attacks on the power of the conservative council of the Areopagus.

Militarily, Pericles first comes to notice as *strategos* around 455, campaigning in the Corinthian Gulf during the First **Peloponnesian War**. In the 440s and 430s he was heavily involved in strengthening the Athenian empire and its grip on the Aegean. In the 440s he led expeditions which restored **Delphi** to **Phocian** control and recovered the island of **Euboea** after its revolt against Athens. In 440-439, after a nine-month siege, Pericles brought Samos back into the Athenian empire after a revolt. During this period he also established **cleruchies**, a colony at Thurii in the west, and in 437/6 a colony at **Amphipolis** in the Chalcidice.

From 443 onward, Pericles was continually reelected *strategos* and was the most influential Athenian politician. **Thucydides** sums up his position during this period as follows: "in what was in name a democracy, the rule was in fact in the hands of the first man" (2.65.9). Pericles played a major role in the outbreak of the Peloponnesian War (2) in 431. Although by refusing arbitration **Sparta** was in breach of the **Thirty Years' Peace**, Athenian activity under Pericles' influence had placed Sparta in an untenable position. The alliance with **Corcyra** in 433 and Athens' activities to improve security in the Chalcidice—especially the measures in **Potidaea**—were major blows to **Corinth**'s position and prestige. The **Megarian** decree of 432 targeted another of Sparta's allies on the Isthmus of Corinth. The growing power of the Athenian empire and the pressure on Spartan allies probably meant that the Spartans saw only two options: go to war to demonstrate that the **Peloponnesian League** under Spartan leadership could, and would, look after its members; or abdicate their position as head of an alliance system. The Spartans chose to go to war.

The Athenian strategy for fighting this war was devised by Pericles. It was an extremely well-designed strategy that capitalized on Athenian strengths and minimized Athenian weaknesses. Pericles' "Island Policy" has been criticized for being too passive—a policy that would avoid losing the war but which could not win it. However, Athens did not need decisive victories to win. Given the pressure on Sparta's allies before the war, Athens could do well simply by showing that Sparta was incapable of really harming Athens and that Athens could raid Peloponnesian territory at will. If Sparta ended the war without a decisive victory, even after several years, Athens would have further strengthened her position in Greece relative to Sparta.

Pericles fell temporarily out of favor in 430 when Athens was reeling from the effects of the plague and was tried and fined. Although reelected *strategos* for 429/8, he died in 429 from the effects of the plague.

PERSEUS (c. 213-166). Eldest son of **Philip V** of **Macedon**; king of Macedon, 179-167. Starting about 183 Perseus was involved in a political struggle with his younger brother, Demetrius, who took a pro-Roman line and was personally very popular in **Rome**. According to the fairly hostile literary tradition, in 180 Perseus persuaded Philip to permit Demetrius' execution, and when Philip died in 179 Perseus succeeded to the throne. He immediately renewed his father's treaty with Rome but also set about restoring some of Macedon's lost power. This, and his role in his brother's death, ensured that the Romans were always suspicious of him.

In 178 Perseus married Laodice, daughter of Seleucus IV (Philopator) of **Syria**. He deliberately attempted to extend his popularity in Greece by supporting the poorer sections of society, something which laid him open to charges of being a social revolutionary. In 176 a Roman commission investigated his activities in Dardania and issued him a warning to ensure he kept to the terms of his alliance with Rome. Roman suspicions were further inflamed by the anti-Perseus propaganda of **Eumenes II (Soter) of Pergamum**. In the summer of 172 the Romans made a series of charges against Perseus, including that he had attempted to kill Eumenes and planned to poison Roman senators. When Perseus refused to submit to the Roman demands, they declared war. Initially successful in this, the Third **Macedonian War**, Perseus was heavily defeated at **Pydna** (1) in 168, captured in Samothrace, and deposed by Rome. He was displayed in Aemilius Paullus' triumph and died at Alba Fucens in 166.

PERSIA. Ancient Persia initially occupied the modern Iranian plateau. The Achaemenid dynasty, which initiated Persian expansion, was originally based in Anshan and Parsa. Under Cyrus the Great and his successors the Persians expanded to conquer Media (550), Lydia (546), Babylon (539-538), and **Egypt** (526-525). Crossing over to Europe, **Darius I** annexed **Thrace** between 512 and 510.

By 500 the Persians controlled the region from the coast of **Ionia** (in modern-day Turkey) to India and from the Black Sea to Egypt and the Red Sea. The empire was an absolute monarchy (although the king sometimes consulted the Persian aristocracy over decisions) with a sound internal organization. Persia was divided into administrative areas known as satrapies (each governed by a satrap who was responsible to the king), which were connected to each other by a fairly well-developed network of roads and an efficient courier system. Literally translated, *satrap* means "protector of the realm." Satrapies were organized right from the start by Cyrus, but it was Darius I's achievement to fix their borders and levy a set amount of tribute from them. The powers of the satrap were very wide—he was governor of a vast territory, but was appointed by and directly responsible to the king. He was ultimately responsible for the defense of the satrapy and conducted the mass levies when mobilization was ordered.

The Persians and their near relations the Medes were the ruling group within the empire, which held many different racial groups in varying degrees of subjugation. The royal family provided a very large proportion of higher officials and army commanders, and the security this provided was considerable. In addition, security was enhanced by the new Achaemenid approach to the administration of conquered areas. They tried to retain as much as they reasonably could of the customary administrative or political systems they found in each new area, and even extended their tolerant approach to native religions. However, in Greek Asia Minor the Persians had to tread particularly warily. The Greeks were less accustomed to domination by a large foreign power and were quick to resent any checks on their liberty. Lydian control had been lax and the Persians encouraged the establishment of tyrannies in the cities to ensure the ruling families would be dependent on Persia for their positions.

By the end of the sixth century the conquest of Thrace suggests that the main direction of Persian expansion was toward Greece. At this stage, Persia presumably saw Greece as a minor frontier state of no particular interest, but this changed when the **Ionian** Greek cities, led by **Miletus**, revolted against Persian rule in 499. Two Greek states, **Athens** and **Eretria** (on the island of **Euboea**), sent ships to aid their Ionian kinsmen during this **Ionian Revolt** and participated in the burning of the king's summer palace at Sardis. The Persians probably saw the actions of Athens and Eretria as unwarranted interference in the internal affairs of the Persian empire, and when the revolt was ended in 493, the Persian king, Darius I, decided to attack Greece.

This led to the **Persian Wars**, which included the **Marathon** campaign of 490 and the Second Persian War, a much larger conflict, which ended in Persian failure in 479 after defeats at **Plataea** (1) and **Mycale** and the destruction of most of their expeditionary force and navy. For much of the rest of the fifth century Persia was on the defensive against an aggressive Athens, the dominant naval power in Greece. At the head of the **Delian League**, Athens encouraged the defection of the Greek parts of the Persian empire and provided protection to anyone who revolted. The dominance of Athens and her league was confirmed in 467/6 with the battle of **Eurymedon River**, a major setback to Persia's attempts to reconstitute her naval power. Some successes, including the suppression of the Egyptian revolt and the destruction of the Delian League expeditionary force there in 454, led to the Peace of **Callias**, which established the relative Athenian and Persian spheres of influence.

Persia never gave up hope of recovering the Greek areas that had revolted and made considerable progress in this regard by assisting **Sparta** in defeating Athens in the **Peloponnesian War** (2) of 431-404. The activities of the two key satraps in the area, **Pharnabazus** and **Tissaphernes**, went a long way toward restoring Persian influence. However, these successes were based on diplomacy and intrigue rather than on solid military power, and the march of the **Ten Thousand** and the subsequent Spartan campaigns (especially under **Agesilaus II**) in Asia Minor in the early fourth century demonstrated the continued Persian military weakness. As long as the Greeks were disunited, Persia could play them off against each other and distract them from the Greeks in Asia Minor. They did so by a combination of bribery and diplomacy (including the use of the **Common Peace** treaties) and the threat of military force. This proved a successful policy for most of the fourth century until **Philip II** of **Macedon** forcibly united Greece. **Alexander the Great**'s rapid conquest of the Persian empire ended the Achaemenid dynasty in 330 and conclusively demonstrated Greek military superiority.

The Persian empire wielded vast resources in both in money and men, and the Medes and the Persians were good fighters. The empire possessed a standing army (drawn predominantly from the Medes and Persians), with the 10,000 Immortals at its center (they were infantry and called the Immortals because when one died he was replaced immediately), and a force of **cavalry**, perhaps of similar size. These were supplemented by mass levies, organized by the satraps, and by **mercenaries**. The commanders of the contingents were also Medes and Persians. The sheer size and racial (and linguistic) diversity of the fully mobilized armies must have made command and control very difficult and posed major logistics problems—exemplified in the Persian defeat in the Second Persian War.

Although the Greek view of the inferiority of **bow**-using Persians compared to the spear-wielding Greeks is oversimplified, it has some basis in truth. The flower of the Persian army was the cavalry and their weapon of choice was the bow. The bulk of the infantry was a mass levy of subject peoples and of indifferent quality; it was simply unable to withstand the Greek **hoplite** or Macedonian **phalanx**. From the late fifth and early fourth century the Persians attempted to make up for their deficiency in heavy infantry by hiring Greek **mercenary** hoplites.

The navy also had a standing component, provided by the Phoenicians and probably also the Egyptians; the marines were Medes and Persians. In times of war, **ships** would be mobilized from other subject peoples, including the Ionian Greeks. Like the army, the commanders of the native naval contingents were Medes and Persians.

PERSIAN WARS. Two closely connected wars fought in Greece and **Ionia** in the early fifth century between **Persia** and various Greek city-states. **Herodotus** is the main ancient source for both these wars.

(1) First Persian War (490). In the late sixth century, Persia had been expanding its control westward to include **Thrace**, but support from **Athens** and **Eretria** to the Greek rebels during the **Ionian Revolt** focused Persian attention on mainland Greece. Incensed by the Greek burning of Sardis, his summer capital, and wishing to remove the possibility of future mainland Greek support to their Ionian compatriots, **Darius I** decided to invade.

In 492 **Mardonius** was appointed to command and ordered to restore Persian control in Thrace and the Chalcidice. The local Athenian ruler, **Miltiades**, had already fled the Chersonese the year before and Mardonius quickly secured Thasos. However, he lost his fleet in a storm off Mount Athos and at the same time suffered a defeat at the hands of the Brygi, a Thracian tribe. By 491, nevertheless, Mardonius had succeeded in establishing Persian control from the Chersonese up to, and including, **Macedon**.

Darius ordered the preparation of a fleet and a seaborne expeditionary force under **Datis** and **Artaphernes** and demanded from the Greek islands and mainland states earth and water—tokens of submission. Many islands, including Aegina, complied, as did some mainland states. The

Athenians, however, executed the Persian envoys sent to them and the Spartans hurled the envoys sent to them down a well, telling them "to take earth and water from there for the king" (Hdt. 7.133). This was seen by the Greeks as patriotic resistance but by the Persians (with considerable justification) as an unforgivable breach of customary **Laws of War** concerning the sanctity of heralds.

Datis and Artaphernes' seaborne force of some 600 warships and transports crossed the Aegean, subduing the islands and levying troops as it went. The Persian force landed on **Euboea**, forcing Carystus to submit, and then besieged Eretria. After a week-long siege, the city was betrayed from within and captured. The Persian force then made the very short crossing to **Marathon** in Attica, a plain which **Hippias**, the exiled Athenian tyrant, advised was suitable for the Persian **cavalry**. The Persian expeditionary force was defeated in battle there with the loss of 6,400 men and retired to their ships. A final attempt to take Athens by quickly sailing around the promontory of Sunium was thwarted when the Greek army conducted a forced march and arrived at Athens first. The Persian fleet sailed home and Athens' reputation was considerably enhanced by its resistance to the Persians. *See also* CLEOMENES I; MEDIZE.

(2) Second Persian War (480-479). Despite the defeat at **Marathon**, **Darius I** decided to attack Greece again. He began preparations for a full-scale land and sea invasion, but this was interrupted by a revolt in **Egypt** in 487 and his death in 486. His son, **Xerxes I**, took up his father's project after the reduction of Egypt in 485. From 483 to 481, a canal was dug across the Athos Peninsula to avoid a repeat of the 492 disaster to **Mardonius'** fleet, and preparations were made for raising a huge army and fleet.

Greek counterpreparations involved creating the **Hellenic League** (1) to resist the Persians, sending spies to **Persia** to gather **intelligence**, and seeking help from **Syracuse** and Crete. In 481 the Hellenic League comprised 31 states, which had agreed to suspend any hostilities among themselves, to accept Spartan command by land and sea, and to punish any Greek state that **medized**. However, support was refused by Syracuse (which was facing a **Carthaginian** attack) and Crete. Many small Greek states—and several important ones, including **Argos**—remained neutral. Apart from the magnitude of the Persian threat, the generally negative attitude of the **Delphic** oracle helped influence some states not to resist.

According to **Herodotus**, the Persian invasion force, commanded by Xerxes himself, was 5 million strong. However, at most it may have been 500,000 strong, including camp followers. The fleet of 1,207 war-

ships and numerous transports operated in parallel with the army, meaning that a purely Greek land defense could in theory be outflanked by sea. However, there were considerable practical difficulties with executing this type of operation, not least that it was rather outside the experience range of fifth-century Persian generals. Landing on an unknown coast and risking being defeated in detail would probably have made the option quite unattractive. When the invasion began in 480, the Greeks met the threat with a joint land and sea response. Their first plan was to deploy an army by sea to defend the pass of Tempe into **Thessaly**, but on arrival they discovered that there were other passes and they had insufficient troops to cover them all. The army consequently withdrew and Thessaly deserted to the Persians.

The first real attempt to stop the Persian march south occurred at **Thermopylae** (1) and **Artemisium**. A 7,000-strong Greek army occupied the pass at Thermopylae, while a fleet, eventually totaling 324 **triremes** and a few **pentecontors**, took station off Artemisium. The Greeks, either reluctant to move so far north or underestimating the Persians (perhaps as a result of **Athens'** victory at Marathon), sent too small an army. It was outflanked and partly destroyed, forcing the withdrawal of the fleet, which up until then had held its own against the Persians. **Boeotia** was left undefended and promptly medized (except for **Plataea** and Thespiae). However, the engagements at Thermopylae and Artemisium did slow down the Persian advance and the Persian fleet was also weakened by considerable losses in storms and action against the Greek fleet.

With the loss of Boeotia, Attica was open to invasion. The Athenians abandoned their city (apart from a small force left on the Acropolis) and joined the Greek fleet at **Salamis**, just off the coast of Attica. The Persians sacked Athens shortly afterward. At this point Greek opinion was divided. The Peloponnesians were putting their efforts into building a fortified wall across the Isthmus of **Corinth** and their naval contingents were in favor of withdrawing there. The Athenians, however, supported by other states north of the wall, wanted a battle at Salamis, where the narrow channel would help neutralize the Persian numerical superiority—as the narrow pass at Thermopylae had done on land. Athenian counsel prevailed, largely due to **Themistocles**, who sent a message to Xerxes prompting him to cut off the Greek withdrawal route. The Greeks now had no choice but to fight, and in the resulting battle the Persian fleet was decisively defeated. Fearing revolts back in the empire, Xerxes withdrew his navy and marched home with a large part of his army. The Greek fleet followed up as far as Andros and then returned to the mainland.

The war was not over, though, as Xerxes left Mardonius in Greece with an army of perhaps as many as 300,000 men. Wintering in Boeotia, this force was still a formidable threat to Greece. In 479 Mardonius invaded Attica again and, when the Athenians refused to join him, again sacked Athens. Meanwhile, the Peloponnesians remained secure behind their wall on the isthmus. It was only when the Athenians, who thought they were bearing an unreasonable share of the hardship, threatened to take Mardonius' proposals seriously that a Peloponnesian force moved north. Linking with the Athenians, this force marched into Boeotia and decisively defeated Mardonius at Plataea. Around the same time a Greek seaborne force under **Leotychides II** destroyed the main Persian fleet and an army at **Mycale**, leading to the detachment of the islands and parts of **Ionia** from Persia. This ended the war and any immediate threat from Persia.

The aftermath of the war saw the Hellenic League carry hostilities to Persian possessions in the Chalcidice, Chersonese, and the Hellespont. This campaign resulted in the transfer of naval command from Sparta to Athens. Although Sparta retained its position as the dominant land power in Greece, Athens' reputation was greatly enhanced and it emerged as the dominant Greek naval power. *See also* DELIAN LEAGUE.

PEZHETAIROI. *See* FOOT COMPANIONS

PHALANX. The term *phalanx* is properly applied to a close **formation** of **hoplites** or of *sarissa*-armed **Macedonian** infantry.

(1) Hoplite Phalanx. This **formation** consisted of ranks of **hoplites**, deployed so that the left-hand half of each man's **shield** protected the right-hand side of the man on his left. In this way, every individual was fully protected, except the man on the extreme right-hand end. This explains why maintaining cohesion was very important and why the shield was regarded as the key item of equipment. The vulnerability of the man on the right-hand end also explains a phenomenon identified in antiquity of phalanxes extending to the right. As the man in this position edged to the right so no enemy faced him, his neighbors also edged right to maintain their protection (Thuc. 5.71). The hoplite phalanx performed best on level ground and had trouble maintaining formation over broken or steep **terrain**.

The width of the phalanx was determined by the terrain and the numbers available, and the depth must also have varied, although it was traditionally 8-16 ranks deep. However, under **Epaminondas** and **Pelopidas**, the **Theban** phalanx was often very deep, with files of up to 64 men (there had been Theban precedents for this—they deployed 50 deep at **Delium** in 424). This gave considerable weight to an attack, especially downhill. It also reduced the frontage and hence the initial **casualties** and was very unpopular with allied contingents, which often felt the Thebans were not pulling their weight—as seems to have happened at **Nemea**. The flanks and rear of a phalanx were vulnerable to attack by **cavalry** or *psiloi* (light troops) and were often protected either by being placed against a river or hills, or by deploying cavalry and *psiloi* to protect them.

(2) Macedonian Phalanx. Technically, the six brigades of **Foot Companions** armed with the *sarissa* and a small **shield**, but ancient sources also generally use "phalanx" to refer to the entire infantry contingent of the **Macedonian** battle line. However, in its more technically correct sense, the phalanx consisted of infantry carrying the *sarissa*. **Polybius** (18.29-33) provides a very useful discussion of the phalanx in his day. The length of the *sarissa* meant that the weapons of the first five ranks protruded ahead of the front rank. The rear ranks carried their *sarissai* upright, at a slant, which provided some protection from missile weapons. Although the depth of the phalanx probably varied, it was traditionally 16 ranks deep. It had a considerable advantage in the reach of its **weapons** compared to the **hoplite** phalanx and was particularly well suited to fixing bodies of enemy infantry in place while the **cavalry** launched the decisive blow. Its main weakness was its need for generally flat and level ground and its relative slowness to maneuver (the fact that at least the first five ranks were hemmed in by *sarissa* shafts on either side of them made turning very difficult). Although generally irresistible against hoplites (but even **Philip II** suffered two early defeats against **Onomarchus**) and Eastern opponents, it proved no match for the more flexible **Roman** formations at **Cynoscephalae** (2) and **Pydna** (1 and 2). Under the *diadochoi* the phalanx retained considerable political power from its Macedonian character—from it (and the cavalry) derived Macedonian political and royal legitimacy.

PHARNABAZUS (PHARNABAZOS). Persian satrap (governor) of Dascylium from around 413 to 370. Pharnabazus has received a good press from **Xenophon**, whose account of his meeting with **Agesilaus II** in 395 portrays him as Persian aristocrat with all the attributes of a Greek gentleman, including a good-looking son (Xen. *Hell.* 4.1.29-40). During the **Peloponnesian War** of 431-404, Pharnabazus was instrumental in getting a **Spartan** naval presence in Asia Minor, and especially the Hellespont, and supported the Spartans against **Athens** at **Abydus** (2), **Cyzicus**, and **Chalcedon**, but encouraged (ultimately unsuccessful) Athenian negotiations with **Darius II** (Thuc. 8.6, 38, 62,

80, 99; Xen. *Hell.* 1.1.6, 1.2.16, 1.3.5-7). He was involved in the death
of **Alcibiades** in 404 and bore the brunt of Spartan campaigns in Asia
Minor, partly because of the intriguing of **Tissaphernes** (Xen. *Hell.*
3.1.9, 4.1.1).

Pharnabazus seems to have been a generally able commander. He
had two major successes against the Greeks by attacking **foragers** instead of the main force—a tactic well suited to the relative strengths and
composition of his and the Greek armies involved. The first was against
the **Ten Thousand** in 401/0, the second in 395 against Agesilaus (Xen.
Anab. 6.4.24; *Hell.* 4.1.17-19). Another of the features of his command
was an interest in the revival of Persian naval power. He was supreme
commander at **Cnidus** with **Conon** in 394 and in later operations (Xen.
Hell. 4.3.11-12, 4.8.1-9). Pharnabazus later married the king's daughter
and was given command of expeditions to recover **Egypt** but was unsuccessful in both 385-383 and 374—despite the presence of **Iphicrates**
on the second occasion (Xen. *Hell.* 5.1.28; Diod. Sic. 15.29.3-4, 15.41-
43). He died around 370.

PHEIDIPPIDES (also PHILIPPIDES). An **Athenian** herald and long-
distance runner sent to **Sparta** to request help when the **Persians** landed
at **Marathon** in 490. He is reputed to have covered the 150-mile (240-
kilometer) trip in two days.

PHEIDON (fl. 670). A king of **Argos** who led a revival of Argive power
shortly after 675. Around 669 he inflicted a heavy defeat on **Sparta** at
Hysiae and in 668 he seized **Olympia** and presided over the Olympic
games. At this time the monarchy at Argos had been reduced to figure-
head status but Pheidon reversed this, which is why he is often referred
to as a tyrant or king-tyrant. His military success against Sparta and his
associated political success may result from the introduction of **hoplite**
tactics at Argos—ahead of his Peloponnesian rivals.

PHILIP II (c. 383/2-336). King of **Macedon**, 359-336. Philip was a remarkable monarch, whose ambition and dedication **Demosthenes** (2)
sums up in the statement: "For the sake of empire and power [he] had
his eye knocked out, his collar-bone broken, his arm and leg injured"
(18 [*On the Crown*], 67).

Philip was taken to **Thebes** as a hostage by **Pelopidas** in 368 or
367, returning circa 365. He came to power in 359 when his brother
Perdiccas III was killed in battle against the Illyrian king **Bardylis I**. On
his accession Philip was under threat from four directions: from the
northeast and east by rival claimants to the throne—Pausanias, sup-

ported by the Thracian king Berisades, and **Argaeus**, supported by **Athens**; from the north by **Paeonians**; and from the northwest by Bardylis, who occupied parts of the kingdom. Philip's actions to deal with these threats are indicative of how he would turn Macedon into the most powerful state in Greece. He neutralized Berisades and the Paeonians by bribery and diplomacy. Argaeus was defeated and eliminated when he landed at Methone, and a treaty was then made with Athens. At the same time, Philip was building up his army. In 358, he took advantage of the death of the Paeonian king to invade Paeonia, decisively defeating the new king and imposing an alliance on him. Philip then defeated Bardylis at an unknown location, probably in Lyncestis, breaking the Illyrian square with his infantry and using his **cavalry** to strike the decisive blow. This reunited Macedon and his success was cemented by a series of strategic marriages, the third (c. 357) to Olympias of Molossia, the future mother of **Alexander the Great**.

This mixture of force and diplomacy, based on a strong army, was typical of Philip's methods as king. Using these, he gradually extended Macedonian control over the Chalcidice and **Thrace**, securing the former with the capture and destruction of **Olynthus** in 348 and the latter with the final defeat of Cersobleptes in 342. In 356 he again defeated the Illyrians and Paeonians, reducing the latter to vassal status. From 358 Philip regularly campaigned in **Thessaly**, reuniting it under his control after his defeat of **Phocis** in 352 (the most likely date for his election as *tagus*, although 344-342 is also possible).

His first involvement in Greece south of Thessaly was to join in the Third **Sacred War** against Phocis. After initial setbacks against **Onomarchus** in 353, Philip defeated him at the battle of the **Crocus Plain** in 352 and forced the Phocians to surrender in 346. Philip increasingly strengthened his position, ending the Fourth **Sacred War** of 340-338 and opening the route south to Athens. The anti-Philip coalition led by Athens and Thebes was decisively defeated at **Chaeronea** (1) in 338. The following year Philip united most of Greece under his leadership with the formation of the **Hellenic League** (2). Philip was assassinated in 336, at the age of 46 and with his preparations for an invasion of Persia well under way.

Philip II's greatest legacy, apart from a secure and powerful Macedon, was the finely honed army that made Alexander the Great's Asian conquests possible. During his detention in Thebes, Philip was heavily influenced by the great Theban generals **Epaminondas** and Pelopidas. This is clearly reflected in his oblique deployment of the **phalanx** of foot and his cavalry charges against the flanks and rear of the enemy infantry. Philip's main innovation in **weaponry** and tactics was the introduction of the *sarissa*, which gave his infantry some advantage over its opponents. He is also credited with the introduction of the wedge-shaped formation into the Macedonian cavalry arm.

However, the major reasons for the success of the **Macedonian army** under Philip was his transformation of it into a well-trained professional army and his inspired leadership. Apart from the Spartans, Philip II was the first in the Greek world to develop a professional army. He made his soldiers far fitter than their opponents through rigorous **training** and **discipline** and inspired his troops' admiration and devotion by participating in the thick of the fighting. A notable feature of his generalship was the speed with which he could move his armies. This often allowed him to take his opponents by surprise before they had time to prepare. Demosthenes remarked: "[Philip] makes no distinction between winter and summer and there is no time of year he chooses to cease from his activities" (9 [*Third Philippic*], 50). *See also* SIEGE WARFARE.

PHILIP V (238/7-179). Son of Demetrius II; king of **Macedon**, 221-179. Philip first made his name in the Second **Social War** of 220-217, fighting successfully against the **Aetolian League**, **Elis**, and **Sparta**. Philip brought the war to a successful conclusion with vigorous campaigns in 218 and 217, during which he **ravaged** parts of **Aetolia** and Sparta, prevented a revolt by three of his generals, and secured **Thessaly**.

In 217-216, he came into conflict with **Rome** in Illyria, and following Rome's massive defeat in the battle of Cannae in 216, Philip allied with **Carthage**, promising to send troops to aid Hannibal in southern **Italy**. However, concerted Roman naval activity in the Adriatic and military action against Philip by Rome's Greek allies, the Aetolian League, prevented this. This First **Macedonian War** involved no contact between Roman and Macedonian ground forces and had petered out by 205. The Romans, though, never forgave Philip for allying with Hannibal in their darkest hour and from this time on regarded him with suspicion. This was an important factor in the outbreak of the Second Macedonian War.

Philip sparked this second war through expansionary naval operations in the Aegean in 202-201 aimed, in alliance with **Antiochus III (the Great)** of **Syria**, against Ptolemaic **Egypt**. His indiscriminate and ruthless use of naval power against shipping in the area alienated **Rhodes** and **Attalus I (Soter)** of Pergamum, who appealed to Rome for help. Philip rejected a Roman ultimatum and was initially fairly successful in keeping the Romans at bay. However, in 197 he was decisively defeated at the battle of **Cynoscephalae (2)** and forced to make peace. Philip's power was temporarily curtailed by having to pay an indemnity, surrender his fleet, and withdraw his garrisons from Greece—which Rome declared free.

Philip largely recovered his position by scrupulous adherence to the treaty and by providing support to Rome against **Nabis** of Sparta, Antiochus, and the Aetolian League. As a reward for this, in 190 Rome eased the financial provisions of the peace treaty and restored some of the Macedonian hostages, including Philip's son Demetrius. From this point onward, Philip focused on reorganizing Macedon and restoring its financial and military power base. His success in this aroused complaints from other Greek states and Roman interference—in 186/5 three commissioners were sent to examine the situation. As a result of this visit and the Roman rulings, Philip seems to have decided that another war was inevitable and set about expanding his frontiers by campaigns in **Thrace** and the Balkans in 184, 183, and 181. These actions, and the execution in 180 of Demetrius, who was popular with the Romans, caused further suspicion. When Philip died in 179 he left his son **Perseus** with a strong kingdom but a legacy of considerable hostility from his neighbors and from Rome.

PHILOPOEMEN (PHILOPOIMEN; c. 253-182). General and political leader of the **Achaean League**; from Megalopolis. As *strategos* of the Achaean League in 208/7, 206/5, 204/3, 201/0, 192/1, 189/8, and 183/2. Philopoemen continued **Aratus'** policy of trying to unite the Peloponnese under **Achaea**. Of good birth and education, Philopoemen first came to attention at the battle of **Sellasia** in 222 when he led an unauthorized **cavalry** charge that saved the day. He spent the next 10 years soldiering in Crete as a **mercenary** leader. Philopoemen became **hipparch** (cavalry commander) of the Achaean League in 210 and by a series of intelligent reforms and intensive **training** transformed what had been an indifferent arm into a very effective force. He killed the commander of the Elean cavalry, Damophantus, in single combat at the Larissus River. As general in 208/7 he reformed the army (according to Plut., *Philopoemen,* 9.1-2, introducing the **Macedonian**-style **phalanx**) and, after eight months intensive training, attacked Machanidas of **Sparta**, defeating and personally killing him at the battle of **Mantinea** (3) in 207. He repelled an attack on Messene by **Nabis**, Machanidas' successor, in 201 and in 200 **ambushed** and destroyed a force of Nabis' mercenaries near Pellene. Losing the election for the post of *strategos* for 200/199, he briefly returned to Crete as a mercenary leader.

As general again in 193/2 Philopoemen blockaded Sparta, forcing it into the Achaean League in 192. For the next few years he controlled

Achaean policy, annexing **Messenia** and **Elis** and in 188 abolishing the Lycurgan system in Sparta. In 183, aged 70, Philopoemen was captured while leading the Achaean cavalry against a Messenian revolt and poisoned. As a young cavalryman, **Polybius** carried Philopoemen's ashes in the state funeral.

Philopoemen was the consummate soldier—demonstrated not only by such victories as Mantinea and his successful reforms but also by his quiet but effective leadership (see Polyb. 11.9-10). However, he was seen by writers in antiquity, brought up on the ideal of the amateur citizen-soldier, as a man who was a little too keen on war and who spent too much time on the martial skills (Plut., *Philopoemen,* 4.5-6).

PHOCION (PHOKION; c. 402/1-318). An **Athenian** *strategos* (general) and statesman; son of Phocus. Phocion was elected *strategos* 45 times in the middle and second half of the fourth century. Born into a wealthy family, he appears to have been a **trierarch** (and possibly a *strategos*) at the battle of **Naxos** in 376. His bravery in this battle, and his initiative in counterattacking the **Spartan** admiral, Pollis, when he was destroying the Athenian left wing, was the key to the Athenian victory. Phocion continued as a protege of **Chabrias**, although his next known command was not until 361, the probable date of an action in Asia Minor during the satrap's revolt against the **Persian** king.

Unfortunately, the state of the evidence means that little detail is known of Phocion's career down to the late 350s. In 351/0 he commanded up to 16,000 **mercenaries** for Persia quelling a revolt on **Cyprus**. In 348 he was joint commander of an Athenian expedition to **Euboea** to counter Macedonian-inspired unrest, and (temporarily) broke the anti-Athenian movement at **Tamynae**, where his handling of the battle demonstrates considerable skill.

From 348-338, Phocion played a prominent role in Athens' military resistance to **Philip II** of **Macedon**'s expansionism. In 343 he apparently prevented **Megara** being handed over to Philip. In 341/0 he captured **Eretria** and expelled the pro-Macedonian tyrant, and in 339 he was one of the *strategoi* who raised Philip's siege of **Byzantium**. Despite a **wound**, Phocion continued to campaign in the area until 338. He was not elected general for 338 and was not present at the battle of **Chaeronea** (1). However, in 322, during the **Lamian War** and aged around 80, he successfully led an Athenian home defense force against a Macedonian landing.

Following Macedon's victory in the Lamian War, Phocion helped negotiate the peace settlement with **Antipater**. Although the embassy was able to gain better terms than the unconditional surrender initially demanded, their acceptance of a Macedonian garrison in **Munychia** paved the way for Phocion's later political demise. In 318, when democracy was restored in Athens, Phocion was condemned to death and executed. His body was cast outside the borders of Attica.

Despite the lack of evidence of many aspects of Phocion's career, he appears to have been an energetic and skilled commander, both on land and sea. A pragmatic politician, he had a reputation for honesty and for acting with Athens' best interests at heart.

PHOCIS (PHOKIS). A region in central Greece, west of **Boeotia** and north of the Gulf of Corinth. In the sixth century most of the cities in the region formed a federation, although **Delphi** maintained its independence from the rest from the First **Sacred War** (c. 595-590). Phocis was frequently under pressure from its neighbors—later Boeotia, but in the sixth century mainly **Thessaly**. **Herodotus** states that the Phocians built the wall at **Thermopylae** to keep the Thessalians out and records successful resistance to two Thessalian attacks. In the first, the Phocians whitened their bodies for a **night attack** on the Thessalians, not only easily distinguishing friend from foe but also throwing the Thessalians into a major panic by their ghostly appearance. On the second occasion the Phocians buried empty jars across the Thessalian line of advance, causing major **casualties** to the Thessalian **cavalry**, whose horses broke their legs in this ancient equivalent of a minefield (Hdt 7.1765, 8.27-8).

Phocis supplied 1,000 **hoplites** to the Greek force defending Thermopylae against the **Persians** in 480 and these were allocated to guarding the Anopaea track. When **Xerxes I** made his decision to outflank the main Greek force, the Phocians were caught by surprise and under a hail of arrows withdrew to high ground and prepared to fight to the death. However, the Persians, with a keener attention to their mission than the Phocians had to theirs, simply ignored them and continued on to attack the main body of the Greek army (Hdt. 7.217-218). Despite the defeat at Thermopylae, the Phocians refused to submit and, with Thessalian help, the Persians thoroughly **ravaged** their country. Some of the Phocians continued to resist, harassing the Persians from mountain refuges. Others, however, were forced to submit and 1,000 Phocian hoplites served in the Persian army at **Plataea** in 479 (Hdt. 8.30, 32-35; 9.17-18, 31).

Following the **Persian Wars** Phocis allied with **Athens** and attacked Doris in 457. A Spartan-led **Peloponnesian League** army forced Phocis to come to terms and return the territory it had captured, although this same army was nearly defeated at **Tanagra** on the way home (Thuc. 1.107-108, 111.1). This alliance with Athens paid off for the Phocians who were left in (temporary) control of **Delphi** at the end of the Second Sacred War because of it (Thuc. 1.112.5). However, in 447 Phocis allied with **Sparta** after **Coronea** (1) and remained loyal to it for the rest of the century, although comments by several Athenian authors indicate they considered there was longstanding friendship between Athens and Phocis (Thuc. 3.95.; Xen. *Hell.* 6.3.1). Phocis was clearly allied to Sparta on the outbreak of the **Peloponnesian War** of 431-404, and it later built **ships** for the alliance to help break Athenian naval supremacy (Thuc. 2.9.2, 8.3.2).

In the early fourth century a (provoked) attack by Phocis on **Locris** helped spark the **Corinthian War** and in 395-394 Phocis provided a base for **Lysander**'s operations and troops to help Sparta at Coronea (2) (Xen. *Hell.* 3.5.3-4, 17; 4.3.15). The Spartans prevented a Theban invasion of Phocis in 374. Phocis also served as a Spartan base for operations against **Boeotia** in 371, prior to the battle of **Leuctra**, and provided troops (including **peltasts**) at the battle itself. At the same time, it was under pressure from **Jason of Pherae** and, after the Spartan defeat at Leuctra, Phocis reluctantly joined with Boeotia (Xen. *Hell.* 6.4.1-3, 21, 27; 7.5.4).

Phocis achieved its greatest power during the Third Sacred War (355-346), which **Thebes** seems to have sparked by its attempts to curb growing Phocian independence. During this war, the Phocians used the treasure at Delphi to hire large numbers of **mercenaries**, with which they defeated Thessaly and exhausted Boeotia. However, Phocis was unable to withstand **Philip II** of **Macedon** and, on losing the war, the region was rendered powerless. Its cities were broken up into villages, a large indemnity was imposed, and Phocis lost her votes in the **Amphictionic League**. From this point on, Phocis was subject to Macedonian (or from time to time **Aetolian League**) control, until freed by **Rome** after the Second **Macedonian War** and re-enrolled in the Aetolian League (Polyb. 18.46-47).

PHOEBIDAS (PHOIBIDAS; d. 378/7). A **Spartan officer** who seized the Cadmea (the acropolis of **Thebes**) in 382. Leading a Spartan force against **Olynthus**, Phoebidas camped outside Thebes, where the anti-Spartan group was dominant. According to **Xenophon** (*Hell.* 5.3.25-31), Leontiades, the leader of the pro-Spartan group in Thebes, persuaded Phoebidas to seize the citadel. **Diodorus Siculus** (15.20.2) and **Plutarch** (*Agesilaus* 23-24), though, suggest that Phoebidas already had secret orders to do so. Ismenias, the leader of the anti-Spartans, was arrested (and subsequently executed) and 300 of his supporters fled into exile. Phoebidas' action was a blatant breach of the King's Peace (**Common Peace** [1]) of 386 and he was fined by the Spartan government. However, as possession of the Cadmea gave Sparta control over Thebes and **Boeotia**, the garrison was retained—despite the criticism this provoked throughout Greece.

Agesilaus II appointed Phoebidas governor in Thespiae in 378/7, after the Thebans had expelled the garrison from the Cadmea and regained their independence. From there Phoebidas regularly and effectively **ravaged** Theban territory with Thespian **hoplites** and a force of **mercenary peltasts**. A Theban army invaded Thespiae in response to these **raids**, but Phoebidas' skillful use of his peltasts severely restricted its movements and forced it to retire in some disorder. While personally leading his peltasts in the pursuit, Phoebidas was killed and his army fled when the Theban **cavalry** turned to fight. *See also* PELOPIDAS.

PHORMIO (PHORMION; d. 428?). An Athenian *strategos* (general) active before and in the early stages of the **Peloponnesian War** of 431-404. He participated in **Pericles**' suppression of the revolt of Samos in 440-439 and in the blockade of **Potidaea** (then in revolt against **Athens**) in 432. The first year of the war proper saw him still campaigning in the Chalcidice, assisted by **Perdiccas II** of **Macedon**. In 430 he commanded a fleet of 30 **triremes** on a successful expedition to Argos in Amphilocia and then led a fleet of 20 ships based at Naupactus to prevent anyone from moving in or out of the Corinthian Gulf.

In 429, in a display of superior seamanship, Phormio defeated a **Corinthian** fleet of 47 ships, sinking an unknown number and capturing 12 (Thuc. 83-84). Shortly afterward, he was forced to engage a fleet of 77 ships, with mixed success. Nine of his triremes were captured in the initial action but the other 11 rallied, routed part of the enemy fleet, captured six ships, and recovered their own nine. That winter Phormio landed in Acarnania and used his *epibatai* (marines) and **Messenians** from Naupactus to arrange affairs in favor of pro-Athenian groups. He sailed back to Athens via Naupactus in spring 428. This is the last we hear of him and he may well have died shortly afterward; his son served as *strategos* in 428/7. Phormio was an experienced and talented naval commander and his actions in 430-429 made a major contribution to the Athenian war effort in the Corinthian Gulf area.

PHYLARCH (*PHYLARCHOS*, pl. *PHYLARCHOI*). The title (literally "tribal leader") of the commander of a military contingent drawn from a *phyle* or tribe, particularly at **Athens** where each *phyle* provided a 100-man **cavalry** squadron. The office was an elected one with the tenure of one year only (although there were no barriers to an individual holding the office repeatedly or in consecutive years).

***PHYLE* (pl. *PHYLAI*).**
(1) Tribe. At **Athens** it was an administrative division of the citizen body, originally based on kinship groups but increasingly artificial from Cleisthenes' reforms in the late sixth century, which created 10 *phylai*. They were used as the basis for the provision of both **hoplite** and **cavalry** contingents to the Athenian army. The numbers of *phylai* were increased from time to time during the Hellenistic period.
(2) At **Athens**, a 100-man tribal **cavalry** contingent or squadron. *See also* PHYLARCH.

PHYLE. A mountainous fortress in Attica, northwest of **Athens**, that controls one of the main routes from **Thebes** to Athens at the pass on Mount Parnes. Phyle was seized in 404 by Athenian democratic exiles under **Thrasybulus** and used as a base in their successful struggle to overthrow the rule of the **Thirty Tyrants**. Phyle's natural and enhanced defenses proved effective when the original 70 exiles were attacked by **cavalry** and infantry from Athens. When their numbers swelled to 700, the exiles attacked and routed the **Spartan** garrison troops and two *phylai* (squadrons) of cavalry that had been deployed against them. The exiles moved into position at night and attacked the enemy camp at dawn. About 120 of the garrison's **hoplites** and some cavalry were killed. From this point on, the Thirty Tyrants' position became increasingly precarious. **Xenophon**'s description of these actions (*Hell.* 2.4.2-7) suggest he may have been present as a member of the cavalry or talked to eyewitnesses. *See also* MUNYCHIA.

PLATAEA (PLATAIA/PLATAIAI). A city in southern **Boeotia** that was a traditional, if much smaller, rival of **Thebes**. At the suggestion of **Cleomenes I** the Plataeans allied with the Athenians circa 519. During the **Persian Wars** they served alongside **Athens** at **Marathon** in 490 and at **Artemisium** in 480. After Artemisium, the Plataeans refused to **medize** and evacuated their city, which the Persians then sacked (along with Thespiae). After the battle of Plataea in 479 (see [1] below) the city was voted special honors and declared inviolable. However, this did not prevent a surprise Theban attack on the city in 431, the first action of the **Peloponnesian War** of 431-404. The Peloponnesians besieged the city from 429 to 427 and, on its surrender, razed it and executed the remaining defenders—although most of the population had been evacuated to Athens before the siege began and half the defenders had escaped during the siege. In Athens the evacuees were given equal rights with citizens until the Spartans rebuilt their city in 382, again as a counterweight to Thebes—which destroyed Plataea yet again in 373. Plataea was restored by **Philip II** of **Macedon** after the battle of **Chaeronea** in 338 and survived into **Roman** times. There were at least two major battles on Plataean territory:

(1) September 479, during the Second **Persian War**. Fought between a coalition Greek force, led by the **Spartan Pausanias** (1), and a **Persian** army commanded by **Mardonius** and Artabazus. According to **Herodotus** (9.30, 32), the Greek force consisted of 110,000 men (38,700 **hoplites**, the rest *psiloi* or light-armed troops). The Persian army was estimated at 300,000, which included an unrecorded but large number of **cavalry** and around 50,000 Greeks. The only extant account of the battle is Herodotus' one, which is unfortunately rather confused. The Greeks were attempting to expel the Persians from Greece and maintain unity in the face of Persian attempts to detach **Athens** from the Greek coalition. Mardonius was apparently aiming to destroy Greek resistance on land.

The battle was preceded by several days of skirmishing, with the Greeks restricted to the foothills by the Persian cavalry. During this period the Persians constantly harassed the Greeks and successfully **ambushed** a Greek supply train, but lost their cavalry commander, Masistius. When the Persians fouled the Greek water supply at Gargaphia, Pausanias decided on a night withdrawal. This went badly as several contingents withdrew past the agreed point and on to Plataea itself, while a Spartan commander, Amompharetus, initially refused to withdraw at all.

The situation at dawn was that the army was split, with some in Plataea, the Athenians and Pausanias' Spartans on the move on different routes, and Amompharetus' contingent following up some distance behind Pausanias. At this point the Persians realized what was happening and launched an attack. However, this was done piecemeal and the Greek hoplites again proved superior, killing Mardonius and routing his infantry. On another part of the battlefield the Athenians were successful against the Boeotian infantry serving with the Persians. Artabazus, who had held his troops back when Mardonius rushed to the attack, retired from the field. The Greeks followed up Mardonius' fleeing troops, breaking into their camp and taking no **prisoners**. Some Greek contingents, though, suffered considerable **casualties** from the **Theban** cavalry during the pursuit phase. This battle ended the Persian invasion of Greece.

(2) 323, during the **Lamian War**. Fought between **Athens** and a pro-Macedonian **Boeotian** army of unspecified size. These Boeotians had been granted **Thebes'** land after its destruction in 335 and were afraid that Athenian success against **Macedon** would lead to the restora-

tion of Thebes and the loss of the distributed lands and associated revenues. They therefore opposed **Leosthenes**, who was marching north against Macedon with a force of around 10,000 **mercenaries**, 5,000 citizen-**hoplites**, and 500 citizen-**cavalry**. Leosthenes defeated the Boeotian force near Plataea and occupied the passes leading north.

PLUTARCH (PLOUTARKHOS; c. A.D. 50-120). A biographer, full name Lucius(?) Mestrius Plutarchus, who wrote on many of the important earlier ancient Greek and **Roman** military leaders. He was born to a prominent family in **Chaeronea** in **Boeotia**, studied philosophy at **Athens**, and traveled in both **Egypt** and **Italy**. He apparently had no first-hand military experience but may have been given civil offices under Trajan and Hadrian. Plutarch wrote many works, including his parallel biographies of famous Greeks and Romans. These *Lives* do contain much valuable information preserved from earlier writers, but they should be read as biography and not history. Plutarch's choice of what to include was not determined by a need to construct an accurate historical picture of his subjects' lives and careers but by his aim to reveal their moral character.

POLEMARCH (*POLEMARCHOS*, pl. *POLEMARCHOI*). "War leader," the title of a military commander in several Greek states. At **Sparta** the polemarch was a subordinate **officer** who commanded a *mora* of troops. At **Thebes** the polemarchs were the military commanders immediately subordinate to the boeotarchs. The polemarch at **Athens** was one of the nine archons (civil magistrates) and traditionally embodied the military powers of the original kings. Despite the institution of the *strategeia* at the end of the sixth century, the polemarch apparently remained the supreme commander of Athenian forces in the field for a short time after that. The polemarch **Callimachus** certainly had the casting vote in the decision to attack the **Persians** at **Marathon** in 490 and held the place of honor on the right of the front rank of the **phalanx**. However, by the mid-fifth century the polemarch was relegated to a ceremonial and judicial role and military command was exercised by the *strategoi* (generals). The office of polemarch also existed at **Mantinea**, **Orchomenus**, and in Crete and **Arcadia** in a variety of periods from the fifth to the second centuries.

POLYBIUS (POLYBIOS; c. 200-118). Author of a history of Greece and **Rome**'s early foreign wars from 264 to 146. Born into a prominent family in Megalopolis, in **Achaea** (his father Lycortas was an associate of **Philopoemen** and had served as *strategos* [general] of the **Achaean**

League), Polybius was destined for a prominent public career. As a young cavalryman he was selected to carry Philopoemen's ashes in his state funeral procession. Polybius was later appointed as envoy to **Egypt** and in 170/69 served as **hipparch** (**cavalry** commander) of the Achaean League forces. After the battle of **Pydna** in 168 he was sent to Rome (in a group of 1,000 Achaeans) as a hostage. While in Rome he became friendly with Scipio Aemilianus and soon became a member of his circle. He campaigned with Scipio in Spain in 151 and was present with him at the destruction of **Carthage** in 146. After much of Greece was made into a Roman province, he acted as a mediator between the Greeks and Romans and helped to reorganize Greece.

Polybius wrote several works, including a manual of tactics, a treatise on Philopoemen (both unfortunately lost), and a 40-book history, now only partly extant, covering from 264 to 146. This is our best source of information on this period. This history was designed for both the statesman and general reader and was intended to explain how Rome had taken over the Mediterranean area and to educate the reader in military and political affairs (1.1-2). Polybios had considerable military experience of his own and in Rome had access to good written sources, both historical works and public records and to oral family tradition. However, as might be expected, his work is pro-Achaean.

PONTUS. *See* MITHRIDATES

PORUS (fl. 326). An Indian prince (his Indian name was probably Paravatesha or Parvataka) who ruled the territory between the Jhelum and Chenab Rivers. He was defeated by **Alexander the Great** at the battle of the **Hydaspes** (Jhelum) **River** in 326. Porus was attempting to prevent Alexander from crossing the river but was outmaneuvered by Alexander, who managed to cross using deception and then decisively defeated Porus. Although completely outclassed by Alexander, Porus fought on bravely until he collapsed from loss of blood from numerous **wounds**. Alexander, impressed by Porus' bravery, left him as client king in his own territory.

POTIDAEA (POTIDAIA). A **Corinthian** colony in the Chalcidice, founded around 600. The city surrendered to the **Persians** at the start of the Second **Persian War**, but after **Salamis** refused entry to the Persian force escorting **Xerxes I** home and subsequently successfully withstood a Persian siege. Potidaea sent 300 **hoplites** to serve against Persia at **Plataea** (1) (Hdt. 7.123, 8.126-127, 9.28).

Potidaea later joined the **Delian League**, while maintaining its links with Corinth—which caused it major problems after Corinth's rupture with **Athens** over the battle of **Sybota**. In 432, when the Athenians demanded it demolish the city walls on the coastal side, provide hostages, and no longer accept Corinthian magistrates, Potidaea revolted against Athens (in conjunction with other cities in the area and supported by **Perdiccas II** of **Macedon**). However, the Athenian demands may only have hastened a revolt that was already planned and which proved to be another major contributing factor in the outbreak of the **Peloponnesian War** in 431 (Thuc. 1.56-59).

The Corinthians sent out a force to assist Potidaea, and when the Athenian expeditionary force sent to quell the revolt arrived, a major battle was fought. The Athenians had 3,000 hoplites, a large but unknown number of allies, and 600 Macedonian **cavalry** provided by Perdiccas, who had recently returned to his alliance with Athens. The size of the rebel force is unknown, but included the 1,600 hoplites and 400 *psiloi* (light troops) sent out from Corinth. The rebel plan was to deploy outside the city and attack the Athenian rear from nearby **Olynthus** once the battle had started. The plan failed because the Athenians placed their cavalry as a blocking force and the contingent from Olynthus was unable to deploy (it made a partial attempt to do so but retired when their allies at Potidaea were fairly quickly routed by the Athenians). The rebels lost 300 men, the Athenians 150. After setting up a **trophy**, the Athenians besieged the city, surrounding it completely with the arrival of an additional 1,600 hoplites under **Phormio** (Thuc. 1.60-65).

After great privation, which saw some of the inhabitants reduced to cannibalism, the Potidaeans surrendered under terms in 430. The siege cost 2,000 talents and the Athenian generals were blamed on their return home for the leniency of the terms accorded to the Potidaeans (Thuc. 2.70). The city was then garrisoned by Athenian **cleruchs** (Thuc. 2.70.4) The cleruchs were presumably expelled from the city at the end of the war, and by 382 it was under the control of Olynthus, although immediately deserting to **Sparta** when it attacked Olynthus and the **Chalcidian Confederacy** in that year (Xen. *Hell.* 5.2.15, 24).

Around 364/3 Potidaea was seized by **Timotheus** as part of Athens' long-term plans to restore its power in the Chalcidice. However, in 356 the city was taken by **Philip II** of Macedon, who killed or enslaved the population and handed it over to the Olynthians as part of his anti-Athenian diplomatic maneuverings in the area (Diod. Sic. 15.81.6; 16.8.5).

Around 316, near the site of Potidaea **Cassander** founded a new city, named Cassandreia, which developed into one of the most promi-

nent and powerful Macedonian cities (Diod. Sic. 19.52.2-3). It was the
seat of Arsinöe, **Lysimachus'** widow, but was seized from her by
Ptolemy Ceraunus (using treachery) in 281/0. With the help of Galatian
mercenaries (Gauls), Cassandreia was seized in 279 by one of its own
citizens, Apollodorus, who established a tyranny (Diod. Sic. 22.5),
which was overthrown by **Antigonus II (Gonatas)** in 276. **Philip V**
used it as his principal naval construction yard (Livy, 28.8). In 169, during the Third **Macedonian War**, Cassandreia successfully withstood a
siege by the **Romans** and **Eumenes II (Soter)** of Pergamum (Livy,
44.11-12). The city passed to Roman control with **Perseus'** defeat.

PRISONERS OF WAR. With some variation according to date and circumstance, prisoners taken in Greek warfare could expect one of several
general fates: execution, **slavery**, ransom, or repatriation. Although the
Greek "**Laws of War**" precluded the execution of those who had surrendered, this was in fact fairly common—particularly in hard-fought wars
and sieges—and the prohibition in any case did not apply to non-Greek
prisoners.

In very broad terms, slavery rather than death was probably more
common down to the **Peloponnesian War** of 431-404, at which time
the pressures of what was essentially total war led to increasing atrocities. The Peloponnesians executed the prisoners taken when **Plataea** surrendered in 427 and the **Athenians** executed all the adult males and sold
the women and children into slavery when Scione surrendered in 421
(Thuc. 3.68.1-2, 5.32.1). That same year the Peace of **Nicias** made provision for the return of prisoners (Thuc. 5.18.7).

The Athenians and allies captured during the **Sicilian Expedition**
suffered a variety of fates: some were hidden by their individual captors
and enslaved, the leaders were executed, and the rest were held under
very poor conditions in the stone quarries in **Syracuse**. Apart from the
Athenians and any Greeks from **Italy** or **Sicily**, they were then sold as
slaves (Thuc. 7.85-87). At least some of the Athenians were apparently
later ransomed or even freed (according to **Plutarch**, *Nicias*, 29, in some
cases because they had impressed their captors with their knowledge of
Euripides' poetry). Toward the end of the war the Athenians decreed they
would cut off the right hand of any prisoners they took in naval actions
and the Peloponnesians executed all the Athenians captured at **Aegospotami** in 405, except one, in retaliation for this decree and the recent
Athenian killings of the crews of two **triremes**.

Again in broad terms, in the Hellenistic period, during the struggles
of the *diadochoi*, there was a fairly good chance that ordinary prisoners
of **Macedonian** origin would be enrolled in the army of the victor,

although more-prominent opponents were often executed (see, for example, the actions of **Antigonus I [Monophthalmus]** after **Gabiene**; Diod. Sic. 19.44). **Philopoemen** was reputed to have regularly spent money ransoming prisoners (Plut., *Philopoemen*, 4.3). At other times prisoners of Greek origin were distributed as military settlers as part of a general process of hellenization (for example, after **Gaza** in 312; Diod. Sic. 19.84.4). The normal fate for prisoners from a state against which a **Sacred War** was declared was death, although this seems to have been interpreted a little flexibly, at least in the fourth century.

PRODROMOI (sing. *PRODROMOS*). **Cavalry** scouts; the word *prodromoi* means "men who run forward." Although any cavalry could be used in a **reconnaissance** role, from the fourth century some states had cavalrymen formally designated as scouts. **Macedonian** *prodromoi* at the time of **Philip II** and **Alexander the Great** were formed into four 200-man squadrons from **Thrace**. They were also known as *sarissophoroi* ("*sarissa*-bearers") and carried the *sarissa* as their primary **weapon**. At **Athens** *prodromoi* are attested around 365, when five may have been part of the establishment of each *phyle* (squadron).

PSILOI (sing. *PSILOS*). Generic term for lightly-armed soldiers, which could include **archers, slingers, peltasts**, or men equipped with nothing more than a dagger or stones. Looked down upon by the **hoplite** class as socially and militarily inferior, for most of the period from the seventh to the mid-fifth centuries, *psiloi* played a secondary role in warfare south of **Thessaly** and east of **Aetolia**. They were normally used to provide flank protection to the hoplite **phalanx** and to conduct much of the **ravaging** of an enemy's territory. However, as *psiloi* were faster and more mobile over hilly country or broken **terrain** than the hoplite, they were also used in border and mobile defense. Although *psiloi* were vulnerable to **cavalry** when caught in the open, their mobility did allow them to operate well with friendly cavalry—increasing the effectiveness of both arms.

　　Psiloi were much more frequently used from the time of the **Peloponnesian War** of 431-404, where they demonstrated their potential against hoplites at **Spartolus** in 429 (in conjunction with cavalry), **Aegitium** in 426, and **Pylos** in 425. Another major success was the destruction of the **Spartan** *mora* by peltasts at **Lechaeum** in 390. Another term for *psiloi* is *gymnetes*. *See also* DEMOSTHENES (1).

PTOLEMY (PTOLEMAIOS). The name used by all pharaohs of **Egypt** descended from the **Macedonian** dynasty of Ptolemy, one of **Alexander the Great**'s generals. The most important militarily are:

(1) Ptolemy I (Soter) (c. 367/6-283/2). A Macedonian noble, the son of Lagus and Arsinöe, Ptolemy ruled **Egypt** from 323 to 283/2 (as king 305/4-283/2). He was a childhood friend of **Alexander the Great** and was one of four of his companions exiled by **Philip II** in 337 during the family turmoil caused by Philip's marriage to Cleopatra. On Philip's death Ptolemy was recalled and made a member of Alexander's seven-man personal bodyguard (Arr. *Anab.* 3.6.5; Plut., *Alexander*, 10). He had a distinguished career under Alexander, commanding 3,000 men at the Persian Gates and leading a large contingent of the Companion and other **cavalry** and a mixed force of infantry on a successful mission to capture Bessus, the assassin of **Darius III** (Arr. *Anab.* 3.18.8-9, 3.29.7-30.3). He subsequently wrote a history of Alexander's campaigns, which is no longer extant but heavily influenced later writers. To a reasonable degree, the picture we now have of Alexander and his campaigns derives from Ptolemy.

Ptolemy became satrap of Egypt in the division of Alexander's empire among the *diadochoi* in 323 and subsequently diverted Alexander's body for burial in Alexandria—a major propaganda coup—which he built upon by a policy of kindness as satrap (Curt. Ruf. 10.10.1, 20; Diod. Sic. 18.14.1-2). In 322 he annexed Cyrene (Diod. Sic. 18.21.6-9) and the following year used his personal popularity with Alexander's **Macedonian** veterans to survive a treason charge brought by **Perdiccas** (2). The next year he defeated Perdiccas' invasion of Egypt by remaining on the defensive and watching Perdiccas' army mutiny and murder him after a disastrous attempt to cross the Nile. Ptolemy played a major part in opposing the crossing, distinguishing himself in the hand-to-hand fighting, and was subsequently confirmed in his satrapy by **Antipater** (Diod. Sic. 18. 33-36, 39.5). In 316 he provided refuge to the future **Seleucus I (Nicator)**, later helping him recover his territory (Diod. Sic. 19.56.1, 19.90.1).

However, Ptolemy was not quite as consistently successful in his conflicts with **Antigonus I (Monophthalmus)** and his son **Demetrius I (Poliorcetes)** over the period 315-301. He joined the anti-Antigonid faction, generally using the war to expand Egypt's territory to include **Cyprus**, **Palestine**, Phoenicia, and most of **Syria** south of Lebanon and Damascus, as well as other places in Asia Minor and the Aegean islands (Diod. Sic. 18.43, 73.2; 19.62.3-5, 78.4-7). In 312 he inflicted a decisive defeat on Demetrius at **Gaza** (Diod. Sic. 19.80.3-86.5). Ptolemy's failure to contribute troops to the coalition forces at **Ipsus** in 301 laid the seeds for future conflict with the Seleucids over who owned Syria. Ptolemy proclaimed himself a king in 305/4, the same year receiving his epithet Soter ("Savior") for assisting **Rhodes** during its famous siege by Demetrius.

In 285 Ptolemy associated his son, **Ptolemy II (Philadelphus)**, with him in the kingship and played little part in affairs between then and his death, probably in 283/2. Although not a particularly successful general, Ptolemy laid the foundations of a strong dynasty and a strong Egypt. Arguably he was the most successful of the *diadochoi*—he was certainly one of the very few to die of natural causes.

(2) Ptolemy II (Philadelphus) (308-246). Joint king of **Egypt** with his father, **Ptolemy I (Soter)**, from 285 and ruled Egypt in his own right from 282 to 247. About 276 his sister Arsinöe returned to Alexandria, displaced Ptolemy's existing wife (also called Arsinöe), married him (giving rise to his epithet Philadelphus—"Sister-loving"), and appears to have had considerable influence over affairs until her death in 270. Ptolemy fought two wars against Seleucid **Syria**, with mixed results.

The First **Syrian War** (274/3-271) confirmed Ptolemy as ruler of most of the Syrian coastline from **Miletus** in the west to the river Calycadnus in Cilicia in the east. Whether or not this represented the status quo before the war, it represented a victory for Ptolemy. The Second Syrian War (260-253) may have started with Ptolemy trying to take advantage of the transition of rule to Antiochus II (Theos) after the death of **Antiochus I (Soter)** in 261. If so, it was a miscalculation, as Ptolemy ended up fighting not only Antiochus but also **Antigonus II (Gonatas)**, who decisively defeated Ptolemy's fleet off Cos in either 261 or 258. This temporarily crippled Ptolemaic naval power and allowed Antiochus to regain **Ephesus** and perhaps to expand his territory at Ptolemy's expense in Cilicia, Pamphylia, Phoenicia, and **Ionia**. Ptolemy made peace with Antigonus in 255 and with Antiochus (probably) in 253. Despite the loss at Cos, Ptolemy generally maintained a strong navy and did make some gains in the Aegean, gaining control of Abdera in **Thrace**, Samos, Samothrace, and Lesbos. His reign included an unsuccessful revolt by his half-brother, Magas, ruler of Cyrene (part of the opening stages of the First Syrian War). Ptolemy also made a major contribution to establishing the well-organized Ptolemaic financial system that provided the wealth on which Egypt's military power was based.

(3) Ptolemy IV (Philopator) (c. 244-205). Son of Ptolemy III (Euergetes); king of **Egypt**, 221-205. The epithet Philopator means "Father-loving." Ptolemy had a troubled reign, which included the Fourth **Syrian War** (219-217). Egypt was eventually victorious in this after the de-

cisive victory over **Antiochus III (the Great)** at **Raphia** in 217. However, it is uncertain how much credit Ptolemy can claim for this, because his chief minister, Sosibius, apparently played a major role in organizing Egyptian resistance. Raphia was won after a rigorous **training** program of native Egyptians in the use of the *sarissa* and the **Macedonian**-style **phalanx**. As a result, Ptolemy recovered southern **Syria and Palestine**. Although the evidence is ambiguous, the role of the native Egyptians at Raphia may to have led to a resurgence of native Egyptian nationalism. Egyptian priests began to take a more independent line, and around 205 the Thebaid revolted and remained independent for about 20 years. This revolt may have contributed to Ptolemy's assassination by Sosibius and another minister, Agathocles, in 205.

(4) Ptolemy V (Epiphanes) (210-180). Joint king of **Egypt** with his father, **Ptolemy IV (Philopator)**, from 210, sole king (as a minor) from 204, and ruled in his own right from 197 to 180. The epithet Epiphanes means "Made Manifest." The death of his father in a palace coup was kept secret for some time and during his minority rule **Antiochus III (the Great)** of **Syria** and **Philip V** of **Macedon** allied to deprive Egypt of its overseas possessions. During the Fifth **Syrian War** (202-195) Antiochus conquered Coele-Syria and **Palestine**, inflicting a decisive defeat on the Egyptians at **Panion** in 200. The flow of conquests was stopped by **Rome**, which ordered Antiochus not to invade Egypt. Rome was also instrumental in stopping the threat from Macedon—Philip V's ruthless use of naval power in 202-201 against Ptolemaic possessions in the Aegean provoked complaints to Rome and led to his defeat in the Second **Macedonian War**. The sole military success stories of Ptolemy's reign were the reconquest of the Thebaid by his general Hippalus in 187/6 and the suppression of a native revolt in 184/3.

(5) Ptolemy VI (Philometor) (c. 186-145). Son of **Ptolemy V (Epiphanes)**, he succeeded to the joint (minority) rule of **Egypt** with his mother Cleopatra I (hence the epithet Philometor—"Mother-loving") in 180, but ruled in his own right from 176 (at which time he married his sister, Cleopatra II) until 170. He exercised joint rule with his brother Ptolemy VIII (Euergetes II) and with Cleopatra II from 169 to 164; he went into exile in 164 but shortly returned to claim sole rule, which he exercised from 163 to 145. His reign was marked not only by internal dissension with his brother but also by rebellions and external threats. If **Rome** had not intervened, **Antiochus IV (Epiphanes)** would probably have conquered Egypt in 169-168 during the Sixth **Syrian War** (which was provoked by Ptolemy's chief ministers). By the time Rome ordered Antiochus out, he had already captured **Cyprus** and most of Egypt. This invasion was initially made with Ptolemy's approval as

part of the internal power struggle with his brother. Ptolemy's return in 163 was also influenced by Rome. Ptolemy gained Alexandria and Cyprus while his brother ruled in Cyrene. In 145 Ptolemy died of wounds received during his victory over Alexander Balas while fighting on behalf of his daughter in Syria.

PYDNA. A Greek colony on the coast of Pieria with a good harbor. Unsuccessfully besieged by **Athens** in 432/1, the city was moved a short distance inland by Archelaus of **Macedon** (c. 413-399), although the inhabitants later moved back to the original site. In 357 the city was taken by **Philip II** and incorporated into Macedon. In 317 Olympias established her royal court there after her coup against Polyperchon. The following year Polyperchon besieged the city, forcing Olympias to surrender. Pydna was the site of two battles during the **Macedonian Wars**.

(1) 22 June 168; the final battle of the Third **Macedonian War** of 171-167, fought between a **Macedonian** army of around 40,000 infantry and 4,000 **cavalry** under **Perseus** and a slightly smaller **Roman** army under Lucius Aemilius Paullus. Perseus initially drew up his army on the plain just to the south of the city and Paullus, advancing without **reconnaissance**, blundered in column of march almost up to it. However, Perseus hesitated too long and by very skillful maneuver Paullus was able to withdraw his army to broken **terrain** before the Macedonian **phalanx** could catch it.

Perseus relocated his camp to prevent the Romans from being resupplied by sea, apparently hoping that lack of supplies would force their withdrawal. However, a few days later, a minor clash between *psiloi* (light troops) caused both generals to deploy their full armies in support, triggering the battle. The Macedonians deployed more quickly and the initial charge of the phalanx caused considerable damage to the Roman army. Perseus, apparently injured, was unable to stop his phalanx pressing too far forward, and rougher ground caused gaps to open in his battle line. Paullus' troops, operating in smaller groups (maniples) penetrated these gaps and broke into the phalanx **formations**. Once this occurred, the superiority of the Roman equipment for close-quarter fighting proved decisive. The Macedonian cavalry, whose horses had never faced them before, were stampeded from the field by the Roman **elephants** and fled to Pydna, along with Perseus.

The Romans took no **prisoners** on the battlefield and the infantry of the line, the phalangites and **peltasts** were annihilated. Around 20,000 Macedonians died and Pydna itself was sacked. This battle effectively destroyed the **Macedonian army** and ended any Macedonian chance of resistance, although the Romans kept campaigning for some time before officially ending the war. Perseus actually performed fairly well in this action and the account of his premature flight and cowardice, apparently based on **Polybius**, should probably be rejected.

(2) 148; the decisive battle of the Fourth **Macedonian War** of 149-48, fought between the pretender to the **Macedonian** throne, **Andriscus**, and a **Roman** army under the praetor Quintus Caecilius Metellus. Few details are known of the battle, but Andriscus was defeated with heavy losses and fled to **Thrace**. Although it took another engagement to subdue him, the extent of the defeat at Pydna effectively put paid to any chance of his success.

PYLOS. A headland in **Messenia** on the west coast of the Peloponnese; modern Navarino Bay. In 425, during the **Peloponnesian War** (2) of 431-404, Pylos was occupied by an Athenian fleet en route to **Sicily**. Although **Thucydides'** account (4.2-4) emphasizes the role of chance, the occupation seems to have been part of a clear plan by **Demosthenes** (1) to regain some initiative for **Athens**. At this stage of the war, the defensive nature of **Pericles'** island policy was clearly apparent and some, at least, at Athens were looking for ways to carry the war more aggressively to **Sparta**. Demosthenes was probably also keen for a success to retrieve his recent failure in Acarnania.

Although initially regarded as a minor problem at Sparta, the occupation of Pylos had an immediate effect on the Spartan army currently **ravaging** Attica. Realizing the danger Pylos posed as a base for ravaging Spartan territory and as a focus for Messenian unrest, **Agis II** led the army home immediately—although bad weather and a shortage of food in Attica also encouraged the decision. The Spartans placed 440 **hoplites** on the neighboring island of Sphacteria and attacked Pylos from both land and sea. However, the Athenian garrison of three **triremes** and crews (two of the five ships originally left had been sent to get assistance), supplemented by Messenians (including 40 hoplites), successfully resisted both Spartan assaults. During the seaborne attack, the Spartan **Brasidas** was **wounded** and lost his **shield**. At this point, 50 Athenian triremes arrived and defeated the 60-strong Peloponnesian fleet in the bay. This not only removed the seaward blockade of Pylos but also trapped the Spartan hoplites on Sphacteria.

The Spartans then sought an armistice, pledging their remaining ships at Pylos and elsewhere in Laconia as security, and offered peace terms to Athens. The Athenians, heavily influenced by **Cleon**, rejected this offer and kept the ships—claiming that the Spartans had breached the terms of the armistice (Thuc. 4.15-23). The Spartans tried to supply their troops on the island by paying helots to swim across with food or to ferry it across in small boats, but only limited quantities could be carried and constant Athenian naval patrols also interfered with this. The lack of water and harbors in the area was also making it difficult for the Athenian force blockading Sphacteria and a decision was made to appoint Cleon to land troops and capture the Spartiates on the island. (Although related from an anti-Cleon perspective, Thucydides' report of the debate in the assembly [4.27-28], gives an interesting insight into the way the Athenian assembly conducted the business of war.)

Cleon took only *psiloi* (light troops) from Athens and once at Pylos adopted a plan already formed by Demosthenes. The Athenians landed 800 **archers**, 800 **peltasts**, and an unknown number of other troops (including hoplites), catching the Spartans by surprise. Demosthenes' plan, influenced by his recent experiences in Acarnania, was to keep his hoplites back and harass the enemy with his *psiloi*. These kept up a continual barrage of missiles from the front and flanks, retiring to safety whenever the Spartan hoplites advanced against them. Without water and in distress from both the missiles and clouds of ash (the island had been burned out in a recent fire), the surviving 292 Spartans surrendered when a contingent of archers worked its way around to their rear (Thuc. 4.29-38).

The Spartiate prisoners were taken to Athens and threatened with execution if the **Peloponnesian League** ravaged Attica again. Ironically, this threat preserved Attica but forced the Spartans to reconsider their whole strategy, and their decision to campaign against **Amphipolis** and the Chalcidice—the only part of the Athenian empire vulnerable from land—turned out to be a much more effective strategy than the annual invasion of Attica. Another major effect of the Spartan surrender on Sphacteria was to demonstrate the potential effectiveness of light troops against hoplites, although few generals other than Demosthenes grasped this until **Iphicrates** repeated the feat at **Lechaeum** in 390. At the time, the surrender caused considerable shock in the Greek world, which had expected to see Spartans fight to the death whatever the odds. Thucydides records that one of the prisoners, on being asked if the ones who had died "were the real Spartans," replied, "A spindle, meaning arrow, would be worth a lot if it could pick out brave men" (4.40.2).

PYRRHUS I (PYRRHOS; 319-272). King of Epirus, 307-303 and 297-272; king of **Macedon**, 287-285 and 274-272. Pyrrhus was placed on the throne of Epirus as a minor but was forced into exile in 303. He joined his brother-in-law, **Demetrius I (Poliorcetes)**, serving with distinction at **Ipsus** and after the battle governing the remains of Demetrius' Greek possessions for him (Plut., *Pyrrhus*, 4). A judicious marriage to Antigone, stepdaughter of **Ptolemy I (Soter)**, allowed him to regain his throne in 297 as joint monarch with his rival Neoptolemus—who was soon executed on the grounds that he was planning to assassinate Pyrrhus. As sole king, Pyrrhus embarked on a consistent and generally successful program designed to strengthen Epirus and secure its independence from Macedon.

Pyrrhus first seized southern Illyria. He then gained from Macedon some frontier areas as well as the regions of Ambracia, Amphilocia, and Acarnania. These were his price for support during a dynastic struggle in Macedon. The death of Antigone allowed Pyrrhus to consolidate his gains by further marriage alliances. The first, to a daughter of **Agathocles**, brought him **Corcyra**. Two more, to a granddaughter(?) of **Bardylis I** and to the daughter of the Paeonian king, secured him useful alliances with tribes on the northern borders of Macedon.

Pyrrhus' relations with Demetrius deteriorated with the death of Deidamea, his wife and Pyrrhus' sister, and with Demetrius' invasion of Macedon in 294. In 291 Pyrrhus invaded **Thessaly**, apparently to divert Demetrius from his siege of **Thebes**, but was quickly expelled, and in 290 Pyrrhus lost Corcyra to Demetrius. The same year, though, Pyrrhus defeated Demetrius' general Pantauchus in battle (killing him in single combat) during Demetrius' invasion of **Aetolia**. This was a major boost to Pyrrhus' reputation and perhaps as early as 288 he **ravaged** Macedon in strength.

Although this led to a peace agreement with Demetrius, in 287 Pyrrhus invaded Macedon from the east working in concert with Ptolemy and **Lysimachus**. Catching Demetrius unawares, he secured the mutiny of Antigonus' army, which proclaimed Pyrrhus king of Macedon. Despite this stunning success, Pyrrhus was not strong enough to hold the whole kingdom and he was forced to divide it with Lysimachus. The details of this are uncertain, but it appears that Pyrrhus received the larger share. Despite another treaty with Demetrius in 287, Pyrrhus seized Thessaly. This was the most territory Pyrrhus ever controlled in Greece—Demetrius' surrender to **Seleucus I (Nicator)** in 285 left Lysimachus free to attack Pyrrhus and expel him from Macedon.

In 281 Pyrrhus accepted an invitation to campaign in **Italy** on behalf of **Tarentum** against **Rome**. With an experienced army and rather reluctant assistance from Tarentum, whose citizens he made liable for conscription and rigorous **training**, Pyrrhus won two battles against the Romans—the **Siris River** in 280 and **Asculum** in 279. Although this caused several Italian tribes to join the coalition against Rome, Pyrrhus crossed to **Sicily** to fight the **Carthaginians** in 278. According to **Plutarch** (*Pyrrhus*, 22) he was discouraged by his irreplaceable losses (which gave rise to the term "Pyrrhic victory") and the apparently inexhaustible source of Roman manpower, as well as attracted by the offer of control of Agrigentum, Leontini, and **Syracuse** if he assisted. Pyrrhus defeated the Carthaginian army in battle and then stormed the stronghold of Eryx. However, his increasingly autocratic methods in the Sicilian cities caused such internal dissension that in 276 he abandoned plans to invade North Africa and returned to Italy, where the Romans had made good most of the ground they had lost in his earlier campaign. With limited support from Italian allies such as the Samnites, who felt they had been left in the lurch by his departure for Sicily, Pyrrhus fought the Romans at **Beneventum** in 275. A surprise night flanking march failed when the contingent became lost and arrived in daylight and the Roman use of **javelins** to deal with Pyrrhus' **elephants** also led to the loss of this battle and his withdrawal from Italy.

Pyrrhus, however, was always on the lookout for new opportunities and in 274 he turned a **raid** of Macedonian territory into a full-scale victory over **Antigonus II (Gonatas)**, briefly regaining most of the kingdom and Thessaly. In 272 he campaigned in the Peloponnese but was repelled outside **Sparta** by temporary entrenchments and the sheer determination of the defenders, who held out until reinforcements arrived. Moving directly from there, Pyrrhus was killed in street fighting in **Argos** while trying to capture the city. In the meantime he had already lost Macedon to Antigonus.

Pyrrhus was an able tactician and leader of men and expended considerable effort on military studies. He wrote a manual of tactics (now lost) and Plutarch preserves a story of Pyrrhus at a dinner party, on being asked which of two flutists was better, answering, "Polyperchon was a good general" (Plut., *Pyrrhus*, 9). However, he was also regarded with some justice in antiquity as a general who did not know when to capitalize on existing success and who often lost what was attainable through overreaching himself (Plut., *Pyrrhus*, 26).

- Q -

QUINTUS CURTIUS RUFUS. *See* CURTIUS RUFUS

- R -

RAIDS. Raiding must have been the earliest form of warfare in Greece, was used throughout Greek history, and was always a major feature of **naval warfare**. Its original aim was undoubtedly the acquisition of **booty**, whether movable property, women, **slaves**, or livestock. However, another common motive was the desire to damage an enemy. Although the introduction of **hoplite** warfare, with its fairly formalized procedure of a pitched battle on the agricultural plain, might be considered to have reduced the opportunity for raiding, this type of operation in fact continued, not only because of the opportunity for acquiring booty but also because it also offered the opportunity for a militarily weaker power to inflict damage on a stronger opponent.

For example, one of **Athens'** responses to **Spartan** attacks on Attica during the **Peloponnesian War** of 431-404 was to mount seaborne raids on the Peloponnese. Unable to face the **Peloponnesian League** forces in a pitched battle, the Athenians landed from the sea, **ravaged** areas, and retired to their ships—any military response was necessarily local and therefore manageable. The raiding was escalated with the Athenian occupation of **Pylos** and Cythera. **Thucydides** states that these raids were so troublesome that the Spartans "contrary to their usual practice" created a force of 400 **cavalry** plus **archers** to help deal with it (4.55.2).

Seaborne raids were particularly effective because of the strategic mobility gained from the navy, but on land considerable tactical mobility could be achieved with the use of cavalry. The mobility and endurance of the horse made it especially suited for raiding and it is not surprising that a solid body of practical experience and theory developed to support this. By the mid-fourth century at Athens, **Xenophon** was describing the techniques involved and stressing the need for rapid movement, combined with stealth where necessary, to enable penetration, damage, and then withdrawal before the enemy could react (*Hipparch.* 7.7-10, 14-15).

The Spartans too used raiding. In the summer of 429 **Brasidas** and other Spartans launched a daring raid on the Piraeus. Reasoning that it would be unguarded because of Athenian naval supremacy, they secretly moved rowers (each carrying his cushion, rowlock thong, and oar) overland across the Isthmus of Corinth, manned 40 **triremes** sitting idle in **Megara**'s harbor Nisaea, and launched a **night attack**. As it turned out, contrary winds and a failure of nerve caused them to strike **Salamis** first and ravage it. This caused a huge panic at Athens and the Peloponnesians were able to withdraw to Nisaea with three captured ships, **prisoners**, and plunder (Thuc. 2.93-94). In 378 Sphodrias mounted a similar daring raid on the Piraeus but by land and in full force. It failed when the army moved too slowly and was spotted in daylight before it reached its objective (Xen. *Hell.* 5.4.20-34).

The second half of the Third **Sacred War** was largely conducted by raiding and the **Aetolian League** made a particular art form of this type of operation. Hellenistic monarchs too made use of raids to damage the enemy and to boost their coffers. **Pyrrhus I** of Epirus' near-conquest of **Macedon** around 274 seems to have originally started as a full-scale raid but was extended in response to favorable circumstances (Plut., *Pyrrhus*, 10.1-2).

Raids were also mounted for very specific purposes. **Alcibiades** used a mounted force, accompanied by hoplites, to raid **Bithynia** to seize the property **Chalcedon** had stored there (Xen. *Hell.* 1.3.3). In 317 **Eumenes of Cardia** launched a raid to capture the baggage of a troublesome satrap (Diod. Sic. 19.23.4). In 480, **Gelon** of **Syracuse** sent cavalry to seize the entrance to a **Carthaginian** camp at Himera and kill its general, Hamilcar (Diod. Sic. 11.21.5). *See also* SOLYGIA.

RAPHIA. The decisive battle of the Fourth **Syrian War**, fought southwest of **Gaza** on the border between **Palestine** and **Egypt** on 22 June 217. **Polybius** (5.79-86) provides a detailed account of the engagement. **Antiochus III (the Great)** led a **Syrian** force of 62,000 infantry, 6,000 **cavalry**, and 102 **elephants** against **Ptolemy IV (Philopator)** and an Egyptian force of 70,000 infantry, 5,000 cavalry, and 73 elephants. The right wing of each army was victorious but Antiochus pursued the enemy left wing too far. Meanwhile the two **phalanxes** at the core of the armies met head on. The Egyptian one (consisting of 20,000 native Egyptians and 5,000 Greeks and which had spent an entire year **training**) destroyed Antiochus' phalanx. Around 10,000 Syrians and five elephants died and 4,000 men were captured. Ptolemy lost only 2,200 men but nearly all his elephants. Antiochus fled the battlefield and shortly afterward made peace. This was the first time for nearly 100 years that the Greek rulers of Egypt had armed their native Egyptian population and this may have caused some resurgence of native Egyptian nationalism. *See also* MERCENARY; TARENTINES.

RAVAGING. Ravaging was an integral part of Greek warfare. Prior to the development of siege **artillery**, it was a key strategy for any state trying to force an enemy to accede to its wishes. Much of the current theory of ravaging derives from the influential work of V. D. Hanson, who corrected an earlier view that it inflicted long term damage on Greek agricultural areas (*WWW*, 3-4, 33-34; *OG*, 14-15, 257, 286, 289, 315). Although this was a necessary correction, Hanson pushed the argument a little too far by claiming that ravaging actually had little effect and the Greeks overestimated the extent of damage it caused. The corollary to this view is his claim that the ideal that one's *chora* (agricultural hinterland) should be *aporthetos* ("untrodden on") by anyone else blinded the

Greeks to reality, causing them to march out in defense of their land—despite the minimal risk of real damage ($WACG^2$, 182-184; WWW, 5-6).

However, the evidence doesn't support this extension of the legitimate (and important) point that ravaging did not cause long-term physical damage to Greek agriculture. For example, Hanson describes the phrase "to turn the countryside into a sheepwalk" as "a common promise of war—if rarely a real consequence of fighting" ($WACG^2$, 10). But in the case of the two passages cited in support of this (Isocrates, 14 [*Plataikos*], 31, and Diod. Sic. 15.63.1), the remark is made in the context of depopulating the countryside—not ravaging it to the point where the only agriculture it can support is grazing (cf. Plut., *Pericles*, 16.5). The overestimation of the effect of ravaging seems to arise from modern interpretations of ancient sources rather than from the sources themselves.

However, although permanent damage was rare, short-term damage was quite serious in its effect on the individual farmer and his family. For many farmers, even some loss would cause problems. Damaged vines and trees certainly had the capacity for quick regeneration and new crops could be planted, but the individual farmer and his family would have nothing to live on in the meantime. Few could afford to run the risk of losing a crop, so all marched out in collective defense of their individual property and families. However, crop damage also had an effect at state level. The loss of even one year's crop was generally seen as a problem (cf. Polyaenus, *Strategemata*, 4.6.20) and the cumulative effects of ravaging several years in a row is likely to have been even more severe. This is illustrated by three passages in **Xenophon** (*Hell.* 5.4.56, 7.2.10, 7.2.17) which suggest that ravaging often did cause sufficient hardship to warrant remedial action on a state level.

This potential for short-term damage was the reason why most Greek **hoplite** states marched out to protect their crops (unless greatly outnumbered, as was the case with **Athens** during the **Peloponnesian War** of 431-404). However, damage to the *chora* could also be reduced by a mobile defense using *psiloi* (light troops) or **cavalry** to restrict enemy movement, and in particular to encourage hoplites to remain in **formation** (Thuc. 1.111.1, 2.100.5; Hdt. 5.63.3-4; Diod. Sic. 15.71.4-5; Xen. *Hell.* 5.3.3). This considerably restricted the area which could be damaged.

The main method of ravaging was almost certainly manual—vines were cut down, crops trampled or cut down, and houses wrecked. Plundering was also part of the process, with troops normally supplementing their **pay** by acquiring whatever **booty** they could. Fire was also used, although it had some limitations—it could not be used extensively if an army was living off the land, the fragmented nature of many Greeks farm plots probably prevented major fires across large areas, and it required fine judgement to arrive at exactly the right time when the crops were ripe enough to burn easily but had not yet been harvested.

RECONNAISSANCE. Although today regarded as a basic military skill, reconnaissance was rather a neglected practice in Greek warfare down to around the end of the Classical period. This was probably because many failed to see the need for it. Most military activity took place in Greece, so local knowledge was a prime source of information about **terrain**, while spies, deserters, and traitors often provided details of enemy troop movements and intentions. The relatively slow-pace of **hoplite** warfare meant that such **intelligence** retained its usefulness for some time.

Several unfortunate experiences arising from a failure to conduct reconnaissance occurred in the first quarter of the fourth century. **Dercylidas** narrowly escaped disaster when his army unexpectedly encountered a Persian army at **Dascylium** in 398/7 and his successor in command, **Thibron**, was killed in 391 while advancing without proper march security (Xen. *Hell.* 3.2.14-20, 4.8.18-19). The **Thebans** and **Spartans** ran into each other at **Tegyra** in 375, neither army apparently using scouts (Plut., *Pelopidas*, 17.1ff.). Incidents such as these, the increasing use of **cavalry** and *psiloi* (light troops), a rise in military professionalism during the **Peloponnesian War** of 431-404, and the increasing frequency of campaigns in Asia Minor led to the development of a higher skill base in reconnaissance in the fourth century.

Xenophon, for example, demonstrates a sound grasp of the theory of reconnaissance and its importance in practice. He criticized **Iphicrates'** reconnaissance of Oneum in 370/69 on the basis that the force was too large for the task, and elsewhere stresses the importance of reconnaissance and the best way to do it (Xen. *Hell.* 6.5.52; *Hipparch.* 4.4ff., 7.9, 8.12-14. His practice of properly reconnoitering his route while leading the **Ten Thousand** in **Thrace** (Xen. *Anab*. 6.3.10-11) demonstrates that he did not just pay lip service to the concept.

Agesilaus II of Sparta also used cavalry scouts in Asia Minor during his campaigns there in the early part of the fourth century (Xen. *Hell.* 3.4.13). The existence of *prodromoi* in the fourth-century **Athenian** and **Macedonian** cavalry also suggest an increase in the practice of scouting. **Alexander the Great** was also scrupulous about conducting reconnaissance. He personally checked the field at **Gaugamela** and is elsewhere recorded as dispatching cavalry or *psiloi* to check on terrain or the enemy location (Arr. *Anab.* 1.12.7, 3.7.7, 3.9.5, 4.30.5-6, 6.20.4, 6.23.2). *See also* MAPS; SELLASIA.

RHIPSASPIS. Throwing away one's **shield** in battle. The standard **hoplite** shield was not particularly suited to individual combat outside the **phalanx** and was fairly heavy. Hoplites fleeing from the battlefield often jettisoned their shield in order to move more quickly. Although fairly common, this was regarded as a sign of weakness or cowardice. *Rhipsaspis* was an indictable offense at **Athens.** *See also* ARCHILOCHUS.

RHODES (RHODOS). A major island in the Aegean, south of Caria (modern Turkey), inhabited in antiquity by Dorian Greeks. The three major settlements on the island until 408 were Lindus, Ialysus, and Camirus. Before this date Rhodes had been a relatively minor contributor to Greek military history. The island is supposed to have sent nine ships to the **Trojan War** (Homer, *Iliad*, 2.653) and in the fifth century was a tributary part of the Athenian empire. Rhodians served (reluctantly) in **Sicily** in 415 and the island revolted against **Athens** in 411 after the defeat of the **Sicilian Expedition,** subsequently providing important support to the Peloponnesian fleet operating in **Ionia** (Thuc. 7.57.6, 8.44, 8.55.1, 8.60.2; Xen. *Hell.* 1.1.2, 1.6.3, 2.1.17).

In 408 the main cities on the island jointly built the city of Rhodes at the northern end of the island, making it the capital. From this point onward, Rhodes developed into an important maritime power which played a much more prominent part in Greek military activities. In 396 Rhodes revolted against **Sparta** (Diod. Sic. 14.79.6) and received **Conon** and his fleet into the city—probably through fear of Sparta's growing influence in Asia Minor, although internal strife between aristocrats and democrats was also a factor (cf. Thuc. 7.57.6; Xen. *Hell.* 4.8.20-24). This internal strife brought Rhodes back into the Spartan alliance in 390 (Diod. Sic. 14.97.1-4) but in 378/7 Rhodes was one of the founding members of the Second Athenian Confederacy. The island followed a generally pro-Athens line down to 364, when a naval expedition of **Epaminondas** brought it over to **Thebes** (Diod. Sic. 15.79.1). Rhodes formally gained its independence from Athens in 355, at the end of the First **Social War** (Diod. Sic. 16.22.2; Dem. 15 [*For the Liberty of the Rhodians*], 26; Isocrates, 8 [*On the Peace*], 16), but later assisted Athens against **Philip II** of **Macedon** (Diod. Sic. 16.77.2).

In the middle of the fourth century Rhodes seems to have fallen under **Persian** influence. A Rhodian, **Memnon,** was one of **Darius III**'s best generals (Diod. Sic. 17.7.2, 17.19.2), and the island supplied ships to the Persian fleet operating against **Alexander the Great**. Despite this, Rhodes surrendered to Alexander without a fight in 332 and was garrisoned (Curt. Ruf. 4.5.9, 4.8.12).

In 323 Rhodes expelled the Macedonian garrison (Diod. Sic. 18.8.1) and for around 150 years maintained its independence and steadily increased its power and prosperity. The basis of this was a strong economy, founded on Rhodes' activities as a free port strategically located on the main grain routes, and a powerful navy—which played an important role in the suppression of piracy in the Aegean. As an island, Rhodes could only be attacked from the sea and this gave it considerable defensive advantages.

Because it refused to support **Antigonus I (Monophthalmus)** against **Ptolemy I (Soter)**, Rhodes suffered an epic siege by **Demetrius I (Poliorcetes)** in 305-304 (Diod. Sic. 20.81-88, 20.91-100). Demetrius' failure to take the city considerably increased its reputation and prestige. Rhodes continued to maintain its independence throughout the third century, although often favorably inclined toward **Egypt**, the source of much of its grain trade. Along with maintaining independence, protection of its trade and commercial interests was consistently a major motivator in Rhodian military activity. In 220-219 Rhodes fought a successful war against **Byzantium** to force it to remove shipping tolls in the Hellespont (Polyb. 4.38, 4.47-52). Around the same time Rhodes went to war against Cretan pirates (Polyb. 4.53) and later against **Philip V** of Macedon. Philip's aid to the Cretans and his aggressive naval expansionism at the end of the third century caused Rhodes to join Pergamum and other states against him (Polyb. 13.4-5, 15.22-23).

An allied naval victory over Philip at **Chios** in 201/0 was followed by a less successful engagement at **Lade** (2), so Rhodes and its main ally, **Eumenes II (Soter) of Pergamum**, appealed to **Rome** for help (Polyb. 16.2-10, 16.14-15, 16.26). This appeal played an important role in the Roman decision to fight the Second **Macedonian War**. Rhodes profited from its assistance to Rome both against Philip and in the subsequent war against **Antiochus III (the Great)**—receiving territory in Caria and Lycia on the mainland opposite as a reward. However, in 167, at the end of the Third Macedonian War, Rhodes fell out of favor with Rome over its lukewarm support and its diplomatic dealings with **Perseus**. As punishment, Rome detached Rhodes' mainland possessions and established Delos as a free port in direct commercial competition with Rhodes (Polyb. 28.2, 29.3-4, 29.19, 29.4-5, 31.7). Both these actions—but particularly the commercial decline caused by the competition from Delos—led to a major reduction in Rhodes' power and independence. Although it regained some favor from Rome for its loyalty

and military contribution during the First **Mithridatic War** (it successfully withstood a siege by **Mithridates VI [Eupator]** in 88), it never
fully recovered. One of the consequences of this was an upsurge of piracy, previously suppressed by the Rhodian navy, that reached crisis
proportions in the early first century. The city's power was finally broken in 43/2, during the Roman civil wars, when it was comprehensively
sacked by Cassius (Plut., *Brutus*, 30; Appian, *Civil War*, 4.72).

ROME. Rome had a major impact on Greek military history from around
the mid-fourth century when it first began subjugating the Greek colonies of **Italy**—although many of these were in the south and were not
threatened until the following century. Prior to that the Greeks had apparently influenced Roman warfare by introducing the **phalanx** and
Cumae had assisted in the Roman struggle for independence from Etruria in the late sixth century. Despite strenuous resistance by some individual cities (notably **Tarentum**, which employed **Pyrrhus I** of Epirus
and others against Rome), by 266 the Romans controlled all the Greek
cities in Italy. In 264 Rome sent troops to **Sicily** to assist in an internal
struggle in Messana (and to protect its interests against **Carthage**), leading to war with **Syracuse**. However, **Hieron II** of Syracuse soon recognized the power of Rome and the following year he allied with them
against the Carthaginians. Hieron's loyal support, which gave the Romans Syracuse as a firm base and source of provisions, played a significant part in Rome's victory on land in the First Punic War (264-241).

During the Second Punic War, Hieron's successors joined Carthage
against Rome after Cannae in 216. Syracuse fell in 211 after a long and
bloody siege, prolonged by the skillful engineering of Archimedes. The
end of the war in 204 led to the incorporation of Sicily into the nascent
Roman empire. The Second Punic War also saw Rome in conflict with
Philip V of **Macedon** (the First **Macedonian War**) who allied with
Hannibal after Cannae. Roman naval supremacy in the Adriatic prevented Philip from assisting Hannibal, and Rome tied him down in a
local war in Greece by allying with the **Aetolian League**.

The First Macedonian War brought Rome into serious military contact with mainland Greece for the first time, and from this point onward
Rome played an increasingly dominant role in Greek affairs. During the
third century a series of wars against Macedon and then **Antiochus III
(the Great)** led to the Roman acquisition of mainland Greece and its
first steps in Greek Asia Minor. The general pattern involved smaller
Greek states appealing to Rome for political and/or military assistance
against more-powerful neighbors. The Romans, acting in some cases
partly through altruism (for example, philhellenism and a desire to assist

Athens appear to have played a part in the outbreak of the Second Macedonian War) but mainly to further their own interests, gradually occupied the whole of the Greek world. The process ended with the defeat
and annexation of Ptolemaic **Egypt** after **Actium** in 31.

Throughout their contact with the Greeks, the Romans proved militarily superior. Their heavy infantry, organized in maniples, proved consistently much more flexible and effective than the Macedonian-style
phalanx from the battle of **Cynoscephalae** (2) in 197 onward. Roman
engineering, organization, and general military ability, backed by an increasing empire and resources, was simply too much for the separate
states of the Greek world, which fell one by one.

- S -

SACRED BAND. The **Theban** *epilektoi*, a force of 300 **hoplites**, established as a standing force in 378. Paid for by the state, this force, reputed to consist of 150 pairs of lovers, spent its time **training** and
achieved the same level of skill as the Spartan hoplites. Founded by
Gorgidas but developed by **Pelopidas**, the Sacred Band was used as the
spearhead of the Theban effort to gain hegemony in Greece in the first
half of the fourth century. It played a major role in the battles of **Tegyra**
in 375 and **Leuctra** in 371 but was annihilated by **Philip II** of **Macedon** at **Chaeronea** (1) in 338.

SACRED WARS. A series of fifth- and fourth-century wars fought for
control of **Delphi** and the **Amphictionic League**. A sacred war involved
the Amphictionic League formally declaring war on a third party over a
religious matter concerning the sanctuary of Apollo at Delphi. A sacred
war was in effect a war without limits. In the First Sacred War, Crisa
was razed, and in the third, **Phocian prisoners of war** were executed.

(1) First Sacred War, c. 595-590. Fought by a coalition comprising
the Amphictiony of Anthela (near **Thermopylae** and dominated by
Thessaly), **Sicyon**, and **Athens**, against Crisa. Crisa controlled access
to the shrine at **Delphi** and levied tolls on visitors, and the war was
fought to free Delphi from this interference. Crisa was defeated, its inhabitants killed or enslaved, the city razed, and its lands left perpetually
uncultivated. Delphi, very probably already a member, was placed under
the protection of the Amphictiony of Anthela, which soon came to be
commonly known as the Delphic Amphictiony or **Amphictionic
League**.

(2) Second Sacred War, c. 449-447. Little is known of this war,
which started when **Phocis** seized **Delphi**. A Spartan expeditionary force

restored Delphi's independence, but almost immediately an Athenian force, commanded by **Pericles**, restored Phocian control. **Athens** and **Sparta** avoided direct conflict with each other so as not to break the Five Years' Truce and Phocian dominance of Delphi seems to have lapsed fairly soon.

(3) Third Sacred War, 355-346. In 356 the **Amphictionic League**, controlled by **Thebes**, fined several prominent Phocians for cultivating the sacred Plain of Crisa and threatened war if the fines were not paid. This seems to have been instigated by Thebes in order to curb Phocian independence from the **Boeotian League**. **Phocis**, led by Philomelus, responded by seeking support from **Sparta** and elsewhere and then seized **Delphi**. A preliminary attack by **Locris** was beaten off and in autumn 355 the Delphic Amphictiony declared war on Phocis.

Using the resources from the sanctuary of Apollo at Delphi the Phocians were able to construct a highly effective army, which included a sizable **mercenary** contingent. **Thessaly** was knocked out of the war after a defeat in the spring of 354 and **Boeotia**, with assistance from Locris, was left to carry it on alone. In autumn 354, Thebes defeated the Phocians at Neon. Although Philomelus died in the battle, the bulk of his forces escaped. Driven into a corner by the Theban policy of killing all **prisoners**, the Phocians regrouped under Onomarchus and continued to resist. In 353 Onomarchus defeated **Philip II** of **Macedon** twice, drove him from Thessaly, and then defeated the Boeotian army and occupied **Coronea**. However, Philip returned to Thessaly in 352 and defeated and killed Onomarchus at the battle of the **Crocus Plain**. The 3,000 prisoners taken may have been executed, but this is not certain. This victory established Philip and not Phocis as the arbiter of Thessalian affairs. Supported by Spartan, **Achaean**, and **Athenian** troops, Onomarchus' brother, Phayllus, occupied **Thermopylae** and Philip decided not to try to force the pass.

From this point on, the war developed into an affair of **raids** and guerrilla operations, resulting in the **ravaging** of considerable areas of Locris, Boeotia, and Phocis. Around this time the Phocians rebuilt **Orchomenus** as a base for ravaging Boeotia. In 352/1 Boeotia and Phocis carried the war on in the Peloponnese, with Phocis aiding Sparta and Boeotia aiding Megalopolis. Phayllus died shortly after this and was succeeded by Phalaecus, Onomarchus' son. In the face of weakened support from Athens, influenced by **Demosthenes** (2), Phalaecus handed over Thermopylae to Philip in 346.

The cities of Phocis were broken up into villages, an indemnity of 60 talents a year was imposed to pay for the treasures taken from the sanctuary, and Phocis lost her votes in the Amphictionic League. The war exhausted both Phocis and Boeotia and considerably increased the power of Philip. Not least of this increase was the transfer of Phocis' votes in the Amphictionic League to Macedon. *See also* LAWS OF WAR.

(4) Fourth Sacred War, 340-338; also known as the Amphissaean War. This war arose from a quarrel between **Athens** and **Thebes** over the premature placement of a **trophy** of **shields** from the battle of **Plataea** (1) on a new temple at **Delphi**. These shields had inscriptions that paired the **Persians** and Thebans as the enemies from whom they had been taken. When the debate at the Amphictionic League meeting in autumn 340 seemed to be going against the Athenians, their chief delegate, Aeschines, countered by stating that the Locrians of Amphissa were illegally cultivating the Plain of Crisa (see [1] above). He refused to take any further part in the proceedings until this had been remedied. The next day an Amphictionic attack on the illegal cultivators was repelled and a sacred war was declared against Amphissa. As Athens and Thebes were wary of open hostility against each other and did not join in, the Amphictiony appealed to **Philip II** of **Macedon**. In 339 Philip led his army into **Phocis** but did not attack Amphissa, instead threatening the route south through Elatea. Athens and Thebes joined together to meet this threat and occupied Delphi and southern Phocis.

The following year Philip gained entry to **Locris**, punished Amphissa, and ended the war with a decisive victory over the Greeks at **Chaeronea** (1). Although this war had started over a religious matter, it had soon been overtaken by the power struggle between Philip and the Greek states. This is reflected in the fairly lenient treatment accorded to the losers. During the war both sides allowed the villages of Phocis to rejoin into cities and the indemnity from the Third Sacred War (see [3] above) was reduced to 10 talents per year.

SACRIFICE. Greek precombat sacrifice fell into two categories: propitiatory, to gain the favor of the gods, and divinatory, to gauge the likelihood of success. Propitiatory sacrifices, known as *sphagia*, could be offered at various times but were most commonly offered immediately prior to the combat commencing. If the offering looked unfavorable (probably due to an unsatisfactory flow, or direction of flow, of the blood), combat would be delayed. An extreme example of this involves **Pausanias** (1) at **Plataea** (1) in 479. The sacrifice was offered as the **Persians** approached but was unfavorable, so Pausanias held his men back, waiting under a hail of arrows, while the *mantis* (seer) continued to sacrifice more victims. Despite the increasing **casualties**, Pausanias did not give the order to advance until the omens proved favorable (Hdt. 9.61-62, 72).

The sacrifice was a blood sacrifice, not a burnt offering, and the **Spartans** at any rate always offered a yearling she-goat when the enemy was sighted (Xen. *Lac. Pol.* 13.8). At **Nemea** this occurred when the two armies were less than a *stadion* (200 yards or 185 meters) apart (Xen. *Hell.* 4.2.20). The Spartans, apparently uniquely, also used to conduct a sacrifice prior to leaving their own territory. If this sacrifice, the *diabateria*, proved unfavorable, it was not unheard of for the expedition to be cancelled. This occurred twice in 419 and again in 416 (Thuc. 5.54.1, 5.55.3, 5.116.1).

Greek armies were regularly accompanied by *manteis* (seers), who conducted divination to determine if the contemplated action would be successful. *Manteis* were of high status and often held the position as a hereditary office (see, for example, Hdt. 9.35; Plut., *Alexander*, 52.1). They were subordinate to the ***strategos*** (general), who made the final decision, and the ancient authors make it clear that only authorized sacrifices should occur, with the general supervising (Xen. *Anab.* 5.6.29; Aen. Tact. 10.4). This was presumably to prevent a loss of morale in the event of bad omens. The Spartans were regarded as particularly religious in the use of sacrifice, but it was also common practice in other armies.

The divinatory sacrifice was known as *ta hiera* (the same term was used in the civilian context), the normal victims were sheep (although goats, calves, and cattle could also be used), and the method of divination was extispicy (examining the entrails, especially the liver). The whole subject is discussed in some detail in volumes 2 and 4 of W. K. Pritchett's *The Greek State at War*. *See also* AGESIPOLIS I; AGIS (1); SEPIA; TAMYNAE.

SALAMIS. An island off the coast of Attica and in 480 site of the major sea battle of the Second **Persian War** (a city in **Cyprus** also shares the same name). Aeschylus' *Persae* is a contemporaneous description of the battle, which is also described in detail by **Herodotus** (8.83-96). Following the withdrawal from **Artemisium**, the allied Greek fleet took station at Salamis. The **Persian** army **ravaged** Attica and sacked **Athens** while the fleet anchored at Phalerum. The Greek fleet, reinforced after Artemisium, had 378 **triremes** and some **pentecontors**. The Persian numbers are usually given as 1,207 triremes but this is probably an exaggeration. This is the total given by Herodotus (7.89) for the start of the expedition but although he states (8.66) that the Persian losses prior to Salamis were replaced by Greek reinforcements, the numbers involved seem far too high (400 were lost in a storm at Sepias, and at least 245 at

Artemisium—excluding unnumbered losses at the final engagement there). **Xerxes I**'s fleet therefore probably numbered around 500 **warships**.

Artemisia advised Xerxes to wait and let the Greek coalition break up from internal dissension—it was already in the process of doing so, as the Peloponnesians wanted to withdraw to the Isthmus of **Corinth**. However, Xerxes decided to attack when informed by **Themistocles** of an impending Greek withdrawal. This was a ruse designed to entice Xerxes to battle and ensure that the Greeks fought at the more favorable site of Salamis rather than at the isthmus. Xerxes dispatched a squadron of ships to block the western exit, occupied Psyttalea (an island between Salamis and the mainland), and blocked the eastern entrance with his main fleet. These preparations took place at night, and the Persian fleet began the action the next day without proper rest.

At dawn the Persians advanced from the east, but their ships lost formation because the channel narrowed and the right wing was able to move much faster than the left. Using local knowledge, the Greek fleet waited until the daily swell rose, which further disrupted the Persian fleet. Launching an attack, they crowded the Persians together even more, causing them to foul each other and generating further confusion. The Persian fleet was routed after some hard fighting and the soldiers on Psyttalea were massacred. The Greek plan and its execution almost certainly derived from Themistocles, rather than from the Spartan admiral, **Eurybiades**.

The battle caused Xerxes' withdrawal from Greece, although he left a large force under **Mardonius** to complete the conquest by land. The Greek fleet followed up slowly and in 478 deployed to Delos and then Samos, leading to the battle of **Mycale**.

Salamis boosted Greek morale (and the reputation of Athens, which had supplied 180 triremes) and allowed a move from defense to offense—both on land and sea. Because of this, it is often regarded as the turning point of the war, although it took the land battle of **Plataea** (1) the following year to expel the Persians from Greece. *See also* ARISTIDES.

SARISSA (pl. *SARISSAI*). A long, two-handed lance, very probably (although not certainly) introduced into some infantry and **cavalry** units in the **Macedonian army** under **Philip II**. It was equipped with a buttspike on the opposite end from the spearhead, providing balance and allowing both ends of the **weapon** to be used offensively. The ancient sources give the length of the *sarissa* around the time of **Alexander the Great** as between 10 and 12 cubits. Using the Greek cubit of 18.25

inches (0.46 meters), this gives a length of 15.2-18.25 feet (4.6-5.5 meters).

The origins of the *sarissa* are uncertain, although it may have developed from the longer spear carried by one type of **peltast**. Following Alexander's successes in Asia, the *sarissa* was gradually adopted in many other Greek states. Over time its length was increased, to as much as 16 cubits (24.3 feet or 7.3 meters). By the second century, when the *sarissa*-armed **phalanx** was defeated by the **Roman** legion, it was a fairly cumbersome weapon, which imposed much greater constraints on the later phalanxes than those under Philip and Alexander. **Polybius** (18.29-33) provides useful information about the *sarissa* and its use in the phalanx.

SCIRITAE (*SKIRITAI*). A unit in the **Spartan army**, probably composed of *perioeci* from the region of Sciritis (Skiritis). This unit apparently had a different status from other units of *perioeci*. The *Sciritae* were tasked with guarding the Spartan camp at night and acting as advance guard on the march. Although this could imply a light or skirmishing role, Thucydides notes that they traditionally held the extreme left of the Spartan battle line, and the description of the role of the 600 *Sciritae* at **Mantinea** in 418 suggests they were **hoplites** (Thuc. 5.67-72).

SCOUTS. *See* RECONNAISSANCE

SCYTHIA. The broad region extending from the north of **Thrace** around the northern edge of the Euxine (Black Sea) to the Caucasus. Its inhabitants, the Scythians (or Scyths), were a generally nomadic people who were regarded as the archetypal barbarians by the Greeks. They were noted for their skills as both riders and **archers**. Scythian **cavalry** and their horses were protected by **armor** (Arr. *Anab.* 3.13.4) and they were reputed to have developed the wedge **formation**, later adopted by the **Macedonians** (Arr. *Tactics*, 16.6; Aelian, *Tactics*, 18.3). They were used as **mercenaries** from time to time in a variety of Greek armies. The **Athenians**, for example, employed Scythian mounted archers in their army from the late sixth to the early fourth centuries and also used Scythian archers as police.

The Greeks had extensive contact with the Scythians from the end of the seventh century, with the establishment of Greek colonies on the north shore of the Euxine. Around 516 the Scythians resisted the invasion of **Darius II** by using mobile defense—they withdrew ahead of the Persians, harassing **foragers** and small parties of troops until Darius gave up and returned home (Hdt. 4.119-132). In the fourth century they extended their control south to the Danube. Their king, Atheas, allied with **Philip II** of Macedon probably around 340, but in 339 failed to fulfill his obligations and Philip invaded. The details of the campaign are unknown, but Atheas was apparently heavily defeated in a battle and killed and Philip captured 20,000 women and children and large numbers of animals (Justin, 9.1.9-3.3). In 331 Zopyrion, **Alexander the Great**'s governor in Thrace and his 30,000-strong army (predominantly Thracian) were in turn defeated by the Scythians (Curt. Ruf. 10.1.44; Justin, 12.1.4, 2.16). Some Scythian tribes served as allies of **Xerxes I** against Alexander.

When united under a strong ruler the Scythians were a dangerous near-neighbor to Greece, but they were more often disunited. From around 250 they were very gradually displaced by the Sarmatian migration south. See also *HIPPOTOXOTAI.*

SECOND ATHENIAN CONFEDERACY. *See* ATHENS

SELEUCID ARMY. Although based on the **Macedonian army** of **Philip II** and **Alexander the Great**, the Seleucid army developed its own characteristics—influenced by local practices and conditions and by warfare against the kingdoms of the other *diadochoi.* A perennial problem was the balancing act between maintaining internal security and possessing an army capable of defending the empire—if native troops were extensively used, the risk of revolt increased exponentially. The two main solutions adopted were hiring **mercenaries** and settling **Macedonians** and Greeks within the empire. Both methods were used, although mercenaries were expensive and sometimes of doubtful loyalty. It has been estimated that at the battle of **Raphia** in 217 soldier settlements allowed the Seleucids to deploy 42,000 heavy infantry of Macedonian/Greek origin, 3,000 light infantry of **Thracian** origin, and 8,000-8,500 **cavalry** of mixed origin (*CAH* 7.1, 265). Native **Syrians** were recruited, but only as auxiliaries—although these formed an important part of the army, protecting the flanks of the **phalanx**.

The *sarissa*-equipped phalanx remained the core of the army and was predominantly drawn from those of Macedonian or Greek origin. Paradoxically, given the greater campaigning distances and the strength of the local tradition, the use of cavalry declined. This probably reflected the more settled conditions of third-century warfare, where large-scale cavalry incursions could no longer win kingdoms. However, it probably also reflected the cost of maintaining cavalry, including the heavier cost of a soldier settlement grant for a cavalryman compared to that needed for an infantryman.

The flanks of the phalanx were protected, as under Alexander, by lighter infantry and cavalry, as well as by **hoplite**-style units such as the *argyraspides*. These seem to have gone over to **Seleucus I (Nicator)** sometime after 312 and remained a unit in the Seleucid army for quite a period after that. The army also used war **elephants** (the Seleucids originally had a monopoly on these because of their links with India), but in general their use declined as countermeasures became more common. Nevertheless, **Antiochus III (the Great)** deployed 54 at **Magnesia**, his last battle against **Rome**, with 20 interspersed in pairs between the phalanx contingents and 16 on each wing.

In terms of **logistics**, the Seleucid army was probably not as efficient as Alexander the Great's. It appears to have had a very large baggage train (perhaps influenced by Eastern practices and the presence of mercenaries), which much have slowed its deployment rate. Conversely, the siege train was retained at a high level of efficiency and benefited from technical improvements to **artillery**, probably emanating from Ptolemaic **Egypt**.

In terms of equipment, not a lot changed, although the phalanx generally seems to have dispensed with body **armor**, apart from a leather cuirass. However, contact with the **Gauls** (Galatians) and **Italians** led to the introduction of the oval **shield**. After Macedon's defeat at **Pydna** (2) in 168, Seleucid Syria joined the rush to remodel at least some of its army along Roman lines. Despite this, it was unable to withstand the sheer numbers and professionalism of the Roman armies, failing to win a single victory against them.

SELEUCIDS. *See* ANTIOCHUS; SELEUCUS; SYRIA

SELEUCUS (SELEUKOS). Common name within the Seleucid dynasty which controlled an empire centered on **Syria** during the Hellenistic period. The most important militarily were:

(1) Seleucus I (Nicator) (c. 358-281/0). Founder of the Seleucid dynasty and ruler from 312 to 281/0. He was a **Macedonian** aristocrat who participated in **Alexander the Great**'s conquest of Asia Minor. One of the *diadochoi*, he held the satrapy of Babylonia, fighting **Eumenes of Cardia**. In 316 hostile action by **Antigonus I (Monophthalmus)** forced Seleucus to flee to **Egypt**. By 312 he had recovered not only Babylonia but also Media and Susiana and in about 309 ended the war with Antigonus by heavily defeating his general, Nicanor, somewhere in the eastern satrapies. Seleucus assumed the royal title about the same

time as the other *diadochoi*, following the lead of Antigonus and his son, **Demetrius I (Poliorcetes)** (Diod. Sic. 20.53).

Despite ceding India to Chandragupta around 304—in return for 500 war **elephants**—Seleucus managed to gain control of much of Alexander's empire, earning the epithet Nicator ("Conqueror"). In 302 he joined the coalition against Antigonus (Diod. Sic. 20.106) and in 301 brought to **Ipsus** around 20,000 infantry, 12,000 **cavalry** (including mounted **archers**), 480 elephants, and 100 scythed **chariots** (Diod. Sic. 20.113.4).

After Ipsus, in which battle his elephants played a major role (Plut., *Demetrius*, 29), Seleucus instead of **Ptolemy I (Soter)** was awarded Coele-Syria, on the basis that Ptolemy had not fought at the battle. Ptolemy, however, had already occupied the territory and Seleucus withdrew, stating that "he would consider later how he ought to deal with friends who decided to encroach" (Diod. Sic. 21.1.5). The issue of Coele-Syria led to continual dissension between Seleucid **Syria** and Ptolemaic Egypt and resulted in several wars (the **Syrian Wars**) after Seleucus' death.

Seleucus married Stratonice, the daughter of Demetrius I (Poliorcetes), but around 295 quarreled with him over the possession of Cilicia. In 285 he captured Demetrius, who had abandoned Greece to try to restore his fortunes in Asia Minor. Seleucus treated his royal captive well (Plut., *Demetrius*, 50).

Seleucus achieved the height of his power in 281 when he defeated **Lysimachus** in battle at **Corupedium** and then invaded Greece, aiming for the throne of Macedon. However, before he could take control, he was assassinated by Ptolemy Ceraunus, who assumed command of the army and Seleucus' position in Europe.

Seleucus was an accomplished general who was prepared to act aggressively. His recapture of Babylonia was achieved by a lightning strike with an initial force of around 800-1,000 infantry and 200-300 cavalry. Using speed and his popularity in the region from his four years as satrap there, he attracted additional troops to his side at every stage of the campaign. After storming the citadel of Babylon, Seleucus defeated Nicanor (Antigonus' satrap in Media) in a surprise night attack, although outnumbered more than 3:1 (Diod. Sic. 90-92; cf. Appian, *Syrian Wars*, 9.54). *See also* SELEUCID ARMY.

(2) Seleucus II (Callinicus) (c. 265-226). Son of Antiochus II (Theos); ruler from 246 to 226. His epithet, Callinicus ("Gloriously Triumphant"), in hindsight was rather ironic. Although Seleucus did well to preserve as much of the Seleucid empire as he did in the face of con-

siderable internal and external pressure, his reign was marked by the loss of considerable Seleucid territory.

On the death of Antiochus II conflict broke out between his two queens, Laodice and Berenice, over whose son should succeed to the throne. Ptolemy III (Euergetes) of **Egypt** invaded **Syria** in support of his sister, Berenice, sparking the Third **Syrian War** (246-241). Ptolemy's army rapidly took Seleuceia in Pieria, secured Antioch, and conquered Cilicia. Ptolemy then advanced on Antioch, but his sister and son were assassinated and he returned to quell a revolt in Egypt. Seleucus recovered most of the lost territory in 245-244 and gained his epithet Callinicus. However, when the war ended in 241 Egypt retained Phoenicia and Seleucia in Pieria, reducing Syria's access to the sea.

During the Third Syrian War Seleucus was forced to set up his younger brother, **Antiochus Hierax**, as ruler in Asia Minor in return for their mother Laodice's financial support for the war. With anyone less ambitious than Antiochus, this arrangement might have been tolerable, but Antiochus allied with the **Gauls** (Galatians) and proved to be a destabilizing influence in the region. In 239 Seleucus attempted to recover his territory from his brother, sparking the War of the Brothers. Although Seleucus had some initial success, he made peace around 236 after his army was decisively beaten by Antiochus and his Gallic allies near Ancyra. This peace ceded Asia Minor north of the Taurus Mountains to Antiochus, but by 228 Antiochus had lost it all to **Attalus I (Soter)** of Pergamum. During Seleucus' reign (probably during the War of the Brothers or the Third Syrian War) Bactria and Sogdiana also apparently became independent and Parthia fell to the Parnoi.

In 228 Seleucus campaigned against the Parthians but was forced to return in 227 to quell a combined revolt by Antiochus and their aunt Stratonice. Seleucus drove Antiochus out of the country and captured and executed Stratonice. The following year he died after a fall from his horse. Seleucus spent almost his entire reign on campaign—he had to fight to secure his throne and against a series of internal and external threats. Although under him the Seleucid empire was considerably reduced, arguably he did well to survive at all. *See also* SELEUCID ARMY.

SELLASIA. Final battle of the **Cleomenic War**, fought in July 222 and described, apparently accurately, in considerable detail by **Polybius** (2.65-70). **Cleomenes III** of **Sparta** was on the defensive against a powerful coalition of **Achaeans** under **Aratus** and **Macedonians** under **Antigonus III (Doson)** and was extremely short of funds since Ptolemy III (Euergetes) had cut off his financial support. In mid-222 Antigonus invaded Laconia, leading an army of 28,000 infantry and 1,200 **cavalry** (comprising the 10,000-strong **phalanx**, 3,000 **hypaspists**, and 300 cavalry from Macedon; 3,000 infantry and 300 cavalry from Achaea; 3,000 **mercenary** infantry; and allied troops from various states). Cleomenes had prepared the various passes into the country for defense and, correctly anticipating which route Antigonus would take, waited for him at Sellasia, where the road to Sparta ran between two hills.

Cleomenes deployed the bulk of his 20,000-strong force (the proportion of infantry to cavalry is unknown) on Evas and Olympus, the two hills on either side of the road, strengthening both positions with a trench and palisade. He also placed a force of cavalry and mercenary infantry on the level ground on either side of the road running between the hills. Impressed by Cleomenes' dispositions, Antigonus spent several days in **reconnaissance** before ordering the advance. He deployed with his Macedonian and mercenary troops opposite Cleomenes on Olympus and placed his hypaspists, Illyrians, and other allies (including 2,000 Achaeans) opposite Cleomenes' brother, Eucleidas, on Evas. The cavalry, 1,000 Achaeans, and 1,000 Megalopolitans equipped as Macedonian phalangites were in the center. His first action was to send his Illyrian troops against Eucleidas, while the cavalry and Achaeans in the center were ordered to hold until they received a signal.

The Spartans' mercenary *psiloi* (light troops) with the cavalry on the level ground next to Evas seized the opportunity to attack the Illyrians in the rear. Caught between Eucleidas' heavy infantry and the mercenaries, the Illyrians were in considerable danger until the cavalry contingent from Megalopolis was persuaded by the young **Philopoemen**, against orders, to charge the enemy cavalry as a diversion. The mercenary *psiloi* abandoned their attack on the Illyrians, who, along with the Macedonian hypaspists, pushed Eucleidas off the hill. Antigonus later personally praised Philopoemen's actions in disobeying orders and ensuring that the attack on Evas succeeded. On the other wing Antigonus, after some very hard fighting, dislodged Cleomenes, causing heavy **casualties** to the Spartans. This battle caused Cleomenes to go into exile and allowed Antigonus to occupy Sparta and end the war.

SEPIA (SEPEIA). A place in the territory of **Argos**, near Tiryns; the site of a major battle around 494 between **Sparta** and Argos, recorded by **Herodotus** (6.76-83). The purpose of the Spartan campaign is uncertain—probably to neutralize Argos as a threat, possibly to force it into the **Peloponnesian League** (although this did not occur). The fact that **Cleomenes I** was (unsuccessfully) prosecuted on his return for not taking the city suggests that some in Sparta were looking for a more permanent conquest.

Cleomenes' army may have been a purely Lacedaemonian force (although this rests on the wording used by Herodotus, which may not bear such precision) and initially advanced by land, via Thyrea, to within three miles (five kilometers) of Argos. Here the **sacrifices** proved unfavorable, so he withdrew to Thyrea and moved his army by sea (which suggests a prearrangement), landing near Tiryns. The Argives deployed to meet Cleomenes and camped at Sepia. According to Herodotus the Argives conformed to the Spartan herald's orders. Cleomenes noticed this and passed a secret message that when the herald ordered the stand-down for the midday meal, the Spartans were to attack instead. The ruse succeeded and the Argives, caught while standing down, were routed. Many took refuge in a sacred grove but Cleomenes lured some out, pretending they had been ransomed, and killed them. He then fired the grove when the rest refused to evacuate it. Cleomenes, later citing religious reasons, did not follow up the victory and returned home.

The battle was a decisive defeat for Argos—which lost 6,000 **hoplites** (a very large number). The scale of the loss caused internal problems, kept the city quiet for a generation, and was used as an excuse for Argos' neutrality in the Second **Persian War** (Hdt. 6.83, 7.148-149).

SHIELD. Several types of shield were used in the course of Greek warfare, predominantly by infantry. The main types were:

(1) *Aspis.* The *aspis* (or *hoplon*) was the central piece of **hoplite** equipment. Up to about 3 feet (95 centimeters) in diameter, it was made of wood with a thin bronze facing and a bronze rim. The wearer slipped his left forearm through a bronze band (*porpax*) in the inside center of the shield and gripped a handle (*antilabe*) inside the rim with his left hand. As the hoplite's left elbow was in the center of the shield, when he faced forward in the battle **formation**, the left-hand half of his *aspis* protected the right-hand half of the man to his left. See plates 1 and 4.

Although the Greek hoplite shield remained essentially unchanged for centuries, it underwent some modification in **Macedon**. From around the last quarter of the fourth century, there is evidence that it became increasingly convex or bowl-like and was reduced in size—while retaining the traditional *porpax* and *antilabe* handle system (Markle, *Hesperia* 68 [1999]: 243-246).

The front of the shield was decorated with a blazon. At the start of the hoplite era, these blazons were apparently chosen by the owner of the shield. However, by the fourth century some states had adopted a uniform blazon, such as the Σ (sigma) of **Sicyon**, the Λ (lambda, for Lacedaemon) at **Sparta**, and the club of Hercules (Herakles) at **Thebes**.

As the *aspis* was central to the integrity of the hoplite line, losing one's shield was a serious matter. The Spartans emphasized a "come back with your shield or (carried) on it" ethic, and at **Athens** throwing away one's shield (*rhipsaspis*) was an indictable offense. *See also* ARCHILOCHUS; BRASIDAS.

(2) Body. A large shield of the **Mycenaean** period. It was often in the shape of a figure eight, with a leather rim and constructed from layers of bull-hide (Ajax's shield had seven) glued together and stretched over a wooden frame. The frame had a horizontal wooden reinforcing piece across the center, which probably also doubled as a hand grip. The shield also had a supporting strap (*telamon*), worn around the soldier's neck (see [9] below). An alternative shape was similar to the rectangular Roman legionary's shield but larger. The body shield is also known as the tower shield.

(3) Cavalry. Shields were not carried by mainland Greek **cavalry** in the Archaic and Classical periods, but coins from **Tarentum** (Taras) in **Sicily** show a military cavalry shield there from around 430. As the Roman republican cavalry regularly carried shields, this may have been a local practice in **Italy** and Sicily.

(4) Dipylon. An Archaic shield, probably descended from the **Mycenaean** "figure-eight" body shield, (2) above, and perhaps used down to the start of the seventh century. Some scholars have argued that the Dipylon shield never existed but was purely an artistic form used to denote a hero or a heroic scene. This use is common in sixth and fifth century art (see plate 1). However, the consistency of the representations, and the existence of very realistic clay models from an eighth-century **Athenian** cemetery, suggest that this shield existed and was used at some time prior to the introduction of the *aspis*, (1) above. The Dipylon shield is also known as the **Boeotian** shield.

(5) *Hoplon*. Another name for the *aspis*. See (1) above.

(6) Macedonian. Members of the **Macedonian phalanx** used a small shield without an offset rim. A *telamon* (strap) over the left shoulder was used to secure it in order to leave both hands free to use the *sarissa*. Two types are recorded, both in use from the mid-fourth to the mid-second centuries and ranging in size from around 27 to 30 inches (70-78 centimeters) in diameter. The first is more bowl-shaped and usually decorated with a large circle (often with a star inside) in the center, surrounded by semicircles with the convex sides facing into the center of the shield. The second is flatter, often represented as undecorated, and assessed by **Asclepiodotus** (*Tactics*, 5.1) as the best shield

for the phalanx (Markle, *Hesperia* 68 [1999]: 246-251). Cf. *aspis*, (1) above.

(7) Oval. During the Hellenistic period an oval shield (like the **Roman** *clipeus*) came into vogue. Attested in art of the third and second centuries, it was long enough to protect from the shoulder to just below the knee and wide enough to cover the bulk of the torso.

(8) *Pelte*. This shield was a wicker shield, usually crescent shaped, originating in **Thrace** and carried by *psiloi* (light troops), mainly to ward off missiles. Worn strapped to the left forearm (in a similar fashion to the *aspis*, [1] above), it provided much less protection than the *aspis*, but was much lighter to carry. This shield was the origin of the name "**peltast**," given to the lightly armed troops who carried it. See plate 2b.

(9) *Telamon*. The name of this shield-type derives from the method of securing it—a strap (*telamon*) over the left shoulder, often in combination with some sort of internal handle, gripped by one hand. It seems to have come in a variety of designs (including the body and Macedonian shields, [2] and [6] above). Although the *telamon* shield did not have as secure a grip as the *aspis*, (1) above, as modified for the Macedonian shield it did allow a soldier to wield a two-handed **weapon** such as the *sarissa*.

(10) Tower. Another name for the body shield. See (2) above.

SHIPS. In addition to the **warship**, the Greeks used ships for a variety of military purposes, principally transport, for both horses and troops, and supply. The **Athenians** introduced specialist horse transport ships for the first time in 430, using old ships (of unspecified type) converted for the purpose, and they regularly took **cavalry** on their seaborne **raids** of the Peloponnese during the remainder of the **Peloponnesian War** of 431-404 (Thuc. 2.56.2, 4.42.1). Horses did not always travel in such comfort, though—**Plutarch** (*Timoleon*, 19.3) records **Corinthians** getting horses across to **Sicily** from Thurii in calm weather by tying their reins to the troop transports and swimming them.

Soldiers could be carried on warships, but—in the Classical period in particular, when the **trireme** was the main vessel—room was limited. As with the horse transports, there is evidence of older ships being converted to troop transports. Towed barges or even rafts were probably also used to move soldiers over shorter distances (Thuc. 26.2.4, 7.7.3).

Requisitioned merchant ships were often used to carry supplies in the Classical period. **Thucydides** states that the supplies for the **Sicilian Expedition** of 415 were "carried on 30 merchant ships loaded with grain and with bakers, stonemasons, and carpenters on board, as well as tools for constructing siegeworks" (6.44.1). These were supplemented by another 100 smaller boats, also requisitioned, as well as private boats traveling with them to conduct trade with the expeditionary force.

SICILIAN EXPEDITION. The Athenian expedition to **Sicily** in 415-13; one of the major factors in the Athenian loss of the **Peloponnesian War** (2) of 431-404. **Athens** had a reasonably long history of contact with the Greek settlements in Sicily and **Italy** and operated a fleet of up to 60 ships in Sicilian waters from 427 to 424. This was ostensibly to help Leontini against **Syracuse**, but according to **Thucydides** (3.86) was really to stop grain from being sent to the Peloponnese and with a view to future conquest. However, in 424 the Sicilians, fearing Athenian ambition on the island, patched up their quarrels and presented a united front in asking the Athenians to leave. The importance of the island to Athens is demonstrated by the fate of the three *strategoi* (generals) involved—when they returned home, they were tried for failing to take control of the island. Two (Pythodorus and Sophocles) were exiled and one (Eurymedon) fined (Thuc. 4.58-65).

In 416 a delegation from Segesta arrived in Athens, seeking help against its neighbor Selinus. **Alcibiades**, still smarting from the failure of his anti-**Spartan** coalition at **Mantinea** in 418, successfully proposed that Athens send an expeditionary force. Thucydides (6.8-26) records the debate in the Athenian assembly when **Nicias** subsequently raised objections to this decision. The new debate did not lead to the cancellation of the enterprise, however—instead the size of the force was increased. Alcibiades, Lamachus, and Nicias were appointed to command the force of 134 **triremes** (100 from Athens, the rest from **Chios** and elsewhere) and two **pentecontors**, 5,100 **hoplites** (including 1,500 Athenian hoplites and 700 citizen-*epibatai* [marines]—the rest were subject, allied, or **mercenary** troops), 30 **cavalry**, 480 **archers**, 700 **slingers**, and 120 *psiloi* (light infantry) from **Megara** (Thuc. 6.43). Thucydides describes it as "the most expensive and best prepared Greek force which had come from a single city up to that time" (6.31.1).

On arriving in Sicily and discovering that the money promised by Segesta to support the expedition did not exist and that there was limited support for their presence, three plans emerged. Nicias argued that they should arrange or force a settlement between Segesta and Selinus and then sail home, making a demonstration of force along the coast to overawe the Sicilian Greeks. Alcibiades considered it disgraceful to return home without accomplishing anything of substance and argued that they should mobilize as much of Sicily as possible against Selinus and Syracuse and then attack them—unless Selinus came to terms with Segesta and Syracuse restored Leontini. Lamachus' plan was to sail straight to Syracuse, the most important Sicilian city, and attack it while it was still reacting to the Athenian arrival. This focused on the Sicilian center of military gravity and offered the best chance of success, but Lamachus was the least influential of the generals and he ended up supporting Alcibiades' plan (Thuc. 6.47-50.1).

Unfortunately for Athens, shortly after this Alcibiades was recalled to Athens to face politically motivated charges of profaning the Eleusinian Mysteries, leaving Nicias—who was not committed to the expedition and who was moreover ill—as the senior general. Alcibiades deserted to Sparta on his way back to Athens and was instrumental in persuading the Spartans to send a force (under a Spartiate, Gylippus) to assist Syracuse. In the meantime the Athenians had tricked the Syracusans into advancing on the Athenian base at Catana, enabling an unopposed Athenian landing at Syracuse itself, and shortly afterward defeated the Syracusans in battle outside the city. However, the Syracusans lost only 260 men because their cavalry protected their withdrawal. The Athenians lost about 50 men and because they were totally outnumbered in cavalry, returned to Catana for the winter (Thuc. 6.62-71).

The following spring (414) the Athenians were reinforced by 250 cavalrymen and 30 *hippotoxotai* (mounted archers), although they had to purchase their mounts locally. With their weakness in cavalry rectified, the army defeated the Syracusans at Epipolae, the high ground that commanded the city, and began constructing forts and a series of walls to blockade the city. The Syracusan response was to build a counter-wall out from the city at right angles to the Athenian wall, in an attempt to cut it off. This was destroyed, and the Athenians also defeated a second attempt to build a counter-wall in a different location, although Lamachus was killed in this action, leaving Nicias in sole command. The Syracusans were now in a dangerous position—local reinforcements had raised the Athenians' cavalry force to 650, their **fortifications** were being constructed very quickly, and the Athenian fleet had sailed into the harbor and was blockading the city from the sea. All of this was inclining them to negotiate a settlement (Thuc. 94-103).

At this point the Spartan Gylippus ran the Athenian blockade and landed at Himera. Along with about 3,000 men (including 1,000 hoplites and 100 cavalry), he managed to cross Epipolae before the Athenians could finish their wall and got into the city. Stiffened by the presence of a Spartiate officer and reinforcements, the Syracusans renewed their efforts to build a counter-wall. They lost the ensuing battle, largely because Gylippus deployed the army between two lines of fortifications, rendering their cavalry ineffective. The next day, however, deployed in more open country, the Syracusan cavalry was able to rout the Athenian left wing, putting the entire army to flight. This initial Syracusan success was sufficient to allow them to complete their counter-wall and prevent the Athenians from completing their wall around the city (Thuc. 7.1-7). From this point onward the Athenians lost the initiative and were continually harassed by the enemy cavalry. The success at Syracuse encouraged the Spartans to reopen the war against Athens on the grounds that Athenian raids on the coast of the Peloponnese were a clear breach of the Peace of **Nicias**. Athens was now faced with a war on two fronts.

Despite this, Athens answered a plea for reinforcements from Nicias, who was suffering from kidney disease and increasingly unable to exercise the responsibilities of command. Eurymedon was sent out with 10 triremes and money in winter 414/3, while **Demosthenes** (1) arrived in spring 413 with 73 triremes, 5,000 hoplites (Athenians and allies), and large numbers of slingers and archers. Nicias had recently lost control of one of his forts, Plemmyrium, and had suffered a naval defeat in the harbor, so Demosthenes' arrival was a major boost to Athenian morale. To capitalize on the fear his arrival caused in Syracuse, Demosthenes planned a daring **night attack** on Epipolae to capture the Syracusan counter-wall and restore the initiative to the Athenians. The attack almost worked, but the Athenian and allied forces fell into confusion in the dark pressing home their initial success and were repulsed with considerable loss (Thuc. 7.16-17, 21-26, 31-44).

The expeditionary force was still quite powerful, though, and Demosthenes, supported by Eurymedon, argued that it should cut its losses and stage a fighting withdrawal. However, Nicias, believing that internal dissension within Syracuse offered a chance of success, refused. Nevertheless, even he was convinced that there was no hope of success when further reinforcements arrived for Syracuse—but decided to delay the withdrawal for 27 days in response to an eclipse of the moon. The delay was disastrous. The Syracusans got wind of the withdrawal and launched a combined land and sea attack on the Athenians. The land attack was not successful, but the Athenian fleet was beaten and Eurymedon killed. When the Syracusans began to block the mouth of the harbor, intending to trap the Athenian fleet, the Athenians had no option but to fight again. The Syracusans inflicted a decisive defeat on the Athenian fleet, ending all hope of a withdrawal by sea. Forced to retire by land, the Athenian army was harassed by the Syracusans, particularly their cavalry, until they were forced to surrender at the **Assinarus River**. Some 7,000 were taken **prisoner**, and Demosthenes and Nicias were executed (Thuc. 7.45-87).

The disaster in Sicily had a major effect on Athens. Athenian aggression in Sicily and toward the Peloponnese had caused Sparta to end its treaty with Athens and renew the Peloponnesian War. At the same time, Alcibiades' defection and his advice to fortify **Decelea** seriously weakened Athens' home front. The loss of so many ships and men in Sicily, coupled with the Syracusan decision to send ships and men to help Sparta and serious revolts in the empire, meant that Athens no longer had much depth to her naval superiority. The loss in Sicily was not the sole reason for Athenian defeat in the Peloponnesian War of 431-404—the plague also played its part, and **Persian** support to Sparta was crucial—but it did give Athens a much smaller margin for error.

SICILY. The island of Sicily was colonized by Greeks in the eighth century (Naxos and **Syracuse** being the two earliest colonies). At this time the island had two groups of local inhabitants—the Sicans, who were restricted to the mountainous areas, and the Sicels who occupied most of the coastal areas prior to the arrival of the Greeks. Greek settlement was largely in the east, as another colonial power, **Carthage**, controlled the west. Most of Sicily's Greek military history (after the original colonization, which involved conflict with the Sicels) revolved around the struggle for dominance between Syracuse and the other Greek cities and the struggle for mastery of the island between the Greek cities (particularly Syracuse) and Carthage. To a large degree, the military history of Sicily is the history of Syracuse. In addition to Carthaginian invasions, the island also had to cope with a major **Athenian** attempt at conquest (the **Sicilian Expedition**, 415-413) and **Roman** invasions in 264 and 213.

Control of the bulk of the island frequently swung between Carthage and the Greeks (especially Syracuse), although neither side was able to gain complete control. The Sicels and Sicans were involved as allies on both sides from time to time. Long-term and serious warfare really developed from the early fifth century, when **Gelon** of Syracuse repelled a major Carthaginian invasion at the same time as the mainland Greeks were repelling the Persians during the Second **Persian War**. Extensive wars were also fought from the end of the fifth century—under **Dionysius I**, Syracuse fought four wars against Carthage between 407 and 367. After some reverses, **Agathocles** of Syracuse (317-289) eventually established Syracusan control over most of the island and restricted Carthaginian influence to the extreme west. However, the Syracusan empire collapsed after his death, leading to instability within the city and on the island in general. Carthaginian influence was ended with the First Punic War (264-241)—after which Rome was the dominant power (although Syracuse remained as an important independent city because **Hieron II** had supported the Romans against Carthage). However, Syracuse chose the wrong side in the Second Punic War and its fall to Rome in 211 led to the island being brought fully under Roman control.

Unsurprisingly, the Greek cities of Sicily maintained Greek-style armed forces. The **hoplite** was the dominant form of infantry, although the Sicilians arguably made more use of *psiloi* (light troops) earlier than many of the mainland Greek states. The emphasis on **cavalry** was certainly greater than in mainland Greece and, in a significant departure from normal Greek practice, some Sicilian Greek cavalry were equipped with **shields**. Sicilian Greek navies were equipped with the **trireme** in the Classical period and again were a more important part of military activity than was the case with many mainland Greek states (Athens, Aegina, and **Corinth** were exceptions to this).

Artillery was first developed in Sicily, under Dionysius I of Syracuse in 399. This and the relative savagery of many Sicilian campaigns perhaps reflect the rather different nature of warfare there. Greek warfare in the Classical and early Hellenistic periods was often settled by a hoplite clash and usually did not involve the permanent acquisition of territory. In Sicily though, the wars between the Greeks and Carthage really aimed at the extermination or expulsion of one side by the other. This was also carried over into inter-Greek warfare on the island with sieges, massacres, and forced resettlement common occurrences, especially under the tyrants. In this respect, Sicilian warfare was closer to the wars waged by Hellenistic monarchs than by city-states of the Classical period.

SICYON (SIKYON). A prominent Peloponnesian city located on a naturally strong plateau northwest of **Corinth**, about two miles (3.2 kilometers) from the southern coast of the Corinthian Gulf. Sicyon was apparently founded from **Argos** and developed into a moderately prosperous maritime power. According to the fourth-century tradition (*FGrH* 105 fr. 2), in the mid-seventh century Orthagoras made his reputation as a soldier, gained command of the Sicyonian frontier guards, and used them to seize power as tyrant. Around 591, under another tyrant, Cleisthenes, Sicyon aided the **Amphictionic League** against Crisa in the First Sacred War. Cleisthenes' reign also saw long-standing hostilities with Argos. Around 555 **Sparta** liberated Sicyon from tyranny, bringing the city firmly into the new **Peloponnesian League**, to which it remained loyal for nearly 200 years.

During the Second **Persian War** Sicyon contributed 15 ships to the Greek forces at **Salamis**, 3,000 **hoplites** at **Plataea** (1), and a force of unknown size at **Mycale** (Hdt. 8.43, 9.28). It supported Sparta throughout the **Peloponnesian War** (2) of 431-404, and remained loyal during the **Corinthian War**, supplying 1,500 troops at **Nemea** in 394 (Xen.

Hell. 4.2.16). Sicyon also mobilized to provide support after **Leuctra** in 371 (Xen. *Hell.* 6.4. 18) but was brought into the **Boeotian** alliance after **Mantinea** (2) in 362. Following this the city began to take a more independent line, and it provided military support to Megalopolis against Sparta in 351.

Sicyon joined the anti-**Macedonian** coalition during the **Lamian War**. It was garrisoned by **Ptolemy I** (**Soter**) in 308—one of his few successes in the Peloponnese—but was captured by **Demetrius I** (**Poliorcetes**) only a few years later. Brought into the **Achaean League** by **Aratus** in 251 (Polyb. 2.43), Sicyon soon became a leading member and the frequent site of league meetings. The city was besieged by **Cleomenes III** in 224, but he was forced to raise the siege and move north to a defensive position on the isthmus when Aratus handed over Acrocorinth to **Antigonus III** (**Doson**). Antigonus wintered in Sicyon in 224/3 and used it as a base from which to recover Tegea and capture **Orchomenus** and Mantinea (Polyb. 2.54). Sicyon's territory was **ravaged** by the **Aetolians** after **Caphyae** in 220 on the outbreak of the Second **Social War** of 220-217 (Polyb. 4.13). With the defeat of the Achaean League by **Rome** in 146, Sicyon, along with other member states, fell firmly under Roman domination.

SIEGE WARFARE. The most common siege technique prior to the fourth century was the blockade. One of the most famous, and longest, was the 10-year siege of Troy in the **Trojan War** which was ended by a famous ruse. For landlocked cities, a siege normally involved surrounding the city with troops or constructing a wall around the city perimeter. The Spartans did this at **Plataea** in 429. **Thucydides'** comprehensive description of this siege (2.71-78, 3.20-24, 52-68) gives a very good idea of the range of options available prior to the use of **artillery**. These included the use of battering rams, building a mound against walls, and mining under them (sapping). In this case, the defenders successfully countered all these measures but were eventually starved into submission by the blockade. For cities with sea access, a siege wall had to be supplemented by naval blockade—as in the Athenian siege of **Potidaea** from 432 to 430. However, sieges (especially ones involving naval forces) were very expensive because of the large numbers of troops required. The two-year siege of Potidaea, for example, cost 2,000 talents—more than five times the annual revenue from the Athenian empire at that time.

Early attempts to speed up sieges by using machinery such as rams or siege towers often had limited success. The Athenian siege engines were burned before they could be brought into use at **Syracuse** in 413

and Spartan siege engines also proved ineffective against the fortress of
Oenoe in Attica in 429. However, some progress did occur. In 427 **Nicias** used ship-mounted towers or ladders to seize towers on the island
of Minoa (Thuc. 3.51). Three years later the **Boeotians** used a primitive
flamethrower to overcome the hastily fortified Athenian position at
Delium (Thuc. 4.100).

The design, construction, and maintenance of siege engines was
very expensive. Because of this, technological developments in siege
warfare did not occur among city-states such as **Athens** or **Sparta** but in
the wealthy Sicilian city of Syracuse. No reliable evidence of the use of
artillery exists before 398, when **Dionysius I**, tyrant of Syracuse, besieged the **Carthaginian** stronghold of Motya in western **Sicily** using
nontorsion catapults. These bolt-firing catapults were successfully used
to kill Carthaginians manning the city walls and the ships blocking the
entrance of the harbor (Diod. Sic. 14.50.4-51.2). However, other methods continued to be used alongside the new catapults. Dionysius also
employed six-story-high towers mounted on wheels and battering rams
at Motya. At the siege of **Mantinea** in 385 (probably before the use of
catapults had really spread to Greece proper) the Spartan king **Agesipolis
I** dammed up the river that flowed through the city. This caused it to
rise above the foundations of the wall and dissolve the sun-dried mud
bricks that formed its upper courses (Xen. *Hell.* 5.2.4-5).

The wealth of **Macedon** under **Philip II** and **Alexander the Great**
enabled the construction and transportation of large siege engines, including battering rams, siege towers, and torsion catapults, further advancing siege warfare. Philip, though, often preferred to take cities by
the use of traitors, and the ancient sources suggest that it was only late
in his reign that he started to use costly siege machines. Philip's earliest
firmly datable use of siege engines is in 340 at Perinthus and **Byzantium** (Diod. 16.74-76; Frontinus, *Strategemata*, 3.9.8). His engineer,
Polyidos of **Thessaly**, seems to have designed battering rams that could
be dismantled, moved, and reassembled at another site. Despite these
improvements, Philip failed to capture either Perinthus or Byzantium.

About this time, the invention of the torsion catapult considerably
enhanced the arsenal of the besiegers, and this **weapon** was put to particularly good use by Alexander the Great. At **Halicarnassus** stone-throwing catapults were apparently used for the first time (although as
antipersonnel weapons). At **Tyre** (see Arr. *Anab.* 2.21-23) the defenses
were elaborate and had to be tackled from both land and sea with a variety of siege techniques, machinery, and catapults. This is the first firmly
attested use of stone-throwing catapults used against walls. Alexander's
sieges of **Gaza**, the Bactrian towns, the Sogdian rock, Chorienes rock,

Massaga in India, and Aornos, all described by **Arrian**, demonstrate a wide variety of challenges and the siege methods used to overcome them.

Further improvements were made to catapults under the Ptolemaic rulers of **Egypt** and Hellenistic siege warfare became a fine art, very different from that of the sixth and fifth centuries. **Demetrius I (Poliorcetes)**, for example, used a nine-story siege tower (*helepolis*, or "citytaker") of 135 feet (41.5 meters) at the siege of Salamis (**Cyprus**) in 306. Its lowest floors held his largest stone-throwing catapults and the middle floors the biggest bolt-firing machines, while the top floors had smaller bolt- and stone-throwing engines. This allowed the large stonethrowers to pulverize the walls while the other artillery cleared the battlements of opposition. This was supplemented by the use of rams. Interestingly, when the situation became desperate, the defenders managed to set fire to the *helepolis* with flaming bolts and the city surrendered only when Demetrius' fleet defeated the force sent to relieve it.

The defenders of Syracuse in 213-211 used a similarly high level of technology (designed by Archimedes) against the **Roman** attackers. Hellenistic siege warfare also apparently influenced Roman practice. As Eric W. Marsden (*Greek and Roman Artillery: Historical Development*, 109-110) points out, Julius Caesar's siege of Avaricum (Bourges) in Gaul in 52 bears remarkable similarities to **Philip V** of Macedon's techniques and equipment at the siege of Echinus in 210. *See also* AGATHOCLES; DELIUM; FORTIFICATIONS; RHODES; SICILIAN EXPEDITION.

SILVERSHIELDS. *See ARGYRASPIDES*; HYPASPISTS

SIRIS RIVER. A battle (also known as Heraclea) fought in the summer of 280 between **Pyrrhus I** of Epirus and the Romans during the war between **Tarentum** and **Rome**. According to **Plutarch** (*Pyrrhus*, 15.1), Pyrrhus' Greek forces consisted of an advance guard of 3,000 men (of unknown type) and a main force of 20,000 veteran heavy infantry, 3,000 **cavalry**, 2,000 **archers**, 500 **slingers**, and 20 **elephants** (although this force had suffered unspecified losses at sea in a storm en route to **Italy**). In addition, he had an unspecified number of troops from Tarentum, albeit enlisted under some pressure. The Roman army, led by the consul Valerius Laevinius, was probably marginally smaller, consisting of two legions and allies, perhaps around 24,000-25,000 strong in all.

Laevinius led his army in good order across the Siris River near Heraclea (on the west of the Tarentine Gulf at the foot of Italy), frustrating Pyrrhus' plan of striking with his cavalry while the Roman army was disorganized by the river crossing. Pyrrhus used his elephants on the wings to drive off the Roman cavalry—whose horses had never seen elephants before and were terrified by them—fixed their infantry with his **phalanx**, and used his cavalry to strike the decisive blow. Plutarch (*Pyrrhus*, 17.4) records two sets of casualty figures: Dionysius', of 15,000 Roman and 13,000 Greek dead, and Hieronymus' of 7,000 Roman and 4,000 Greek dead. Hieronymus' figures are the more likely—with **casualties** of slightly over 50 percent, Pyrrhus would have been unable to continue the war. Even with 4,000 dead (including quite a few of his trusted **officers**), Pyrrhus regarded it as an expensive victory. Pyrrhus captured the enemy camp, ravaged to within 40 miles (64 kilometers) of Rome, and secured the defection of several Roman allies, including the Samnites and Lucanians. *See also* ASCULUM; BENEVENTUM.

SITALCES. *See* PERDICCAS; THRACE

SITERESION. See PAY

SLAVES. Ancient Greek society was a slave-owning society, and new slaves were often generated by warfare and **raids**. This involved enslaving not only foreign opponents—the inhabitants of a conquered city or town in Greece itself were sometimes enslaved, particularly from the fifth century onward. The practice became more common under the Hellenistic monarchs and the **Romans**, when the practice of selling defeated enemies or rebellious subjects into slavery was used as a measure to encourage others to remain passive.

However, the use of slaves in warfare, and particularly combat, was generally peripheral. Their employment was limited because the risk of desertion and the danger that they might turn on their masters caused an understandable reluctance to arm slaves—except in times of grave crisis. To meet major shortages of rowers prior to **Arginusae** in 406 the **Athenians**, for example, had to call on slaves (along with other normally exempt groups such as metics [resident aliens] and the **cavalry**) to man their fleet. However, in recognition of the general lack of incentive for slaves to perform well in combat, the Athenians freed en masse those slaves who fought in this battle.

Otherwise, slaves apparently normally accompanied soldiers on campaign as attendants or baggage carriers (*skeuophoroi*) to care for and carry their equipment, cook for them, and undertake a variety of other tasks such as **foraging** or caring for the **wounded**. At times these slaves may also have acted as baggage guards or even been involved on the fringes of battle, throwing stones or other impromptu missiles. Captured slaves, or those who did desert to the enemy, were usually simply regarded as **booty** and sold.

SLINGERS. The sling is attested as a military **weapon** as early as **Homer** (*Iliad* 13.712-716; cf. 12.156, 279-285), but the general lack of evidence after this means we cannot really gauge its use down to the sixth century, although it is mentioned in **Archilochus** (poem 3) and occasionally represented on Archaic period vases.

The evidence is much more extensive for the Classical and Hellenistic periods. For example, **Gelon** of **Syracuse** included 2,000 slingers in his offer of troops to aid the Greeks in the Second **Persian War** (Hdt. 7.158). In large numbers they could generate sufficient firepower to be very effective. Acarnanian slingers caused considerable trouble to the **Spartans** and their allies in both 429 and 389 (Thuc. 2.81; Xen. *Hell.* 4.6.7). Slingers also served under **Demosthenes** (1) at **Pylos**/Sphacteria among the various types of missile troops that proved so effective against the Spartan **hoplites** (Thuc. 4.32). At the Himeras River in 311, 1,000 slingers from the Balearic Islands drove back a determined Greek assault on the **Carthaginian** camp (Diod. Sic. 19.109.1-3). Other notable slingers were the **Rhodians**, who in the late fifth century could apparently outrange Persian slingers and most Persian **archers** (Xen. *Anab.* 3.4.15-17). The **Athenians** took 700 Rhodian slingers with them as part of the **Sicilian Expedition** in 415 (Thuc. 6.43).

The sling consisted of a central pouch for the amunition with two three-foot (one-meter)-long thongs attached. The thongs were probably usually leather, gut, wool, or twisted horsehair, but other materials could be used. The ammunition could be stones, clay shot, or cast lead bullets—the last having the advantage of a regular size and weight. Numerous finds of lead slingshot bullets survive from the Hellenistic period, some stamped with the names of the ruler whose army used them. These bullets are almond shaped and generally weigh around 1.5-2.0 ounces (55 grams). **Diodorus Siculus** (9.109.2) states that the Balearic sling bullets at the Himeras River could penetrate **armor** and that the slingers were so skillful because they practiced from an early age. This mirrors Greek experience. The best slingers were drawn from the wilder and more remote parts of Greece, where the skill must have been in continual demand for the protection of livestock from predators. Ranges of around 435 yards (400 meters) could be attained, with the bullet probably traveling in excess of 60 miles (100 kilometers) per hour.

SOCIAL WARS. The name given to wars between former allies (from the Latin for allies, *socii*).

(1) First Social War (357-355). A war between **Athens** and **Chios**, Cos, **Byzantium**, **Rhodes**, and others who revolted against the Second Athenian Confederacy. Unrest had been growing among the allies for some time over Athens' increasingly high-handed actions—for example, **Chares'** support for an oligarchic coup at **Corcyra**, the establishment of **cleruchies** on Samos (by **Timotheus**) and **Potidaea** (technically legal, but disturbing), and the financial exactions of Athens' underresourced generals. Athens' concentration on **Amphipolis** had also allowed interference with the allies by **Jason of Pherae** and **Thebes**. This unrest was aided by Mausolus of Caria, who seems to have helped foment the revolt and provided naval assistance to the rebels.

The Athenians were defeated in the first action, a land and sea battle at Chios, in which **Chabrias** was killed. This setback caused other allies to revolt and the rebels ravaged Lemnos and Imbros. In 356, reinforced from Athens, Chares threatened Byzantium but was defeated in a naval battle at **Embata** (or Embatum) when his colleagues **Iphicrates**, Menestheus, and Timotheus failed to join him. Chares was then appointed to sole command, but lack of money forced him to hire his fleet and troops out to the satrap Artabazus, who was in revolt against **Persia**.

The war ended in mid-355 when Artaxerxes III of Persia threatened to supply the rebels with 300 ships unless Athens recalled Chares. Athens did so and in the subsequent peace negotiations recognized the independence of Chios, Cos, Byzantium, and Rhodes.

(2) Second Social War (220-217). A war of the **Hellenic League** under **Philip V** of **Macedon**, allied with the **Achaean League**, against the **Aetolian League**, which was supported by **Sparta** and **Elis**. The war was started by the depredations of Aetolian privateers operating in the Aegean with the tacit approval of the Aetolian League. In 220 an Aetolian force under Dorimachus and Scopas invaded and ravaged **Messenia**, defeating a small force led out against them by **Aratus** in a battle at **Caphyae**. The Aetolian League officially disowned the attacks as the actions of private citizens but threatened the Achaean League if it interfered. The Achaeans appealed to Philip, who joined them in war against **Aetolia**. In 219 an Aetolian force under Scopas invaded Macedon and destroyed the religious center of Dium. A successful coup at Sparta by supporters of **Cleomenes III** resulted in its defection to the Aetolian side, although Cleomenes and his band of supporters died in **Egypt** that same year.

Sparta and Elis declared war on **Achaea**, which was ravaged by Aetolian forces based in Elis. In autumn, after a successful campaign in

Dardania, Philip defeated Elis. In 218, after an unsuccessful naval attack on Cephallenia, Philip countered an Aetolian invasion of **Thessaly** by a lightning attack on Aetolia. He destroyed the Aetolian federal center of Thermum, returned to the Peloponnese and **ravaged** Sparta, and then crushed an attempted revolt by three of his generals. Peace was made in 217 after Philip secured Thessaly and Phthiotis, threatening Aetolia. The war demonstrated that the Achaean League was no match for Aetolia without outside help, somewhat curtailed Aetolia's power, and firmly established the young Philip V as a power in the region.

SOLYGIA (SOLYGEIA). A village and mountain southeast of **Corinth**, site of a major battle between the **Athenians** and Corinthians in 425, during the **Peloponnesian War** of 431-404. **Thucydides** (4.42-43) provides a fairly detailed account of the action, which involved a large-scale seaborne landing. The Athenian force, carried in 80 **ships** and horse transports, consisted of 2,000 **hoplites** and 200 **cavalry**, with an unspecified number of allied troops. It was commanded by **Nicias** and two other *strategoi* (generals) and was part of the ongoing Athenian program of raids on the coast of the Peloponnese in retaliation for the **Peloponnesian League**'s annual invasions of Attica. The size of the Corinthian force is not given, but it was half of their army and had no cavalry with it.

One Corinthian general occupied the village of Solygia with a *lochos* of hoplites while the other attacked the Athenians, who had sailed undetected during the night. The Athenian right wing bore the brunt of the first assault. It was a hard-fought battle, with the Athenian right wing pushing the Corinthian left back twice (in between being beaten back to the beach when the enemy left wing was reinforced). Eventually Athenian persistence and the enemy's lack of cavalry took its toll and the Corinthians were routed, suffering the heaviest loss (including their general) on their right wing. The Athenians set up a **trophy**, recovered their dead (about 50), and withdrew. Approximately 212 Corinthians died—despite the lack of a determined pursuit by the Athenians, who presumably did not want to get too far from their ships and whose cavalry horses may not have fully recovered from their voyage. Nicias had to make a truce in order to recover the bodies of two of his dead who had accidentally been left behind. This was an act of considerable piety on his part, as it was a formal admission that the Athenians did not control the battlefield and therefore could not claim the victory (Plut., *Nicias*, 6).

SPARTA. The main city in Lacedaemon (Lakedaimon), an area within Laconia (Lakonia) in the southeast Peloponnese. Sparta had established domination over the other Lacedaemonian cities by about the end of the eighth century, reducing them to *perioeci* (lit. "dwellers around"), subject allies who governed themselves but with no independent foreign policy or political input into ruling Lacedaemon. Other inhabitants of Lacedaemon were reduced to serflike status as helots, who worked the land, providing food to maintain the Spartiates, each of which was allocated a block of land to provide his sustenance. The neighboring territory of **Messenia** was reduced to the same status in the second half of the eighth century, after the First **Messenian War**. This allowed the Spartans considerable leisure, but the threat of helot revolt meant that this had to be spent in military **training**. The ever-present threat of revolt also tended to make the Spartans cautious about deploying large numbers of troops at any distance from home. It was probably also a major motive in the formation of the **Peloponnesian League**, a defensive alliance that could be used against internal as well as external threat.

    Sparta was a conservative state, with a balanced constitution supposedly deriving from the early lawgiver, Lycurgus. Although in theory the people's assembly (the *apella*), comprised of all male Spartans over age 30, was the ultimate source of authority, Sparta was ruled by a dual monarchy. The kings commanded the army in the field, but from the sixth century one of the five annually elected ephors accompanied them on campaign. The kings were also assisted by a council of aristocratic elders, the *gerousia*. The system had a lot of checks and balances and this gave Sparta a considerable level of political stability.

    From the seventh century to the early fourth century Sparta was arguably the dominant Greek military power. This was due not only to its political stability but also to the Spartan system of education (the *agoge*), which was designed to achieve one thing—the production of skilled **hoplites**. The rigorous training started from the age of seven and distinguished the Spartans from their fellow Greeks until other states created bands of *epilektoi*, or elite troops, in the fourth century. It is a measure of the success of the Spartan hoplite that the city did not have permanent defensive walls until the second century (although temporary defenses were set up to repel specific attacks in the third century).

    Largely unchallenged (except by **Argos**) in the sixth century, at the start of the fifth century Sparta held the supreme command against the **Persians** in the Second **Persian War**, both on land and sea. This illustrates Sparta's high military reputation as they were not a naval power and the Athenians had provided nearly half of the allied Greek fleet. Following the successful defense of Greece in 480-479, Sparta's position was twice challenged by **Athens**. The First **Peloponnesian War** ended with the prewar status quo restored. Although Sparta won the next war, the Peloponnesian War (2) of 431-404, it demonstrated deficiencies in Sparta's system. The *agoge* was extremely well designed for producing good hoplites but tended (quite deliberately in many ways) to produce a conservative soldier, obedient to the laws but not necessarily suited to the less hoplite-centered warfare of the late fifth century.

Sparta maintained its dominant position in Greece for nearly the first 30 years of the fourth century. Its professional hoplites and judicious use of diplomacy, which ensured the support of Persia in several of the **Common Peace** treaties of the era, maintained Spartan power against both **Thebes** and a resurgent Athens. However, in 371, the Theban defeat of Sparta at **Leuctra** shattered the myth of Spartan invincibility. More importantly, subsequent Theban-led invasions of the Peloponnese broke up Sparta's alliance system and detached Messenia from Spartan control. Sparta never really recovered from this—despite fairly frequent strenuous attempts to regain her former greatness (or at least to preserve her independence from new powers in the Peloponnese, such as the **Arcadian League** and the **Achaean League**). After the battle of **Chaeronea** in 338, **Philip II** of **Macedon** detached more border territory from the city, further weakening it.

Sparta was also affected by a steady decline in the number of full citizens, from around 8,000 in 480 to around 2,000 in 371 and 700 in 242. This was partly because of the loss of territory and partly because of the increasing concentration of land and wealth in fewer hands. During the third century Sparta was firmly relegated to the position of a second- or third-rank power in Greece, despite the efforts of reformist monarchs such as **Agis IV** and **Cleomenes III** to restore the earlier "Lycurgan" constitution. After Cleomenes' defeat at **Sellasia** in 222, Sparta was captured and a pro-Macedonian government installed. Continual quarrels with its neighbors brought Sparta to the attention of **Rome** and resulted in Roman intervention to force Sparta to give up territory in 195. The assassination of **Nabis** in 192 essentially marked the end of an independent Sparta. It was forced to join the Achaean League (unsuccessfully attempting to secede in 189-188) and subsequently formed part of the Roman province of Achaea. *See also* DISCIPLINE; SPARTAN ARMY.

SPARTAN ARMY. Many of the details of the **Spartan** army are obscure and contradictory, partly because of the Spartan penchant for military secrecy (cf. Thuc. 5.68.2) but also because of a lack of precision in the ancient authors. For example, "Lacedaemonians" is used to mean both

Spartiates and the inhabitants of the whole of Laconia (Lakonia), and the term *lochos* is used both as a technical term, meaning a company-sized group of soldiers, and as a general term, denoting any group of soldiers. Because of this, there is no agreed modern opinion on the organization of the Spartan army, even in the Classical period, for which we have the most information. Lazenby's *The Spartan Army*, Sekunda's *The Spartan Army*, and the appendix in Anderson's *Military Theory and Practice in the Age of Xenophon* together provide a good idea of the different theories, as well as a detailed discussion of the evidence and its problems.

Although some have argued that the Spartan army was a very conservative organization that remained basically unchanged over several hundred years, it seems more likely that it underwent periodic revision. There may have been two such reorganizations in the Classical period, one between the Second **Persian War** and the First **Peloponnesian War** (perhaps caused by the **casualties** in the earthquake and subsequent helot revolt [Third **Messenian War**] in 464) and one following the Spartan defeat at **Leuctra** in 371. These reforms help to, but do not entirely, explain the differences between our two main sources for Spartan military organization, **Thucydides** and **Xenophon**.

In general terms the army consisted of **hoplites**, provided by Spartiates, *perioeci* (local subject allies), and even freed helots (cf. Thuc. 5.67.1); **cavalry**; and light troops. Other specific units known are the **Hippeis** (kings' bodyguard) and the **Sciritae** (a force of *perioeci* with special duties). By 425 it seems that *perioeci* outnumbered Spartiates in the hoplite force, and this seems to have remained the case during the fourth century (Thuc. 4.20.2; Xen. *Hell.* 6.4.15). The Spartan army also mobilized craftsmen for campaigns, had a fairly well organized commissariat, and took more care than most Greek armies in the security and organization of their camps. Camp security was aimed at both the enemy threat and the internal helot threat (Xen. *Lak. Pol.* 11.2, 12.1-5).

The Spartiate hoplites were highly regarded, largely because they were prepared for hoplite warfare from the age of seven by their **training** system, the *agoge*. Even on campaign Spartiates engaged in gymnastic exercises to maintain fitness and morale (Xen. *Lak. Pol.* 12.5-6). However, their system of subunits within larger **formations** and subordinate commanders at each level also assisted their performance on the battlefield (cf. Thuc. 5.66.2-4). Xenophon (*Lak. Pol.* 11.5-10) points out that this training and the command structure meant that Spartan hoplite **phalanxes** could easily carry out maneuvers that instructors of hoplite warfare elsewhere thought were very difficult. The Spartan superiority in this area was demonstrated at **Nemea** in 394, **Corinth** in 392, and (in retrieving an error) at **Mantinea** in 418, although they proved unsuccessful at Leuctra in 371 when outmaneuvered by the **Boeotians**.

Prior to the fourth century the largest internal division within the Spartan hoplite force mentioned in the ancient sources was the *lochos*. According to Thucydides (5.68.3) the Spartans deployed seven *lochoi*, plus the *Sciritae*, at the battle of Mantinea in 418. From his account each *lochos* had 512 men and contained four *pentecostyes* (each of 128 men). Each *pentecostys* contained four *enomotai*, each of 32 men. This gave a total of 3,584 men at Mantinea and a possible total of 6,144 hoplites, if the full army consisted of 12 *lochoi*.

Xenophon, however, uses the term ***mora*** (pl. *morai*) for the largest subdivision of the Spartan army—at least for the period down to Leuctra, after which he uses the term *lochos*. From his description of the organization (*Lak. Pol.* 11.2) and his account of Leuctra (*Hell.* 6.4.12), the Spartan army in the early fourth century consisted of six hoplite *morai* (and the same number of cavalry *morai*). Each *mora* had 576 men and was dived into four *lochoi* (each of 144 men). Each *lochos* consisted of two *pentecostyes* (each of 72 men), which were further divided into two *enomotai* (each of 36 men). This accords with his statement that the Spartan *mora* destroyed at **Lechaeum** in 390 was about 600 strong (Xen. *Hell.* 4.5.11-12) and gives a total hoplite force of 3,456.

Reconciling the two accounts at this stage of our knowledge is impossible. Thucydides may have used the term *lochos* instead of *mora*, and there are contradictions between his description of the organization at Leuctra and of the battle itself. In addition, the existing text at 11.2 of Xenophon's *Lak. Pol.* may be corrupt (the original may have had two *lochoi*, not four) and some (notably Lazenby, *The Spartan Army*, 7-9) have argued that the total of the Spartan hoplite force in both Thucydides and Xenophon is far too low.

According to Thucydides (4.55.2) Sparta first raised a force of cavalry (400 strong) in 425/4 in response to **Athenian** seaborne raids. However, the Spartans must have had horsemen in the distant past—as reflected in the name of the *Hippeis*, the kings' bodyguard. The cavalry always took second place to the hoplite force and was never highly regarded. Xenophon describes it as of "very poor quality" at the time of the battle of Leuctra in 371 because the rich paid for the horses, but the men, drawn from those "least physically fit and least keen to win honor," were only provided with their mounts and equipment on mobilization at the start of each campaign (*Hell.* 6.4.10-11). At Corinth in 392 **Pasimachus**, the Spartan cavalry commander, chose to dismount his force and fight on foot rather than on horseback (Xen. *Hell.* 4.4.10).

The organization of the cavalry is also rather uncertain, but in the fourth century apparently consisted (like the hoplites) of six *morai*. Each *mora* may have been around 120 men strong and was commanded by a **hipparmostes**. This gave a total of 720 cavalrymen, which included both Spartiates and *perioeci*. The whole cavalry force was probably commanded by at least one **hipparch** (Xen. *Lak. Pol.* 11.4; *Hell.* 4.4.10, 4.5.12, 5.2.4).

Psiloi (light troops) were generally drawn from helots (despite a reluctance to arm them) or provided by allies and/or mercenaries. A force of **archers** was raised, along with the cavalry, in 425/4 but the light troops at Nemea, **Coronea**, Leuctra, and the **Tearless Battle** were **mercenaries** or allies (Xen. *Hell.* 4.2.16, 4.3.16, 6.4.9, 7.1.32).

While the evidence from the fifth and fourth centuries is problematic, there is almost no information on the Spartan army of the third and second centuries. After Leuctra, with the loss of **Messenia** and the **Peloponnesian League** Sparta was never again dominant in Greece. The Spartans, however, clearly maintained skills as drillmasters and tacticians and were in some demand as mercenary leaders. Examples include **Archidamus III** in Crete and **Italy**, Acrotatus in **Sicily**, Cleonymus (uncle of **Areus I**) in Italy, Leonidas in Syria, and the **Xanthippus** who helped the Carthaginians during the First Punic War (Diod. Sic. 16.62.4-63.1, 19.70-71.5, 20.104-105, 23.14.1-2; Plut., *Agis*, 3.6; Polyb. 1.32-34). Under **Cleomenes III**, the Spartans converted to the *sarissa* and the **Macedonian**-style phalanx (Plut., *Cleomenes*, 11.2) and achieved some local successes, ended by their defeat at **Sellasia** in 222. By the time of the battle of Mantinea (3) in 207, the Spartan army was thoroughly Hellenistic—its core was the *sarissa*-equipped phalanx, supported by **Tarentines** (light cavalry), **artillery**, and mercenary light troops (Polyb. 11.11-12).

SPARTOLUS (SPARTOLOS). A city in the Chalcidice, near **Olynthus**. It was the site of a major battle between the **Athenians** and Chalcidian rebels in 429 during the **Peloponnesian War** (2) of 431-404. **Thucydides** (2.79) provides an account of this engagement (which is also mentioned in Isaeus, 5 [*On the Estate of Dicaeogenes*], 42). An Athenian army of 2,000 **hoplites**, 200 **cavalry**, and an unspecified number of *psiloi* (light troops) **ravaged** the country around Spartolus in an attempt to force it into surrender. However, the inhabitants, reinforced from Olynthus, marched out and gave battle (Thucydides does not specify their numbers). The Athenians defeated the Chalcidian hoplites but their own cavalry and *psiloi* were defeated by their opponents. At this point more **peltasts** arrived from Olynthus, giving fresh heart to the Chalcidian cavalry and light troops, who now pressed the Athenian hoplites. In a classic demonstration of the vulnerability of unsupported hoplites to cavalry and *psiloi*, the Chalcidians attacked with **javelins**, retired when the slower hoplites charged out, and attacked again. The Athenians were driven back on their baggage and then routed—largely by the enemy cavalry. All three Athenian *strategoi* (generals) and 430 of their men were killed.

SPHACTERIA (SPHAKTERIA). *See* PYLOS

SPHAGIA. See SACRIFICE

SPY. *See* INTELLIGENCE

STRATEGOS (**pl.** *STRATEGOI*). Originally a general, leader of an army, the term later came to also include major naval command and was used as a title of officeholders in a variety of Greek states. As an office, it is attested at Tegea and **Eretria** in the fourth century and Aegina in the second century. The senior official of the **Achaean League** was the *strategos*.

Athens had a board of 10 *strategoi*, elected annually. Although an individual held the *strategeia* (office of general) for only one year, there was no barrier to reelection, and it was common for prominent Athenians to be *strategoi* for extended periods. During the campaign season *strategoi* were allocated to command land or naval expeditions, with two or three often appointed to the same expedition. For example, **Alcibiades, Nicias**, and Lamachus held joint command of the **Sicilian Expedition** in 415, and eight *strategoi* commanded at **Arginusae** in 406. The generals were accountable to the people and could be fined, exiled, or even executed if their performance was held to be unsatisfactory. Six of the eight generals at the battle of Arginusae were executed for failing to rescue Athenian sailors from the sea, and **Timotheus** was fined 100 talents for failing to support **Chares** at **Embata**. Fear of punishment probably encouraged caution—Nicias' decision making in Sicily was certainly affected by this, according to **Thucydides** (7.8, 11-15).

In the fifth century, *strategoi* were often politicians as well as military leaders—the *strategeia* was the most important elected office in the state. This worked well with experienced men but occasionally led to problems such as **Cleon's** mishandling of the action at **Amphipolis** in 422. In the fourth century, *strategoi* were more likely to be military specialists than politicians, but were often hampered by Athens' lack of resources. Generals such as **Chabrias**, Chares, **Iphicrates**, and Timotheus often had to go to extraordinary lengths to raise money for campaigns, including hiring their men out as laborers or **mercenaries**.

STRUTHAS (fl. 390). A **Persian** sent by Artaxerxes II to take control in **Ionia** in 392. He favored **Athens** over **Sparta,** provoking a Spartan expeditionary force under **Thibron** in 391. At the head of a large force of **cavalry**, Struthas surprised Thibron near **Ephesus**, killing him and large numbers of his men. He was less successful against Diphridas, Thibron's successor, who captured Struthas' daughter and son-in-law.

SYBOTA. An island off the east coast of **Corcyra**, site in 433 of a naval battle between Corcyra and **Corinth**. **Thucydides** (1.24-55) provides a detailed account of the battle and the diplomatic maneuvering that preceded it. The battle arose from a quarrel between Corcyra and its mother city, Corinth, over internal strife in Epidamnus, one of Corcyra's colonies. Different factions within Epidamnus had appealed to both Corcyra and Corinth and, worried by the imbalance in power with Corinth, Corcyra had made a defense alliance with a rather reluctant **Athens** (despite Corinth's understandable, but technically incorrect, belief that this was a breach of the **Thirty Years' Peace** which had concluded the First **Peloponnesian War**).

The Corcyraean fleet had 110 **triremes**, supported by 10 Athenian ships under instruction to avoid contact with the Corinthian ships unless Corcyra was directly threatened. The Corinthian fleet had 150 triremes—90 from Corinth and the rest provided by **Elis** (10), **Megara** (12), Leucas (10), Ambracia (27), and Anactorium (1). The Corcyraean fleet anchored at Sybota, the Corinthians on the mainland opposite. After a night approach by the Corinthians, the two sides met at dawn off Sybota.

Thucydides describes the fighting as "rather old-fashioned" (1.49.1), with all ships carrying **hoplites**, **archers**, and **javelin**-equipped *psiloi* (light troops). The implication here is that both sides intended to fight using their marines, rather than fight a maneuver battle using the ram—as Thucydides' description of the course of the battle confirms. He states that maneuver was restricted because of the number of ships involved and that it was fought more like a land battle, with the marines engaging each other from the stationary ships. The Corcyraean left wing routed the Corinthians opposite them and, pursuing them to land, began plundering the enemy camp. Already outnumbered and now deprived of the 20 ships involved in the pursuit, the remaining Corcyraean ships were hard-pressed. The Athenian contingent, which had initially hung back and avoided contact now became increasingly involved and once the Corcyraeans were forced back and the Corinthians maintained the pressure, the Athenians directly engaged Corinthian ships.

In the aftermath of the battle the Corinthians sailed around killing the men in the water (including, inadvertently, some of their own) and then recovered their own disabled ships. Now late in the day, the Corinthian fleet then formed up again and sailed toward Corcyra, to be met by the surviving Corcyraean ships and the Athenians, determined to prevent a landing. The *paean* was sung by both sides, but the Corinthians then spotted 20 new Athenian triremes arriving to help and retired. The next day both fleets put out to sea, but the Corinthians, worried by the arrival of the fresh Athenian contingent and the thought that Athens might now consider itself at war and attack them in full force on the return voyage, parleyed with the Athenians instead of fighting. On receiving advice that the Athenians did not consider themselves at war with Corinth and that they could sail anywhere they liked except Corcyra, the Corinthians sailed for home.

Both sides claimed victory and set up **trophies**—the Corinthians had won the initial battle, had taken 1,000 **prisoners**, and had sunk 70 Corcyraean ships; on the other hand, the Corcyraeans had sunk about 30 of the Corinthian fleet, plundered their camp, recovered their own dead and the disabled ships without asking for a truce, and the Corinthians had refused battle the next day. Although the battle was fairly inconclusive, thanks to Athenian support for Corcyra, it added greatly to the tensions between Athens and Corinth and helped lead to the outbreak of the Peloponnesian War (2) of 431-404.

SYRACUSE. The largest and most important city on **Sicily**. Syracuse was a Dorian Greek colony, founded by **Corinth**, in 734. It was a natural defensive site, with the best harbor on the east coast of Sicily. The original settlement was on the island of Ortygia, which was soon connected to the mainland by a causeway and served as Syracuse's citadel. Originally an aristocracy, Syracuse became a democracy in the early fifth century but was most commonly ruled by tyrants (interspersed by democratic and occasionally oligarchic government). Under these, Syracuse from time to time controlled large portions of Sicily, mainland **Italy**, and even places in the Adriatic. Its history was marked by conflict with other Greek cities on Sicily and, especially, with **Carthage**.

The expulsion of the aristocracy, the *gamoroi*, led to the takeover of the city by its first tyrant, **Gelon** (c. 485-478). By an aggressive military policy, which involved resettling in Syracuse people from conquered cities, Gelon boosted Syracuse to the position of the premier Greek city in Sicily. It was to Gelon that the mainland Greeks unsuc-

cessfully appealed for help at the start of the Second **Persian War**—he
offered aid but under unacceptable conditions and was anyway distracted
by a major Carthaginian invasion, which he decisively defeated outside
Himera in 480 (Hdt. 7.153-157). The middle part of the century saw the
city continue to cement its position in Sicily, despite internal dissension
and even civil war that resulted in the establishment of a democracy
(Diod. Sic. 11.67-68, 72-73, 76).

It was under this democracy that Syracuse successfully resisted the
Athenian Sicilian Expedition, launched in 415 and destroyed outside
Syracuse in 413 (Thuc. 6.1-7.87). As a result, Syracuse entered the **Peloponnesian War** of 431-404 on the side of **Sparta** and under **Hermocrates** played an active role in the naval campaigns in **Ionia** and the
Hellespont that led to the destruction of Athenian naval power and Spartan victory (Thuc. 8.26; Xen. *Hell.* 1.1.26, 1.2.8-10).

Toward the end of the Peloponnesian War Syracuse was again
threatened by Carthage, now allied with Segesta. In a series of successful
campaigns between 409 and 406 the Carthaginians took Selinus, Himera, and Agrigentum and gave **Dionysius I** the opportunity to seize
control in Syracuse, where the inhabitants lacked confidence in their existing generals. As tyrant from 406/5 to 367 Dionysius ended this war
on terms favorable to Carthage and, after crushing an internal revolt, began a program of ruthless expansion. During his reign **artillery** was invented and in 398 he used it in an attack on the Carthaginian city of
Motya. The resulting war was fairly close run—Syracuse survived a
siege mainly because of an outbreak of plague in the Carthaginian army;
the Carthaginians made peace in 392 on terms favorable to Syracuse.
From 391 to 387 Syracuse was involved in a campaign in southern Italy
that brought much of it under Dionysius' control—resistance essentially
collapsed after the Italian defeat at **Elleporus River** in 389. Two more
wars against Carthage followed (around 382-375 and 368/7) with mixed
success—Syracuse was forced to cede some territory in 375 after a defeat
at Cronium, but was doing well in the second war when Dionysius died.
Under Dionysius, Syracuse's **fortifications** were extensively improved—not only in quality but also in the area included within them.
Syracuse was now the best-protected city in Sicily and its defenses included areas of land that could be farmed during a siege.

Dionysius' death was followed by a period of considerable internal
instability. After 10 years of calm under his son **Dionysius II**, an attack
by the exiled **Dion** effectively split Syracuse—Dion controlled most of
the city; Dionysius' men, the citadel of Ortygia. The situation was so
unstable that some of the citizens appealed for help to their mother city,
Corinth. In 344 **Timoleon** was sent over and fairly rapidly stabilized the

situation, restoring Syracuse's population by immigration and settling its constitution. Syracuse enjoyed a return to its former prosperity and a period of relative peace and tranquility, ended by the reign of **Agathocles** (tyrant, 317-307; king, 307-289).

Under Agathocles Syracuse successfully fought a coalition of Sicilian cities despite their being led by the Spartan prince Acrotatus. Syracuse was left in control of most of the island, except Heraclea, Himera, and Selinus, which remained under Carthaginian control (Diod. Sic. 19.65, 19.70-72). In 312 the Carthaginians and a mixed alliance of Sicilians (including Syracusan exiles) attacked Syracuse. Agathocles' defeat at the Himeras River in 311 caused the defection of many subject and allied states and led to a Carthaginian siege of Syracuse (Diod. Sic. 19.102-110). Agathocles answered this with an invasion of Africa and in turn besieged Carthage. Although ultimately unsuccessful—eventually losing his entire force in 307/6—Agathocles came very close to capturing Carthage. During the siege, Syracuse had held out and Agathocles had been able to restore the situation in Sicily, defeating a bid by Acragas to lead a coalition of cities against him. Despite the African disaster, Agathocles was also able to reestablish Syracusan control over most of the island (Diod. Sic. 20.3-90). From around 300 Agathocles extended Syracuse's power in Italy and the Adriatic, forming a marriage alliance with **Pyrrhus I** of Epirus, giving him **Corcyra** as a dowry, capturing Croton, and fighting the Bruttians (Diod. Sic. 21.1.4, 2-4, 8; Plut., *Pyrrhus*, 9).

With Agathocles' death in 289, Syracuse and its empire again fell into anarchy, reversed by the rule of **Hieron II** (275-215). His rule involved struggles against Campanian mercenaries based in Messana—initially unsuccessful, although he won at least one major victory. In 264 he made the mistake of initially allying with Carthage against the Roman expeditionary force sent to Messana on the outbreak of the First Punic War. However, a Syracusan defeat by the Romans led him to reconsider and for the rest of the war Syracuse was a loyal ally of **Rome** (Polyb. 1.8-11, 16). With the Roman victory over Carthage in 241, Rome controlled Sicily but left Syracuse independent as a reward for the major assistance it had provided. Under Hieron Syracuse's defenses were further strengthened (under the direction of Archimedes), confirming its position as the best-fortified Sicilian city.

Unfortunately for Syracuse, Hieron's successor, Hieronymus, reversed his policy and defected to Carthage during the Second Punic War. Syracuse was besieged by the Romans under Marcellus in 213-211. When the city eventually fell, it was comprehensively sacked and then incorporated into the province of Sicily.

Syracuse was besieged on several occasions, the two most famous being by the Athenians in 415-413 (during the Sicilian Expedition) and by the Romans in 213-211. The first siege was prior to the invention of siege artillery and essentially focused on walls of circumvallation and counter-walls. The second is notable for the contribution Archimedes made to the defense and the difficulty the Romans had in taking the city. It fell in several stages, the first, resulting from the Roman scaling of a poorly guarded section of wall, led to the capture of most of Epipolae and the surrender of Tycha and Neapolis. The second stage occurred soon thereafter, when the commander of the important fort of Euryalus surrendered it. The Romans then survived a major Carthaginian relief effort (which was decimated by illness) and finally took the stronghold of Ortygia when a **mercenary** officer betrayed it. The rest of the city surrendered (Polyb. 8.5-9, 37; Livy, 24.33-34, 25.23-31).

The Syracusan armed forces were organized along Greek lines, with a **hoplite** core, supported by **cavalry** and *psiloi* (light infantry). However, there was a greater emphasis on the use of cavalry than was often the case in mainland Greece, at least in the Classical period. Gelon used his cavalry as the mainstay of his defense against Carthage in 480 and the Syracusans again extensively used their cavalry in the mobile defense role and in pitched battles during the Athenian expedition to Sicily. Although its hoplites were initially no match for their more experienced Athenian opponents, under the leadership of the Spartan **Gylippus** they were eventually victorious. The Syracusan navy was generally of good quality, as it demonstrated in the latter stages of the Peloponnesian War. Apart from this, Syracusan military activity was characterized by the frequent use of mercenaries, the development and use of artillery, and extensive fortifications to protect the city—all products of the city's considerable wealth.

SYRIA. In antiquity this consisted of Syria proper and Coele-Syria ("Hollow Syria"), although the terms are often used rather loosely. The former extended from the Amanus Mountains (a branch of the Taurus range) in the north to the River Eleutherus in the south; Coele-Syria comprised the Phoenician coast and inland Lebanon and was often also regarded as including northern or even all of **Palestine**. As possession of Coele-Syria was in continual dispute between Seleucid Syria and Ptolemaic **Egypt** the definition of the border tended to change with circumstances.

Prior to **Alexander the Great**'s conquest of the region in 332, Syria was a satrapy (province) of the **Persian** empire. After Alexander's death it was allocated to Laomedon, who was ejected in 319-318 by **Ptolemy I (Soter)**, who subsequently contested its ownership with

Antigonus I (Monophthalmus). In 301 the victors over Antigonus at **Ipsus** granted Syria to **Seleucus I (Nicator)** as a reward for his efforts at the battle—and as punishment of Ptolemy, who had previously been granted it but had failed to show up at Ipsus. However, Ptolemy already had Coele-Syria and Seleucus chose not to fight for it. The Ptolemies later claimed that their possession of it was formally confirmed under the terms of an agreement with Seleucus in return for remaining neutral during his war with **Lysimachus.** The Seleucids denied this and the region was fought over in a series of wars (the **Syrian Wars**) from 274/3 to 168. Coele-Syria (or large parts of it) basically remained under Ptolemaic control until the Fifth Syrian War (202-195) when **Antiochus III (the Great)** captured it, extending Seleucid control to the borders of Egypt. The Seleucids were still in possession of Coele-Syria when **Rome** intervened in 168 and ordered **Antiochus IV (Epiphanes)** to withdraw his troops from Egypt. After Antiochus' death in 164, Seleucid control of Syria gradually but relentlessly declined—despite a temporary resurgence under **Antiochus VII (Sidetes),** who reigned from 139 to 129. Many parts of Syria became independent or quasi-independent and in 83 Tigranes II of Armenia annexed it. His defeat by Gnaeus Pompeius Magnus (Pompey) led to the incorporation of Syria into the Roman empire in 64-63. *See also* SELEUCID ARMY.

SYRIAN WARS. A series of wars fought in the third century between Seleucid **Syria** and Ptolemaic **Egypt,** generally for control of the border area of Coele-Syria. The grounds for the conflict lay in the agreement made in 303 between **Ptolemy I (Soter), Cassander, Lysimachus,** and **Seleucus I (Nicator)** that Ptolemy would receive Syria for his assistance in the war against **Antigonus I (Monophthalmus).** However, his failure to send troops to the battle of **Ipsus** in 301 led to the allocation of Syria to Seleucus instead. That same year Ptolemy occupied most of Syria south of Lebanon and Damascus, and in 286 he completed his acquisition of Phoenicia when he gained **Tyre** and Sidon from **Demetrius I (Poliorcetes).** The Ptolemies subsequently claimed that Seleucus had ceded Phoenicia, **Palestine,** and Coele-Syria to Egypt in 282 in return for remaining neutral during his war with Lysimachus. The Seleucids denied this and claimed all of Coele-Syria down to the Egyptian border. Modern interpretations of these wars often differ considerably because the sources are fairly scanty until **Polybius'** account of the Fourth Syrian War. *See also* SELEUCID ARMY.

(1) Syrian War of Succession (c. 280/79). A clash between **Antiochus I (Soter),** who was trying to secure his inheritance after the death of **Seleucus I (Nicator),** and **Ptolemy II (Philadelphus).** Details are

obscure but it appears that Ptolemy extended **Egyptian** influence in parts of southern and western Asia Minor. Antiochus was mainly preoccupied with securing northwestern Asia Minor.

(2) First Syrian War (274/3-271). Fought between **Syria**, under **Antiochus I (Soter)**, and **Egypt**, under **Ptolemy II (Philadelphus)**. In 275 Antiochus secured the alliance of Magas, Ptolemy's half-brother and ruler of Cyrene. Magas revolted and marched on Alexandria, but was forced to retire to quell a revolt in his rear before Antiochus and Ptolemy commenced hostilities in 274. Antiochus himself was placed under great pressure in the north, where a Ptolemaic fleet attacked Cilicia, pirate fleets were hired to attack other coastlines, and the **Gauls** probably also launched raids against his possessions. Antiochus was forced to make peace in 271, with Ptolemy confirmed as ruler of most of the Syrian coastline from **Miletus** in the west to the River Calycadnus in Cilicia in the east. This may simply have been the position prior to the outbreak of war, but even so represented a victory for Ptolemy.

(3) Second Syrian War (260-253). Fought between **Syria**, under Antiochus II (Theos), supported by **Antigonus II (Gonatas)**, and **Egypt**, under **Ptolemy II (Philadelphus)**. Little is known of this war, but it may have started with Ptolemy trying to take advantage of the transition of rule to Antiochus II after the death of **Antiochus I (Soter)** in 261. Antiochus apparently made considerable gains in Cilicia, Pamphylia, and **Ionia**. He also regained **Ephesus** (lost to the Ptolemies circa 262/1), liberated **Miletus**, and secured all of Phoenicia north of Sidon. Most of these conquests were made possible by Antigonus' crippling of Egypt's hitherto dominant fleet off Cos circa 258. Ptolemy made peace with Antigonus in 255 and (probably) in 253 with Antiochus—to whom he married his daughter Berenice Syra. Antiochus had to divorce his first wife, Laodice, to marry Berenice but may have gained the revenue from Coele-Syria as a dowry from Ptolemy.

(4) Third Syrian War (246-241), also known as the Laodicean War. Fought between **Syria**, under **Seleucus II (Callinicus)**, and **Egypt**, under **Ptolemy III (Euergetes)**. This arose from the settlement of the Second Syrian War, which caused a conflict between Queens Laodice and Berenice over whose son would succeed on the death of Antiochus II (Theos). Ptolemy sent forces to assist his sister Berenice. They took Seleuceia in Pieria, secured Antioch, and conquered Cilicia. In spring 246 Ptolemy advanced on Antioch, securing Syria on the way and gaining the adherence of the eastern satraps. On his arrival he not only found that Berenice and her son (whose name is unknown) had been murdered but also had to return to Egypt to put down a revolt.

In 245 the Syrians under Laodice and her son, newly crowned as Seleucus II, went on the offensive. In spring 244 Seleucus recovered the allegiance of the eastern satrapies and most of Syria down to Antioch. The war ended in 241 with Egypt keeping Phoenicia and Seleucia in Pieria but the boundaries otherwise as they had been before the war. Seleucus was left with the problem of his younger brother, **Antiochus Hierax**, who had been set up as ruler in Asia Minor in return for Laodice's financial support to Seleucus during the war.

(5) Fourth Syrian War (219-217). Fought between **Syria**, under **Antiochus III (the Great)**, and **Egypt**, under **Ptolemy IV (Philopator)**—although his chief minister, Sosibius, played a major role. The war began with Antiochus' capture of Seleucia in Pieria and his seizure of much of Phoenicia. He was held up at the fortress of Dura and, worried by **Achaeus'** activities and believing a false story that Ptolemy had a large army at Pelusium, accepted a four-month truce with the Egyptians. They used this time to build up an army consisting of **mercenaries** and, very unusually, large numbers of native Egyptians. In 218 Antiochus concentrated on pacifying southern Syria and conquering northern **Palestine**. When he advanced south to the border with Egypt in 217 he was decisively defeated at the battle of **Raphia**. Ptolemy recovered southern Syria and Palestine but left Antiochus with Seleucia in Pieria, ratifying this with a peace treaty. One of the results of the war may have been a reassertion of the native Egyptian population against the Greek rulers, stemming from their participation at Raphia. Several uprisings occurred and the Egyptian priests apparently began to take a more independent line.

(6) Fifth Syrian War (202-195). Fought between **Syria**, under **Antiochus III (the Great)**, and **Egypt**, under **Ptolemy V (Epiphanes)**. In 203/2, tempted by Egypt's weakness, **Philip V** of **Macedon** and Antiochus agreed to dismember its overseas possessions. Antiochus rapidly captured Coele-Syria in 202-201 and in 200 inflicted a decisive defeat on the Egyptians at **Panion**. By 199 he had secured Sidon but was then told by **Rome** not to invade Egypt. He agreed—perhaps because he had no intentions of invading anyway—and concentrated on pacifying Coele-Syria and raiding Ptolemy's coastal possessions in Anatolia. In 197 he invaded Pergamum and in 196 crossed to **Thrace** and began rebuilding Lysimacheia. Warned off again by the Romans, he outmaneuvered them diplomatically in 195 by arranging peace and a marriage alliance with Ptolemy. This peace confirmed Antiochus' possession of Coele-Syria and temporarily prevented further Roman action, although they remained suspicious of him.

(7) Sixth Syrian War (169-168). Fought between **Syria**, under **Antiochus IV (Epiphanes)**, and **Egypt**, under **Ptolemy V (Epiphanes)**. The immediate cause of this war is uncertain, with some sources blaming Syria and others Egypt. The dispute was referred to **Rome**, but when no firm answer was received, Antiochus invaded Egypt in 169, occupying most of it except Alexandria. When Ptolemy decided to negotiate with Antiochus (his uncle), the Alexandrians proclaimed his younger brother Ptolemy VIII (Physcon) and his sister Cleopatra II rulers. Antiochus withdrew but returned the following year, capturing **Cyprus**. The war ended when Rome, freed from the distraction of the Third **Macedonian War**, ordered Antiochus to evacuate Egypt and Cyprus in 168 and he did so.

- T -

TACTICAL WRITERS. *See* AELIAN; AENEAS TACTICUS; ARRIAN; ASCLEPIODOTUS; MILITARY MANUALS; ONASANDER; POLYBIUS; XENOPHON

TAGUS (TAGOS). The chief civil and military magistrate of **Thessaly**. Although the position was not always filled, once elected the *tagus* apparently held power indefinitely. The Aleuadae controlled the office at the time of the Second **Persian War** but lost it after the **Persian** defeat. The office was reactivated by **Jason of Pherae** in 374. It carried significant military power and, according to **Xenophon**, when Jason was *tagus* he commanded "a force of 6,000 **cavalry** and a **hoplite** army of over 10,000" (*Hell.* 6.1.14). The office came under the control of **Philip II** of **Macedon** around 352 (although he possibly did not assume it until 344-342) and was subsequently also held by **Alexander the Great**.

TAMYNAE (TAMYNAI). A town on the east coast of **Euboea**, site in 348 of a battle between the **Athenians**, supporting Plutarch, tyrant of **Eretria**, and rebels, possibly supported (or at least encouraged) by **Philip II** of **Macedon**. The Athenian force of perhaps 3,000 men comprised both infantry (including the *epilektoi*) and **cavalry**, in addition to **mercenaries** commanded by Plutarch. The Athenian *strategoi* (generals) were **Phocion** and the otherwise unknown Molottus.

Phocion marched his army across Euboea to meet the rebels, led by Callias of Chalcis. Awaiting reinforcements, and with unfavorable omens from the prebattle **sacrifice**, Phocion did not march out against the rebels. Plutarch interpreted this as cowardice and charged out with his mercenaries, and the Athenian cavalry followed suit—apparently against Phocion's orders. This attack was repulsed and the rebels followed up, but lost cohesion as they did so. At this point Phocion announced that the sacrifices were favorable, ordered the **phalanx** to receive the fleeing mercenaries, and led the *epilektoi* against the disorganized rebels. The battle was won when the Athenian cavalry regrouped and charged the enemy. This battle demonstrated Phocion's coolness under pressure and helped bring Euboea back under Athenian influence—although Athens recognized its independence that same year.

TANAGRA. A city in **Boeotia**; site of a major battle fought in the summer of 457 during the First **Peloponnesian War**. A **Peloponnesian League** force of 11,500 (including 1,500 **Spartans**) fought a 14,000-strong Athenian army which included 1,000 **Argives**, contingents from other allied cities, and a force of **Thessalian cavalry**. The **Spartan army** was returning from northern Greece where it had forced **Phocis** to restore territory to Doris and had reestablished the **Boeotian League** under **Thebes** as a counterweight to **Athens**. Unable to return via the Corinthian Gulf because of an Athenian fleet there nor via the Isthmus of Corinth because of Athenian troops operating in **Megara**, the Peloponnesians advanced toward Athens. They were met by the Athenians at Tanagra in Boeotia, near the Athenian border. The battle was close-run but the Peloponnesians won when the Thessalian cavalry allied to Athens deserted during the battle. This victory allowed the Peloponnesians to return home via the Megarid, but shortly afterward Boeotia was lost to Athens at the battle of **Oenophyta**. *See also* CIMON.

TARENTINES. A type of light **cavalry**, probably equipped with **javelins**, used from the end of the fourth century down to at least the second century. They are first attested under **Antigonus I (Monophthalmus)**, who had 2,200 at **Gabiene** in 316 and under **Demetrius I (Poliorcetes)** at **Gaza** in 312. From this point on, they regularly appeared in the armies of Hellenistic monarchs and Greek city-states and leagues and were in wide use in the late third and early second centuries. Although the original type may well have come from **Tarentum** (Taras) in **Italy**, almost from its first appearance the term apparently described a type of soldier rather than his city of origin. Livy's claim (35.28.5) that they were cavalrymen who led a second horse into battle is almost certainly mistaken—leaping from one horse to another during combat seems intrinsically unlikely, as well as extravagant in horseflesh. **Mercenary** Tarentines were common, although citizen-Tarentines are also known. *See also* MANTINEA (3).

TARENTUM (TARAS). A colony of **Sparta**, founded in the late eighth century and located on the instep of the heel of **Italy**. Tarentum (modern Taranto) had the best natural harbor in southern Italy and a good agricultural hinterland. Its strong links with Sparta and the Lacedaemonian features of the city support the story that it was founded by the illegitimate children born to the wives of Spartans who had served away from home for a long period during the First **Messenian War**. Treated as less than full citizens, they posed a political and social threat to Spartan stability and were sent overseas after attempting a revolution (Strabo, 6.3.3; Diod. Sic. 15.66.3).

Little is known of Tarentum's early history, but the city appears to have gained control over some of the surrounding areas, at the expense of the native tribes (Paus. 10.10.3). However, around 473 the Tarentines suffered a very heavy defeat at the hands of the Messapians. **Herodotus** states of this action: "nobody has ever heard of so great a slaughter of Greeks" (7.170)—the number of Tarentine **casualties** is unknown, but 3,000 of their allies from Rhegium died. Although the magnitude of this disaster probably held it back for some time, the defeat of Croton by Locri and Rhegium around 450 left Tarentum as the most powerful Greek city in southern Italy. In 433/2, for example, after a war with Thurii, it founded its own colony, Heraclea, which later became the headquarters of an Italiote confederacy (Diod. Sic. 12.23, 36). Despite its origins, Tarentum essentially took a neutral stand in the **Peloponnesian War** of 431-404, although refusing supplies to the **Athenian** fleet en route to **Sicily** in 415 (Thuc. 6.44.2).

During the fourth century, Tarentum's response to the pressure of its Italian neighbors was to hire a succession of Greek **mercenary** leaders and troops. This policy met with mixed success. **Archidamus III** of Sparta was defeated by the Messapians in 338. **Alexander I of Molossia** defeated the Lucanians in 334, but then quarreled with his employers and was acting independently when he was killed in 331 (Diod. Sic. 16.63.1; Livy, 8.24). Cleonymus of Sparta not only quarreled with the Tarentines, but in 303 was also defeated by the Italians (Diod. Sic. 20.104-105).

In the early third century Tarentum came into conflict with the expanding power of **Rome**, war breaking out in 282/1. The Tarentines called in **Pyrrhus I** of Epirus, who had some spectacular early successes against Rome but became distracted by a campaign in Sicily and neglected the war in Italy. Pyrrhus withdrew from Tarentum in 275 and the garrison he left behind surrendered to Rome in 272. From this point on, Tarentum was under Roman control and had no independent action—with the exception of the period 213-209, when it was in revolt and assisting Hannibal. The city was sacked when the Romans recovered it in 209 (Polyb. 8.26-36; Livy, 25.8-10, 27.12, 27.15-16).

Tarentum lent its name to a type of light cavalry; the **Tarentines**.

TAXIARCH (*TAXIARCHOS*; pl. *TAXIARCHOI*). Generally, the commander of a *taxis*, an infantry subunit of varying size attested for **Athens**, **Sparta**, **Macedon**, Hellenistic **Egypt**, and elsewhere. At Athens the taxiarch was elected; in most other *poleis* (city-states) or regions it was an appointment. The word could be used rather more loosely to indicate a subordinate commander in other arms, for example, the navy (Xen. *Hell.* 1.6.29, 35).

TAXIS (pl. *TAXEIS*). A body of soldiers, normally infantry, although the term could also be used of **cavalry** and even of **ships**. A *taxis* could be of varying sizes and, unless a number is specified, it is not always clear how big a group is meant. At **Athens** a tribal regiment of **hoplites**, theoretically 1,000 strong, was called a *taxis* (Lysias, 16 [*For Mantitheus*], 16), but a *taxis* of the **Ten Thousand** was 200 strong (Xen. *Anab.* 6.5.11). In the writings of the later tacticians a *taxis* was a company-size unit of 128 men (Asclepiodotus, *Tactics*, 2.8; Aelian, *Tactics*, 9.3). *See also* FOOT COMPANIONS.

TEARLESS BATTLE. Fought just north of **Laconia** in summer 368 between the Spartans (aided by troops sent by **Dionysius I** of **Syracuse** and **mercenaries** paid for by **Persia**) and the **Arcadian League**, which was trying to become the dominant power in the Peloponnese. The Arcadian army broke and ran before the impact of the **phalanxes** and suffered heavy **casualties** from the **cavalry** in the ensuing pursuit. **Sparta**, which was suffering a manpower crisis, was delighted that no Spartans had died in the battle and this gave rise to the name "Tearless Battle."

TEGYRA. An encounter battle in **Boeotia** in 375 during the struggle for hegemony over Greece. A **Spartan** army and a **Theban** force, both apparently marching without scouts, met on the road from **Locris** to **Orchomenus** (Boeotia). The smaller Theban force launched itself at the Spartans and cut its way through to safety. This was another indication, like **Coronea** (2) in 394, that the Theban **hoplite** was starting to overtake the long-standing supremacy of the Spartan hoplite. This shift in military power was confirmed by the Theban victory over Sparta at **Leuctra** in 371.

TEN THOUSAND. A force of Greek **mercenaries** created by Prince **Cyrus the Younger** of **Persia**, ostensibly for use in his satrapy but actually used in his unsuccessful attempt to overthrow his brother, Artaxerxes II, king of Persia. Drawn principally from the Peloponnese, but also from **Boeotia**, **Thessaly**, and the Chersonese, the Ten Thousand were accompanied by the **Athenian Xenophon**, who recorded their history in the *Anabasis* (called, in the Penguin series, *The Persian Expedition*). Although successful on their wing at **Cunaxa** in 401, Cyrus' death in that battle left the Ten Thousand stranded in the Persian heartland. Following the battle, their leaders were seized at a meeting and executed by the Persians. The Ten Thousand elected new leaders, including Xenophon, and marched approximately 2,000 miles (3,200 kilometers) across enemy territory to return to **Thrace**. Initially greeted with hostility, in 399 Sparta employed around 6,000 in **Thibron**'s campaign against Persia in Asia Minor. The success of the Ten Thousand led to an increasing Greek contempt for Persian military power and inspired calls for a panhellenic war against Persia which that have helped influence **Philip II** of **Macedon**'s decision to attack Persia. *See also* CLEARCHUS.

TERRAIN. Terrain affects numerous (if not most) aspects of land warfare, too many to consider in a dictionary. However, its major influence in ancient Greek warfare primarily derived from the difficulties hilly, rough, or broken terrain caused for maintaining a close-ordered infantry **formation** (e.g., the **hoplite** or the **Macedonian phalanx**) and for riding a horse without stirrups. The main effect of this was on both the development and the employment of troops.

As might be expected, very hilly regions, such as Acarnania, developed the more mobile *psiloi* (light infantry) rather than hoplites—although poverty (in turn influenced by the lower fertility of mountainous regions) was also a factor here. **Thessaly**, the area with the largest expanse of plain in Greece, was the premier Greek **cavalry** power, although both **Boeotia** and **Athens**, which had some plains and were relatively wealthy, also had creditable cavalry. The location of the best agricultural land on flat ground, coupled with the needs for the phalanx to operate in the same sort of terrain, heavily influenced the pattern of hoplite warfare throughout its existence. Greek exposure to the rather different terrain of Asia Minor led to the development of **reconnaissance** and an increased emphasis on cavalry in the fourth century.

Terrain also had a major influence on battlefield tactics. Hoplite phalanxes often secured their flanks against rivers or hills, as did **Alexander the Great** and the generals of the Hellenistic period. **Xenophon** (*Hipparch.* 7.11) also advocates attacking with cavalry an enemy whose army had partly crossed a river, and this was a feature of several Greek battles, including the **Crimisus River** in 341 or 339 and the action against the **Gauls** at the Spercheius River in 279.

THEBES. Located on the eastern end of the fertile plain of **Boeotia**, Thebes was the major city in Boeotia and a first-ranking city in Greece as a whole. Much of Thebes' history is characterized by its efforts to dominate Boeotia, its general competition with its southern neighbor, **Athens**, and its brief dominance of Greece during the fourth century. Thebes was inhabited in the **Mycenaean** period and is often mentioned in **Homer** (although not as a participant in the **Trojan War**). Because of its location on a plain and its proximity to **Thessaly**, Thebes always possessed a useful **cavalry** force. However, in the fourth century Thebes became famous for developing **hoplite** warfare to its highest point.

The city's first appearance in a historical military context is in 519 when **Plataea** separated from the **Boeotian League**. At the time, Thebes was apparently dominant in the league, or attempting to be so, and Plataea allied with Athens (a **Spartan** suggestion, according to Hdt. 6.108). This resulted in a war with Athens in which Thebes was defeated and lost territory to Plataea, Hysiae, and Athens (which probably gained Eleutherae at this time). From this point onward, Thebes generally followed a hostile policy toward Athens and in 506 participated in the unsuccessful Spartan-led coalition attack on it. Although they captured Hysiae and Oenoe, the Thebans were defeated with heavy loss, including 700 men captured (Hdt. 5.77).

The Thebans participated on the Greek side in the early stages of the Second **Persian War**, sending 500 men with the Greek expedition to Thessaly in 480 and 400 men to **Thermopylae** (1). According to **Herodotus** (7.233), the Thebans abjectly surrendered to the **Persians** at Thermopylae. However, this story may well result from Thebes' later **medizing** and the hostility this aroused in Greece. With the loss of Thermopylae, Thebes was open to the full force of the Persians and, like Thessaly before it, medized—an action that colored Greek views of the city from then on. Theban hoplites and cavalry fought against the Greeks at Plataea and the cavalry particularly distinguished itself. After the Persian defeat, the Greeks besieged Thebes, forcing it to surrender and executing those most prominent in the medizing party. Collectively the Greeks never really forgave Thebes for what was seen as an act of treachery against the Hellenes.

Thebes lost its dominant position in Boeotia for some time after the war (perhaps to **Tanagra** or possibly **Orchomenus**) but recovered it during the 450s, partially because of Spartan support designed to make

Thebes a useful counterweight to the growing power of Athens. The Thebans (and other Boeotians) fought against Athens in the First **Peloponnesian War**, initially unsuccessfully. They lost the major battle of **Oenophyta** in 457 and Boeotia was dominated by Athens until the Athenian defeat at **Coronea** (1) in 447. Thebes was apparently preeminent in the restored Boeotian League and housed the federal treasury.

Thebes triggered the **Peloponnesian War** (2) of 431-404 with a surprise attack on Plataea before hostilities had been declared. It played a major role in this war, joining in the annual **Peloponnesian League** invasions of Attica and contributing cavalry to make up for the Peloponnesian deficiency in this arm. Thebes led the Boeotian League forces to victory against Athens at **Delium** in 424, ending Athens' attempt to return to the land-based strategy of the First Peloponnesian War. Thebes' hostility to Athens was such that in 421 it refused to sign the Peace of **Nicias**, instead arranging a seven-day truce, which was continually renewed. It also joined enthusiastically in the resumption of the war in 413 and provided considerable assistance in the garrisoning of **Decelea** and the **ravaging** of Attica. At the end of the war Thebes argued unsuccessfully for the destruction of Athens and fell out with her ally Sparta over this issue. It subsequently supported the overthrow of the **Thirty Tyrants** and the restoration of democracy in Athens.

Still leading the Boeotian League and occasionally in temporary alliance with Athens, during the fourth century Thebes actively worked against Sparta. It played an important role in starting the **Corinthian War** (395-387), during which it fought against Sparta. In the final stages of the battle of Coronea (2) in 394 the Thebans forced their way through the Spartans, giving some hint of their future hoplite expertise. Despite this, the King's Peace (**Common Peace** [1]), which ended the war in 386, left Thebes in a vulnerable position—isolated and unable to sign on behalf of Boeotia. Spartan garrisons were placed in several Boeotian towns and in 382 the Cadmea (the citadel of Thebes) was seized by **Phoebidas** and Thebes lost its independence. With Athenian assistance the Theban exiles recovered their city in 379 and embarked on war with Sparta. The war ended in 371 when Athens made peace with Sparta and Thebes was again left isolated by the Restatement of the King's Peace (Common Peace [2]). This time, however, Sparta's attempt to enforce the provisions of the peace met with a decisive defeat at the battle of **Leuctra**.

In the years preceding Leuctra, Thebes, under the guidance of **Pelopidas** and **Epaminondas**, had developed new techniques for hoplite warfare. These were based around the professionalism of the **Sacred Band** (the Theban *epilektoi*) and involved combining an unusually deep

phalanx with an oblique advance designed to disrupt and dislocate the enemy phalanx. Using these techniques, Pelopidas and Epaminondas permanently ended Spartan military supremacy. Thebes now became the dominant power in Greece. It invaded the Peloponnese, freeing **Messenia** and assisting in the establishment of the **Arcadian League**—both aimed at crippling Sparta. Thebes also established reasonably complete control over Thessaly to its north. The death of Pelopidas in Thessaly in 364 was a setback and the death of Epaminondas at **Mantinea** in 362 was an even worse one. These deaths, along with the uniting of Athens and Sparta and other Greek cities against Thebes, considerably curtailed its power.

Theban influence on **Euboea** (aimed at Athens) was greatly reduced in 358 and Thebes was unable to deal with **Phocis** during the Third Sacred War (355-346). This war established **Philip II** of **Macedon** as the arbiter of northern Greek affairs, and he fairly regularly outmaneuvered Thebes in the following period. In 338 Thebes was on the losing side at **Chaeronea** (1), and became subject to Macedon. The Cadmea was garrisoned and Thebes lost its position as head of the Boeotian League. On Philip's death in 336 Thebes joined in the abortive Greek revolt and, although treated leniently, revolted again the following year when **Alexander the Great** was reported killed in the north. Alexander reacted rapidly, taking the city by assault. Its fate was decided by the **Hellenic League**, which decreed the city should be destroyed. Alexander razed the city (except for Pindar's house). The site remained uninhabited until **Cassander** rebuilt it in 315. However, it never regained its former position. In 292 Thebes allied with **Pyrrhus I** of Epirus and the **Aetolian League** and revolted against Macedon but was recaptured in 291 by **Demetrius I (Poliorcetes)**. In 229 it was annexed by the Aetolian League.

Thebes' walls were destroyed by the **Romans** in 146 and it was incorporated into the empire. Following its support to **Mithridates VI (Eupator)** in the First **Mithridatic War** (88-85), Lucius Cornelius Sulla confiscated half of the Theban territory, reducing it to little more than a village.

THEMISTOCLES (THEMISTOKLES; c. 528-460). A member of a minor aristocratic family in **Athens**, Themistocles was largely responsible for turning Athens into a major naval power prior to the **Persian Wars**. Elected eponymous *archon* in 493, he was a *strategos* (general) at **Marathon** in 490. In 483 he persuaded the Athenians to spend a large surplus from the silver mines at Laurium to expand the navy from 70 to 200 **triremes**. Under his command, this fleet was the largest contingent in the allied Greek fleet and formed the backbone of the Greek naval defense against the Persian invasion of 480-479. **Herodotus** (8.19 and 58ff.) gives him full credit for the plans adopted by the Greek fleet at **Artemisium** and **Salamis**, although **Eurybiades** was in command. This may be pro-Athenian bias, but the story is supported by **Thucydides** (1.74.1), Eurybiades cannot have been an experienced admiral, and Athens provided around half of the entire Greek fleet. Under these circumstances, Herodotus is quite likely to be correct. Themistocles had less influence in 479 and appears to have played only a minor role in the subsequent campaigns in the Hellespont and the transfer of naval command from Sparta to Athens. He was ostracized circa 471 and went into exile in **Argos** and the Peloponnese. Accused by **Sparta** of involvement in anti-Spartan and pro-Persian plots, he fled to **Persia** where he later died—perhaps around 460. He was renowned for his cunning and trickery and the literary tradition is generally hostile to him, although Thucydides is an exception to this. *See also* ARISTIDES; CIMON.

THEOPOMPUS (THEOPOMPOS).

(1) Eurypontid king of **Sparta** from circa 720 to circa 670. He led the Spartans to final success in the First **Messenian War**.

(2) Historian from **Chios** (b. c. 378). A pupil of the Athenian orator Isocrates, his work is represented by fragments of his *Hellenica* (a history of Greece continuing from where **Thucydides** left off) and the *Philippica* (a world history, beginning in 359). It is possible that Theopompus was the author of the *Hellenica Oxyrhynchia*, a fragmentary history which appears to continue Thucydides, but current opinion generally discounts this.

THERMOPYLAE (THERMOPYLAI). A 50-foot (15-meter)-wide defile on the eastern coastal road from northern Greece to central and southern Greece, where the mountains come very close to the sea (the modern pass is much broader than the ancient one because of changes to the coastline). A natural defensive position, it was the site of several battles:

(1) 480, during the Second **Persian War**. A Greek army of around 7,000 men from the Peloponnese, **Boeotia**, **Phocis**, and **Locris** occupied the pass in conjunction with a Greek fleet taking station at **Artemisium**. Together, they were to prevent **Persian** land and naval forces from moving south. The force seems to have been more in the nature of an advance guard. **Sparta** sent only 300 men (possibly the *Hippeis*), led by **Leonidas I**, who held overall command of the Greek army, claiming they could send no more because of a religious festival (the Carneia). This may indeed have been the reason, but it is also possible that they

underestimated the Persians because of the battle of **Marathon**. Other
states were celebrating the Olympic festival and did not expect the pass
to be forced so quickly. However, the later **fortification** of the Isthmus
of Corinth, and Spartan reluctance to leave its protection, suggests that
Sparta was not prepared to commit large numbers of troops so far north
of the Peloponnese.

Thermopylae was a natural defensive site, improved by a wall and
too narrow for the Persians to deploy their **cavalry**. The Persians
(probably a maximum of 500,000 strong, including noncombatants)
were held up for several days, proving no match for the Greek **hoplites**
in the narrow pass. Then, guided by a local man, **Ephialtes**, the Persians outflanked the Greek force using a little-used mountain track, the
Anopaea path. Leonidas had placed a contingent of Phocians to guard
this track but they were brushed aside. Leonidas did get some warning,
though, and released most of his force before the Persians arrived. With
his 300 Spartans, 700 Thespians, and 400 Thebans (whom **Herodotus**
rather maliciously states were forced to remain), Leonidas fought to the
last. All the Spartans and Thespians were killed, but the bulk of the
Thebans surrendered. The immediate result of the battle was to open the
route south—the Greek fleet had to withdraw when the accompanying
army was destroyed—leading to the defection of Boeotia and the evacuation and destruction of the city of **Athens**. Despite this, the gallantry of
the defenders and the heavy **casualties** they inflicted on the Persians
raised Greek morale and boosted Sparta's reputation. *See also*
DIENECES.

(2) 323, during the **Lamian War**. A **Macedonian army** of 13,000
infantry and 600 **cavalry**, led by **Antipater**, and accompanied by the
Macedonian fleet, marched south against a Greek force led by **Leosthenes**. Already outnumbered because of recent large drafts of Macedonian
troops to Asia Minor, Antipater was defeated at or near Thermopylae
when his Thessalian cavalry deserted to the enemy. Unable to withdraw
far, he retired to **Lamia**, in **Thessaly**, where he was besieged by
Leosthenes.

(3) 191, during the war between **Antiochus III (the Great)** and
Rome. Antiochus had invaded Greece in 192 and, lacking any real local
support and awaiting reinforcements from Asia, he fortified the pass of
Thermopylae. Livy (36.15-19) provides an account of the battle, at least
in part based on a no longer extant section of **Polybius**. Antiochus had
10,000 infantry (which included a *sarissa*-equipped **phalanx**, **archers**,
slingers, and troops equipped with **javelins**), 500 **cavalry**, and an unknown number of **elephants** (Livy [36.19] cites these numbers from
Polybius in preference to much larger numbers given by other histori-

ans). He fortified the pass at Thermopylae with a wall and ditch and secured the mountainous inland flank with 2,000 **Aetolian** allies (half of what he had expected would be available). When the Romans arrived he deployed his army in front of the rampart, with the missile-equipped *psiloi* (light troops) on the hilly right flank, his phalanx in the center, and the elephants and associated infantry on the left, or sea, flank, with the cavalry behind them.

The Roman expeditionary force consisted of 20,000 infantry, 2,000 cavalry and 15 elephants (Livy, 16.14), and was led by the consul Marcus Acilius. Having dispatched 2,000 infantry against each of the three Aetolian positions in the mountains, Acilius launched a frontal attack at dawn. After some hard fighting, Antiochus' force was pushed back behind the rampart and then broke and ran when one of the Roman flanking contingents managed to penetrate an Aetolian position and outflanked the **fortifications**. Although the Roman pursuit was initially slow because of the wall and the narrow pass, the next night (having beaten off an audacious Aetolian attack on their camp) Acilius sent his cavalry on ahead and it overtook many of Antiochus' troops. The Roman **casualties** (150, with another 50 killed in the Aetolian attack on the camp) seem suspiciously low compared to those reported for Antiochus' troops—only the king and 500 men survived. As a result of this battle Antiochus evacuated Greece, but he was followed up by the Romans and at **Magnesia** in Asia Minor was again defeated.

THESSALY. A region south of **Macedon** and north of **Boeotia**, famous for its **cavalry** for most of antiquity. The region contains extensive plains and preserved an aristocratic way of life, both of which helped in the development of its cavalry arm. Thessaly dominated northern Greece in the sixth century, partly through its control of the **Amphictionic League**. By the mid-fifth century, though, following the region's **medizing** in the Second **Persian War**, the cities of Thessaly had become strong enough to challenge the feudal aristocrats and to play an important part in the region's history. The late fifth and early fourth century saw the rise of dynasts, particularly in Pherae, which led to considerable internal strife within the region. Thessaly was briefly united under **Jason of Pherae** but came under Macedonian control under **Philip II**. Following the **Lamian War** Thessaly essentially ceased to play an independent role in Greek affairs. Apart from the sixth century, Thessaly was never a major force in Greek military history, because of a lack of heavy infantry and because the region was rarely united.

In the sixth century Thessaly was administratively divided into four tetrarchies, each of which was further divided into *kleroi*. Each *kleros* provided 40 cavalrymen and 80 **peltasts** (Rose, *Aristotelis Fragmenta*, fr. 498—with the manuscript reading of "**hoplites**" amended to "peltasts"). These troops, in conjunction with those (mainly cavalry) raised from the large feudal estates, formed the forces available to the Thessalians. The large number of cavalry was particularly impressive for a period when most Greek states only had very small numbers. **Herodotus** (5.63-64) records, for example, that in 511-510 the Thessalians sent 1,000 cavalry to assist the Pisistratids at **Athens** against **Sparta**. The Thessalians were largely responsible for the defeat of the first Spartan invasion led by Anchimolius, but were later unsuccessful against a larger force under **Cleomenes I**. In 476/5, when Athens' entire cavalry force was only around 300 strong, one large landowner, Menon of Pharsalus, was able to raise 200-300 cavalry for a campaign against the **Persians** (Demosthenes, 23 [*Against Aristocrates*], 199).

The Thessalians joined the Persians during the Second Persian War when the Greek expeditionary force sent to help defend them returned home on discovering that it was too small to cover all the passes into Thessaly (Hdt. 7.172-174). Thessaly provided an unknown number of soldiers on the Persian side at **Plataea** (Hdt. 9.31) and after the war was unsuccessfully attacked by the **Hellenic League** (Hdt. 6.72), which had sworn to punish medizers.

For the rest of the fifth century, Thessaly drifted in and out of alliance with Athens. Around 462/1, following their rupture with Sparta over Mount **Ithôme**, the Athenians allied with Thessaly and **Argos** (Thuc. 1.102.4) to remedy their deficiencies in cavalry and hoplites. However, the Thessalian cavalry deserted from Athens at **Tanagra** in 457, causing the loss of the battle. Around 455 an Athenian army sent to aid one Thessalian faction had some minor success but was largely restricted to its camp by the Thessalian cavalry acting in a mobile defense role (Thuc. 1.111.1). In 431, at the start of the **Peloponnesian War** (2), the Thessalians sent cavalry to assist Athens resist the first **Peloponnesian League** invasion (Thuc 2.22.2-3). These contingents were apparently organized on a *polis* (city-state) basis, not on the old tetrarchy system, but there is no evidence of when this structural change had taken place. The organization of the Thessalian forces on this basis continued down to at least the late fourth century as **Alexander the Great** had *polis*-based Thessalian cavalry contingents in the army he led against Persia (Arr. *Anab.* 3.11.10). **Thucydides** does not mention any further Thessalians deployed to assist Athens during the Peloponnesian War, but they were instrumental in causing the failure of the Spartan-led colony at Heraclea, which threatened **Euboea**, and generally hindered Spartan attempts to move troops to the Chalcidice (Thuc. 4.102.2, 5.13.1).

However, at least some Thessalians aided **Brasidas'** activities in the north (Thuc. 3.92-93, 4.78, 4.108.1), demonstrating one of the perennial barriers to Thessalian power—a lack of united action.

The tyranny of Lycophron in Pherae (c. 406-390) divided Thessaly. He may have based his tyranny on a popular and anti-aristocratic movement (if Xen. *Hell.* 2.3.36 is linked with his rise to power)—certainly he was opposed by the aristocratic elements of Thessalian society, especially the Aleuadae, based in Larissa. This division and the later activities of Jason (tyrant of Pherae c.385-370) encouraged the intervention of outside powers—Sparta, Athens, and **Thebes**. Jason had gained control by 374, but was assassinated shortly afterward (in 370). Thessaly again degenerated into internal anarchy, which allowed the **Boeotian League** to extend its influence over the region—although from around 361 opposed fairly successfully for a time by Athens. On the outbreak of the First **Social War** in 357, Boeotia was able to reassert its control over Thessaly and in late 355 used this control to secure the votes of the Amphictionic Council to declare war against Phocis (Third **Sacred War**). Thessaly was initially an ally of Boeotia, but was defeated in 354 and shortly afterward fell under the control of Philip II of Macedon. In 352 Philip decisively defeated **Onomarchus** and reunited Thessaly. Although Thessaly attempted to regain its independence during the Lamian War, from 352 it was essentially under Macedonian control (although the level of control initially varied from time to time). It was one of the regions contested by the *diadochoi*, and later by **Pyrrhus I** of Epirus, and parts of it fell under **Aetolian** control in the third century. **Rome** detached Thessaly from Macedonia in 196 and it became part of the Roman province of Macedonia in 148.

The Thessalian cavalry of the classical and early Hellenistic periods was marked by its generally high quality and its use of a diamond- or rhomboid-shaped formation. Thessalian cavalry often wore heavy boots, breastplate, and **helmet**, and as a group carried a mix of the thrusting spear and **javelins**. It performed well at the battle of the **Crocus Plain** in 352 and at **Gaugamela** in 331 (Diod. Sic. 16.35.5; Arr. *Anab.* 3.15.3) but failed to prevent **Agesilaus II** from crossing Thessaly in 394 and was defeated by **Pelopidas** around 364 (Xen. *Hell.* 4.3.5-8; Plut., *Pelopidas*, 32.2).

THIBRON (d. 391). A **Spartan officer** sent twice to **Ionia** to campaign against the **Persians**. In 400/399, with a force of 500 Peloponnesians, supplemented by troops from **Athens**, the Ionian Greek cities, and the **Ten Thousand**, he had some minor successes against **Tissaphernes**, capturing Magnesia. He was replaced by **Dercylidas** and exiled on his return for allowing his troops to plunder allied territory. Later, he was recalled from exile and sent to Ionia again in 391. He was surprised and killed near **Ephesus** by a force of **cavalry** under **Struthas**, who on a number of previous occasions had observed Thibron's poor march discipline.

THIRTY TYRANTS. The name given to a group of oligarchs who ruled **Athens** and fought a civil war against the democrats from 404 to 403. The Thirty came to power at the end of the **Peloponnesian War** of 431-404, with the support of **Sparta** and many of the upper classes who believed that they had borne an inordinate burden during the closing stages of the war. They had supported the war financially through taxes such as the *eisphora*, their property in Attica was **ravaged** more regularly following the Peloponnesian occupation of **Decelea**, and they (or their sons) had served in the **cavalry** by day and manned the walls of the city as **hoplites** at night. In short, they considered that they and not the navy had been the defenders of the state. A significant proportion of the 1,000-strong cavalry corps appears to have supported the overthrow of the democracy. To secure their position, the Thirty installed a Spartan garrison in Athens (Xen. *Hell.* 2.3.13.4).

The accession of the Thirty led to a civil war when 70 democratic exiles under **Thrasybulus** occupied **Phyle** with **Theban** support (Xen. *Hell.* 2.4.2; Diod. Sic. 14.32.1). The cavalry and the Spartan garrison actively, but unsuccessfully, attempted to dislodge this group. A dawn attack by the democrats caught the oligarchic troops by surprise—120 hoplites and three cavalrymen were killed. This, and the increasing hostility to their extremist actions within Athens, alarmed the Thirty, who decided they needed a safe refuge. They used the cavalry to secure Eleusis, expelling the inhabitants and executing all the adult males (Xen. *Hell.* 2.4.2-10; Diod. Sic. 14.32.4-33.1). Late in 404 Thrasybulus and his men defeated the oligarchs at the battle of **Munychia** and the Thirty retired to Eleusis (Xen. *Hell.* 2.4.10-24; Diod. Sic. 14.33.2-5). A near-success by the democrats against the Spartan garrison and dissension between **Pausanias** (2) and **Lysander** led to the Spartans withdrawing their support from the oligarchs and effecting a reconciliation between the parties in Athens. By 401 the Thirty had lost all power, with all the oligarchs either dead, permanently exiled, or, in some cases, reconciled with the democracy, which accepted them back under an amnesty (Xen. *Hell.* 2.4.28-43; Diod. Sic. 14.33.5-6).

THIRTY YEARS' PEACE. The peace treaty which ended the First **Peloponnesian War** of 459-445. The terms are notoriously obscure (there is no coherent account of it in any ancient source, only scattered remarks in **Thucydides** [1.35, 40, 44-45, 67, 78, 140]). It was to last for 30 years, determined peace on the basis of the status quo, specified that differences should be arbitrated, and allowed any neutral state to join either alliance. Other possible clauses were the guarantee of free trade and (much less likely) the guarantee of the autonomy of allies. Breaches of this peace led to the outbreak of the Peloponnesian War (2) of 431-404.

THRACE. An area to the northeast of **Macedon**, covering from the Danube to the Aegean and from the Black Sea to (approximately) the River Strymon. In antiquity, the Thracians were Indo-Europeans who lived in villages and whose society was tribal. Although the occasional ruler succeeded in temporarily uniting parts of Thrace, the Thracians were perennially divided along tribal lines and generally failed to take any united action. The Thracians had no written records, so all our knowledge of them comes from non-Thracian writers (including **Herodotus**, **Thucydides**, **Xenophon**, **Diodorus Siculus**, Strabo, and **Arrian**) and from archaeology.

The Greeks regarded the Thracians as barbarians but considered them very useful as **mercenaries**. The area produced good light infantry and **cavalry** (it was particularly known for its white horses and in Greek tradition was the intermediary between the **Scythians** and the Macedonians for the introduction of the wedge-shaped cavalry **formation** into Greece). Arms and **armor** varied from region to region, but there are differences between the southern and northern parts of Thrace. The latter was more closely aligned with Scythian practices and equipment, while the southern part was more subject to Greek influence. Armor was probably generally the preserve of the nobility, although the practice of wearing **helmets** and other items of body armor seems to have become more common in the south from the fourth century.

The core of the Thracian armies were the **peltasts** and cavalry, supplemented by other *psiloi* (light infantry) such as **javelin**-equipped troops (more lightly equipped than the peltast but otherwise probably very similar), **archers**, and **slingers**. The general deficiency in heavy infantry was sometimes remedied by hiring Greek **hoplite** mercenaries—even Seuthes, a minor Thracian dynast (not to be confused with Seuthes I, below), employed 6,000 of the **Ten Thousand** on their return from **Persia**; Xen. *Anab.* 7.2ff.). However, the nature of Thracian forces meant that skirmishing and **ambush** were the generally favored tactics.

Thracians are mentioned in **Homer**'s *Iliad*, but the first real contact between Greeks and Thracians occurred in the seventh century with

Greek colonization of the Chalcidice and neighboring coastline. Late in the sixth century the **Athenians** established a strong foothold in the area when **Miltiades** became ruler of the Chersonese (Hdt. 6.34ff.), but were driven out by the Persians, who established loose control over Thrace (Hdt. 4.118ff., 6.40ff.). Some Thracians served in **Xerxes I**'s army that invaded Greece in 480 (Hdt. 7.75), but after the battle of **Plataea** (1) Greek control was soon reestablished over the coastal regions of Thrace.

Following the **Persian Wars**, the Thracians had to contend with further Greek expansion in the south. The founding of **Amphipolis** was seen as a major threat and earlier, around 465, near Drabescus, a combined army of Thracians annihilated a strong Athenian force pushing further north into Thrace (Thuc. 1.100). At this time the Odrysian king Teres had succeeded in uniting a large portion of Thrace under his rule, and this was maintained under his son Sitalces, who supported Athens at the start of the **Peloponnesian War** of 431-404 (Thuc. 2.95ff.). In 429 Sitalces led an army against Macedon and the Chalcidians, who had revolted against Athens. It was around 150,000 strong and consisted of about one-third cavalry and the remainder *psiloi* (Thuc. 2.98; Diod. Sic. 12.50). Sitalces was killed in 424 fighting another Thracian tribe, the Triballi, and was succeeded by his nephew, Seuthes I, who brought the Odrysian kingdom to its greatest power (Thuc. 2.97, 4.101). Thrace played little part in the remainder of the war other than providing a ready source of mercenaries to both sides (e.g., Thuc. 5.6, 7.27).

After the death of Seuthes I late in the fifth century, the kingdom disintegrated, with semipermanent struggles between the parts (Xen. *Anab.* 7.2.32, 7.3.16, 7.5.1; *Hell.* 4.8.25). It is at this time that Seuthes employed the remnants of the Ten Thousand to restore him to power. This internecine warfare allowed the Athenians to regain control of the Chersonese (although they lost most of it again to Cotys I, who ruled from 383 to 359) and by 339 **Philip II** of Macedon had reduced the Odrysians to a subject state. Thracian troops served in **Alexander the Great**'s expedition against Persia (Arr. *Anab.* 3.12.4) but the region was involved in a rebellion against him (Curt. Ruf. 10.12.43).

After Alexander's death, the Odrysian king, Seuthes III, fought **Lysimachus** with mixed success but retained the independence of the core of his kingdom. In essence, Thrace was now a loose Macedonian protectorate, although by no means subjugated. Despite the presence of several Macedonian garrisons, parts of the region, under numerous small dynasties, maintained varying levels of independence. The political fragmentation, however, further weakened Thrace, and in 279 Gallic invaders overran large parts of the region. One group of **Gauls** was able to temporarily establish a kingdom in the Chersonese that survived until the end of the third century. Political instability and disunity also hampered resistance to **Rome** after the fall of Macedon and substantial parts of the Thrace were incorporated into the Roman empire in 168 and 129.

THRASYBULUS (THRASYBOULOS; d. 388). An **Athenian** military leader and statesman. As a **trierarch** in the fleet at Samos in 411 he led the democratic resistance against the oligarchic revolution at Athens and was elected *strategos* (general) by the sailors. He commanded a contingent of 20 ships at **Cyzicus** in 410. Thrasybulus campaigned successfully along the **Thracian** coast in 407 and was active in the Hellespont in 406, serving as a trierarch at **Arginusae** in 406. **Xenophon** involves him (perhaps unfairly) in the failure to recover the Athenians whose ships had sunk in this battle, and in the subsequent execution of six of the *strategoi* involved (*Hell.* 1.7). He was exiled by the **Thirty Tyrants**, but seized **Phyle** and then the Piraeus and played a major part in the restoration of the democracy.

In 389-388, during the **Corinthian War**, Thrasybulus commanded 40 ships in the Hellespont. After fairly successful operations, including defeating and killing the **Spartan** governor of Methymna on Lesbos, he crossed to the mainland, exacting contributions (or **ravaging** the territory of those who did not comply) in order to **pay** his troops. At Aspendus he was murdered by locals angered by the fact that, even though they had paid the contribution, their property had been looted by Thrasybulus' men. *See also* BOOTY; THRASYLLUS.

THRASYLLUS (THRASYLLOS; also THRASYLUS/THRASYLOS; d. 406). An Athenian general and politician active during the **Peloponnesian War** of 431-404. A **hoplite** at Samos in 411, he was associated with **Thrasybulus** in leading the democratic resistance to the oligarchic revolution at **Athens**. Thrasyllus was elected *strategos* (general) by the fleet and was one of the naval commanders at the Athenian victory at **Abydus** (2) (Cynossema) in 411. From then on he served in the Hellespont and **Ionia**. In 409 he lost a land battle at **Ephesus** but defeated a Peloponnesian naval contingent off that city shortly afterward. He was associated with **Alcibiades** in the command at **Chalcedon**. Along with five of his fellow generals, Thrasyllus was executed after the battle of **Arginusae** in 406 over the failure to rescue those Athenians whose ships had been sunk.

THREE HUNDRED. Most probably the *epilektoi* (elite troops) of **Elis**. They are attested circa 365, in conjunction with the Elean **cavalry**, putting down a democratic revolution in Elis. They stormed the Acropolis, which had been occupied by the democrats, and exiled about 400 of them (Xen. *Hell.* 7.4.15-16).

THUCYDIDES (THOUKYDIDES; c. 460-400). Author of the main source for the history of the **Peloponnesian War** of 431-404. Thucydides, son of Olorus, was a well-to-do **Athenian** whose family had mining interests in **Thrace**. He caught and survived the plague at the start of the Peloponnesian War and in 424 served as a *strategos* (general). His failure to prevent the Spartans from capturing **Amphipolis** in that year (he was absent at the nearby port of Eion) led to his exile from Athens until the general amnesty in 404 at the end of the war. He states that he began collecting information for his history of the Peloponnesian War from the very start of that war and that his exile gave him the opportunity not only to spend considerable time writing it but also to acquire information from non-Athenian participants in the war. Despite this, the work is incomplete, ending (in mid-sentence) in 411.

Thucydides was a thoughtful historian with a solid understanding of military affairs. He was an eyewitness to many events of the war and in general seems to have been a careful and intelligent observer. However, he did hold aristocratic and antidemocratic views (believing that the educated upper classes were better suited to decide military and political matters than the ordinary man in the street) and these do seem to have influenced his account of the war. In particular, he has been accused of strong bias against the democratic politician **Cleon**. *See also* BRASIDAS.

TIMOLEON (d. c. 335/4). A Corinthian general; son of Timodemus. He is famous for his liberation of Greek **Sicily**. Little is known of his early life except that he was a brave soldier and killed his own brother, Timophanes, who was setting himself up as tyrant of **Corinth**.

In 345 Syracusan exiles requested assistance from Corinth, **Syracuse**'s mother city. The tyrant **Dionysius II** was ensconced in Ortygia (the island fortress of Syracuse), while Hicetas, tyrant of Leontini, held the rest of the city and was allied with the **Carthaginians**, who were clearly preparing for an attack on the Greek areas of Sicily. In 344 Corinth sent Timoleon to Sicily with seven **ships** (including four **triremes** and three ships from other cities) and 700 **mercenaries**. Having dodged a Carthaginian fleet sent to intercept him and welcomed only by Andromachus, tyrant of Tauromenium, Timoleon soon defeated Hicetas in battle outside Adranum by attacking his much larger army at dinnertime. The victory gained him further support and induced Dionysius to surrender Ortygia to him. This was a major breakthrough as it provided

Timoleon with a considerable quantity of arms and money—both of which he lacked. With these, plus reinforcements from Corinth (2,000 infantry and 200 **cavalry**), he began systematically deposing the tyrants in all the Greek cities (except Tauromenium). However, he also had to plunder parts of Carthaginian Sicily to gain **booty** from which to **pay** his mercenaries. He razed the fortress on Ortygia and established a democratic constitution in Syracuse and took similar measures elsewhere. He also encouraged settlers to come from mainland Greece.

In 341 (or possibly 339) Timoleon saved Greek Sicily from a major Carthaginian invasion by heavily defeating a much larger enemy army at the **Crimisus River**—probably with the support of Hicetas of Leontini (Diod. Sic. 16.77.5, although not mentioned in **Plutarch**'s *Timoleon*). Shortly afterward, Timoleon successfully renewed his war against Hicetas and executed him. The publicity Timoleon generated in Greece about the Crimisus River further boosted Greek immigration and within a few years of his victory, Timoleon had stabilized the whole of Greek Sicily and increased its population considerably. The highest figure for the number of settlers in the literary sources is 60,000 but archeological evidence suggests this may be too low.

Around 337 Timoleon, who was going blind, retired from public life, but he stayed in Syracuse, dying there a few years later. Timoleon's achievements are remarkable (particularly considering the small size of his force and the strong possibility that it was sent out with rations but no money for pay) and he undoubtedly restored and revitalized Greek Sicily. However, the extant sources for him (Plutarch and **Diodorus Siculus**) are based on extremely favorable accounts of his rule (including the historian Timaeus, son of Andromachus of Tauromenium) and are rather uncritical. Timoleon was clearly not averse to the use of trickery, unscrupulous diplomacy, and killing, but the benefits of his work in Sicily ensured he was seen there as a spotless and altruistic liberator.

TIMOTHEUS (TIMOTHEOS; c. 415-354). An **Athenian** *strategos* (general) active in the second quarter of the fourth century; son of **Conon**. He was elected *strategos* on the foundation of the Second Athenian Confederacy in 378/7 and was reelected on numerous occasions after that. **Chabrias'** expedition to northwest Greece and Timotheus' decisive victory over the Spartan navy at Alyzia in 375 gained important new members for the confederacy (including **Corcyra**). In 374 his restoration of exiles on Zacynthus broke the newly made peace with **Sparta**. The following year he was given command of a 60-**trireme** fleet to Corcyra against the Spartan Mnasippus. However, this expedition was delayed by a lack of funds and shortage of crews, which led Timotheus to cruise the islands raising money, and he was recalled in disgrace to stand trial. This problem was a perennial one at the time for Athenian generals, who were frequently sent out without sufficient money. In another case Timotheus had to issue emergency copper coinage for soldiers to buy food and later redeemed these with traders for silver (Isocrates, 15 [*Antidosis*], 120). Although Timotheus was acquitted in 373/2, he left Athens and served as a **mercenary** general for **Persia** against **Egypt**.

From 366 Timotheus was prominent in the more aggressive Athenian policy within the Athenian confederacy and against Persia. The following year he seems to have exceeded his instructions to help Ariobarzanes, a Persian satrap in revolt, without breaking the King's Peace (**Common Peace [1]**). Actions such as his establishment of a **cleruchy** on Samos and a visit to Erythrae in Asia Minor (in the Persian empire) helped bring about the First **Social War** of 357-355. In 356 he was fined 100 talents for his failure to support **Chares** at the naval battle of **Embata**. He went into exile and died in 354.

TISSAPHERNES. Persian commander of the coastal satrapies (provinces) of Asia Minor, 413-408 and 401-395. In 412, during the **Peloponnesian War** (2), he adopted a policy of partial support for **Sparta**. This was encouraged by **Alcibiades** and aimed at weakening both Sparta and **Athens** and freeing his coastline from Athenian influence and interference. Tissaphernes was responsible for negotiating the Persian-Spartan agreement of 412/1 and ending the revolt of Amorges in Caria in the same year (Thuc. 8.5, 18, 28, 37, 46). In 408 he was demoted to command of Caria, perhaps because of his brother's conspiracy against **Darius II**, but also because of a change of Persian policy under **Cyrus the Younger** to one of wholehearted support for Sparta (Xen. *Hell.* 8-9).

Tissaphernes ended up in open military conflict with Cyrus and in 401 alerted Artaxerxes to Cyrus' attempt to overthrow him (Xen. *Anab.* 1.1.6-8, 1.2.4-5). Following the battle of **Cunaxa**, Tissaphernes was instrumental in the arrest and execution of the leaders of the **Ten Thousand** (Xen. *Anab.* 2.5-6). Restored to his command of the coastal provinces, Tissaphernes was faced with the attacks of **Thibron**, **Dercylidas**, and then **Agesilaus II** of Sparta. Although for a time managing to divert these attacks into the province of his subordinate satrap, **Pharnabazus**, Tissaphernes eventually had to face Agesilaus and was heavily defeated by him near Sardis in 395. This defeat led to his subsequent murder by Artaxerxes (Xen. *Hell.* 3.1.3, 9; 3.4.1-25). Tissaphernes was a fairly able politician but not so able a military commander, although the great disparity between the quality of his local troops and the Greeks under Agesilaus and others rendered his task of defending his provinces extremely difficult.

TRAINING. Although training is now regarded as fundamental to the success of any army, in most periods of Greek warfare training was non-existent or limited. Soldiers generally learned on the job and there were no staff colleges or training academies. An **officer** would normally receive no specialist training, learning basic skills by serving in the ranks of the infantry or **cavalry** and then graduating to command. However, in the **Macedonian** and Hellenistic armies, aspiring young officers could gain good preparation for command by serving on a general's staff or in the Royal Companions. Preparation for command was therefore essentially by on-the-job experience and talking to experienced practitioners, perhaps supplemented by the reading of **military manuals** or history, rather than by formal training.

Training in individual military skills was also limited. This particularly applied to *psiloi* (light troops), who acquired skills as an **archer** or **slinger** in civilian life as shepherds or hunters. At **Athens** in the fifth century there were professional teachers of the use of **hoplite** weaponry, although Socrates (or Plato) at least rather looked down upon them (*Laches*, 178a-179c, 183c-184b). From the fourth century the *ephebeia*, an Athenian public training scheme for youths aged 18-20 provided some training in light infantry skills and later on in the use of **artillery** and in horsemanship. The Athenian cavalry also undertook **weapons** training. **Xenophon** was a strong advocate of training, especially for the **javelin**, which was apparently quite difficult to throw from horseback (Xen. *PH*, 8.10; *Hipparch.* 1.6, 21, 25; 6.5). There are several vase paintings showing cavalrymen throwing javelins at a **shield** on a pole and this was an event (the *aph' hippou akontizon*) in public games for which prizes were awarded. **Sparta**, though, was the greatest exception to the general lack of individual training. The whole Spartan *agoge* or education system was designed to produce good hoplites and involved both physical exercise and skill at arms.

The general view seems to have been that being in a hoplite **phalanx** actually required little training. Aristotle comments on this and adds that what made the Spartan hoplites superior was not the quality of their training but the fact that they trained and the others did not (*Politics*, 1297b18-22, 1338b24-9). In addition, the burden of training could be avoided by hiring **mercenaries** with the requisite skills. Nevertheless, collective training was better catered for than individual training, particularly from the end of the fifth century and some training did occur—not only at Sparta. The Athenians from time to time held reviews of their hoplites, and these almost certainly involved some maneuver. Nevertheless, for most of the Classical period the Spartan phalanx was the only one that could execute more than the most basic maneuvers in battle. When necessary, training was sometimes conducted after mobilization—as **Agesilaus II** did in Asia when raising his army against the **Persians** (Xen. *Hell.* 3.4.15-18).

Extensive experience, such as that possessed by the **Ten Thousand**, could produce similar results to the Spartan-style individual and collective training but the average Greek hoplite could not hope to emulate this until the fourth century. In many states this period saw the rise of standing elite hoplite forces, *epilektoi*, and these did engage in training, which led to improved performance. Training was also necessary when new organizations or practices were introduced, such as the *sarissa*-equipped Macedonian phalanx—as occurred in the **Achaean League** under **Philopoemen** in 208/7. Prior to the battle of **Raphia** in 217, **Egypt** spent a whole year raising and training its phalanx, with impressive results.

Cavalry and **naval warfare** were usually regarded as requiring more training than hoplite warfare. The Athenian cavalry seem to have engaged in more elaborate training than their infantry counterparts, not only in weapons use but also in maneuvering in **formation**. The *anthippasia* (a mock cavalry battle) was a feature of Athenian public festivals and prizes were awarded for the winning squadron. Skilled rowers and steersmen were required for **naval warfare**, particularly with the **trireme**. Prior to the battle of **Lade** in 494, the **Ionians** embarked on a good training program but abandoned it as being too onerous (Hdt. 6.11-12). In 373, **Iphicrates** used the long voyage to **Corcyra** to conduct exercises and prepare his fleet up to a high level of efficiency (Xen. *Hell.* 6.2.27-32). In 431 **Pericles** dismissed the threat of the Peloponnesians developing a navy for several reasons, including the Athenian control of the seas, which would prevent the Peloponnesians from training to achieve the required standard (Thuc. 1.142.6-9).

Training generally did not sit easily with the concept of the volunteer citizen-soldier who was called up as required. When it did occur down to the end of the Classical period (other than at Sparta), it was usually because of the initiative of a vigorous commander or resulted from the pressing needs of a particular circumstance. Although increasingly common from the fourth century, training was still fairly often conducted on an ad hoc basis, with the emphasis on preparation by on the job experience.

TRIERARCH (*TRIERARCHOS*; pl. *TRIERARCHOI*). Commander of a **trireme**; at Athens, a citizen who equipped, maintained, and normally captained, for a 12-month period, a trireme provided by the state. Holding the trierarchy was a liturgy (i.e., a public duty) that an individual could be asked to perform only once in any three-year period. The system worked well in the fifth century but was in some difficulty in the fourth when a decline in wealth and willingness to undertake expenditure coincided with the need for a higher level of military preparedness and a quick response time for crises. *See also* NAVAL WARFARE; OFFICERS.

TRIREME. The trireme (Greek *trieres* [pl. *triereis*], "three-rowing") was the main fighting **ship** of the Greeks and most of their opponents from the late sixth to the late fourth centuries (see figure 6). Used primarily as fighting ships, versions also served as troop or horse transports. From the late fourth century they were superseded as frontline ships but continued to be used as transports, for **reconnaissance**, and to tow troop ships or supply ships.

About 20 feet (6 meters) wide and 130 feet (40 meters) long, the trireme was propelled by three rows of oars on each side (see figure 7a), supplemented by sails on a mainmast and a foremast. However, before combat the masts were usually stored at a suitable site on nearby land. An average speed of around 6 knots was an unhurried traveling pace, while an average speed of around 7.5 knots is attested as a fairly fast pace over distance. Speeds of up to 10 knots could probably be achieved for short periods during combat. On trips the trireme was normally beached at night to allow the crew to eat and sleep. Because of this and its overall structure, it was not a deep-sea vessel and tended to remain within striking distance of the shore. Trireme routes therefore tended to either follow the coast or move from island to island.

The Athenian trireme carried a complement of 200 men, comprising 170 oarsmen, 16 other crew (including captain, steersman, etc.), and a complement of 14 *epibatai* (marines) (10 **hoplites** and four **archers**). However, troop-carrying triremes could carry up to 40 hoplites. Each oarsmen was designated by his position in the three rows as either a *thranites* (top row), *zygios* (middle row), or *thalamios* (bottom row).

The trireme was equipped with a bow-mounted bronze ram, which was used to hole enemy ships. As the trireme had positive buoyancy, if holed it did not sink to the bottom but floated at or just below surface level until it broke up or was recovered. This explains why no trireme wrecks have been discovered, although some examples of rams have been found. The trireme, in the hands of an experienced and well-trained crew, was a very effective **warship**. Its considerable speed and maneuverability combined with the well-designed ram posed a formidable threat to other vessels. It began to be replaced from the second half of the fourth century by a variety of bigger warships, including the four, five, six, seven, eight, nine, and ten (no ship bigger than a ten is attested in combat, although they were built). *See also* NAVAL WARFARE; PENTECONTOR; TRIERARCH.

TROJAN WAR. Although the details of the Trojan War preserved in **Homer** and other authors' works are clearly fictitious, the epic story of a 10-year siege may be based on a real incident. Levels VI and VIIa of the site (Hisarlik) in northwest Asia Minor plausibly (but not certainly) identified as Troy did suffer violent destruction around 1270 and 1190, respectively—about the time the story is set. The broad outlines of the epic fit the current state of our archaeological knowledge and make sense, but should not be pressed for detail. The two Homeric poems, the *Iliad* and *Odyssey*, deal, respectively, with a very small section of the siege and with **Odysseus'** return home. The remaining parts of the story are supplied from a variety of later authors, whose works often differ in detail.

   The essential version is that Paris, son of Priam, king of Troy, abducted Helen, wife of Menelaus of **Sparta**. Enraged, the Greeks launched an expedition, led by **Agamemnon** of **Mycenae**, and laid siege to Troy for 10 years. As **Thucydides** (1.11) very logically points out, the Greeks must have won an initial victory in order to establish a camp outside Troy, but the length of the siege necessitated troops being drawn from the combat to growing food to supply the besiegers, ironically extending the siege because the Greeks could not deploy all their combat power. Troy held out for 10 years, with the assistance of allies (and various gods) and despite the death of their champion **Hector** at the hands of **Achilles**. Finally, the quarrel among the gods was resolved, with those supporting Troy forced to give way. The city was taken by a ruse, suggested by Odysseus, the Greeks feigning withdrawal and leaving behind a large wooden horse filled with men. The Trojans took the horse into the city and during the night the Greeks released themselves and opened the gates to the rest of their army. The city was sacked with great bloodshed, but Aeneas escaped and later founded **Rome**.

TROPHY. A victorious Greek army normally set up a trophy (*tropaion*) on the battlesite to commemorate its victory. The most common type was composed of **weapons** and **armor** stripped from the enemy and attached to a tree that had been cut back to a rough cross shape or to a pillar. A series of traditions generally governed their use, although there were some departures from these. The trophy was set up at the point where the battle swung in favor of the victors, or on a nearby headland or shore in the case of naval actions. In addition, they were not supposed to be restored or renewed—the **Thebans** were criticized after the battle of **Leuctra** in 371 for setting up a permanent monument (plate 4) to their victory over **Sparta**. The trophy signified that the victor had gained control of the battlefield, usually indicated by the losing side having to ask for a truce to recover its dead. In theory, once set up, trophies were inviolable, but occasionally where the result of the battle was disputed or there had been a delay between the battle and setting it up, a trophy was destroyed (e.g., Thuc. 8.24.1). Away from the battlefield, though, permanent trophies were often set up in cities to commemorate victories. From Leuctra onward, particularly under the Hellenistic monarchs, permanent trophies became more common on battlesites.

TYRE. A major seaport in Phoenicia (modern Lebanon). It, along with the other Phoenician cities, provided a significant contingent to the **Persian** fleet from around 539. The Phoenician fleet assisted in the Persian conquest of **Egypt** and a naval contingent from Tyre served alongside other Phoenician squadrons in the invasion of Greece during the Second **Persian War** of 480-479. During the fifth and fourth centuries the Phoenician squadrons were the mainstay of the Persian fleet. The city of Tyre itself was built on a naturally strong defensive position, an island about half a mile (1,700 meters) offshore.

In 332 **Alexander the Great** besieged Tyre as part of his operation to destroy the Persian navy by capturing its ports. Two other major Phoenician centers, Byblus and Sidon, had already surrendered to him, but Tyre, because of its position and its large amount of **artillery**, was generally regarded as impregnable and decided to resist. The seven-month siege of the city (January to August 332) is regarded as one of Alexander's major successes in **siege warfare**.

Alexander first constructed a mole out to within artillery range of the city and under the protection of two large hide-covered towers, bombarded the wall opposite with stone missiles. This wall was about 160 feet (50 meters) high and constructed of large stone blocks. Despite its strength the defenders were sufficiently worried to launch a sortie using a fireship, which destroyed the two towers. Alexander's response was to widen the end of the mole and build even more towers and catapults to batter the walls. At the same time a boost in his naval strength by the arrival of 80 ships from the Phoenician cities of Aradus, Byblus, and Sidon and from elsewhere allowed him to mount stone-throwing engines in ships and to probe the city walls from every direction. The city walls were first breached on the southern side, facing Egypt, but an amphibious assault was repelled. Alexander brought even more ship-mounted catapults to bear on the weakened section and when a sufficient length of wall had been brought down, personally led his troops ashore from other ships using drawbridges. As was often the case, the length and difficulty of this siege led to considerable bloodshed when the troops stormed the city. Many of the survivors were sold as **slaves**.

The loss of Tyre was a fatal blow to Persian naval power, opened up Egypt to invasion, and removed much of Alexander's worries about Persian interference back in Greece. This siege is the first attested case of stone-throwing machines being employed against walls rather than personnel, a significant development in siege warfare. It also demonstrates the power of contemporary siege techniques and engines when used imaginatively and marks an advance on the siege tactics Alexander had displayed at **Halicarnassus** in 334. Tyre soon recovered from Alexander's sack and remained an important maritime city. It was subject to various Hellenistic monarchies until it regained its independence (along with other Phoenician cities) in the mid-second century. Tyre was one of the first states in the region to make a treaty with **Rome** and was peacefully absorbed into the Roman empire.

TYRTAEUS. A seventh-century **Spartan** poet and general whose inspirational martial poetry is credited with stiffening Spartan resistance in the dark days of the Second **Messenian War**. Tradition from at least the time of **Pausanias** (3) has it that Tyrtaeus' poems played a major part in reinvigorating the Spartan war effort and inspiring victory. However, this is not justified by anything in the extant poetry, which mentions the Spartan conquest of **Messenia** in the First Messenian War but whose other military references do not actually mention Messenia and could apply to any other seventh-century wars. Tyrtaeus' poetry also praises *eunomia* (good order) as an important Spartan discipline and describes the constitution that helped give the early political stability which was an important feature in Sparta's ability to dominate the Peloponnese.

- W -

WAR OF THE BROTHERS. *See* ANTIOCHUS (5); SELEUCUS (2)

WARSHIP. Early Greek warships were not really specialist military vessels. They developed from, and often doubled as, cargo or personnel transports. They had a single row of oars on each side, supplemented by

a mast and sail, and had no deck. The **Homeric** poems refer to both 20-
and 30-oared **ships**. Gradually the **pentecontor** became dominant as the
pirate vessel or warship of choice and by the eighth century an advanced
version, the two-level pentecontor (bireme), had developed as a fairly
specialist warship. It was fast and maneuverable and its bronze ram was
effective in ship-to-ship combat. Further specialization occurred with the
development of the **trireme** in the second half of the sixth century, reputedly first built in Greece at **Corinth**.

The trireme was a highly effective warship for the time and rapidly
became the mainstay of Greek fleets. Although larger warships were
built from circa 399, the trireme only really began to lose its importance
late in the fourth century, because of a shift toward larger ships mounting catapults and sizable contingents of *epibatai* (marines). Their size
and increased protection made ramming less effective and boarding difficult from smaller ships. While giant ships (ranging from the eleven to
the forty) were built by the Hellenistic monarchies from the third century, these were used largely as ceremonial vessels or as demonstrations
of the power and wealth of their owners. No ship larger than a ten is attested in combat. Excluding the trireme and pentecontor, which are
listed separately, the range of ships which made up Greek navies from
around the mid-fourth century are given below (the lighter ships first).
See also NAVAL WARFARE.

(1) *Hemiolia* (pl. *hemioliae*). A fast, light ship, attested from the
early fourth century as a favored vessel of pirates. From the mid-fourth
century the *hemiolia* also appears in Greek navies, presumably mainly
for **reconnaissance** or dispatch purposes. It probably had 50 oarsmen,
distributed in one and a half rows on each side, with the half row amid-
ships either above or below the full row.

(2) *Trihemiolia* (pl. *trihemioliae*). A fast, light ship, attested from
the end of the fourth century, particularly in the navies of **Rhodes** and
Egypt. Bigger than the *hemiolia*, the *trihemiolia* probably had three
rows of oars on each side—two full rows and one half row. There may
have been 24 men in each full row and 12 in the half row, giving 60
men on each side, a total of 120 rowers. The full ship's crew may have
been 144 men. The *trihemiolia* was used as a warship and as a raider.

(3) Liburnian. Probably weighing around 14.75 tons (15 tonnes)
and powered by 50 oarsmen, this was a fast, light ship (although not
quite as fast as a *hemiolia*). The Liburnian was developed by the Liburni
of Illyria, who practiced piracy in the Adriatic. It was probably config-
ured with either one row of oars or one and a half rows of oars on each
side and used particularly for **reconnaissance**.

(4) Four (*tetreres*). Based on the **pentecontor** and invented (perhaps by the **Carthaginians**) around the middle of the fourth century. The four probably had 90 oarsmen, operating two rows of oars on each side of the ship, with two men per oar. The four played an important role in Greek **naval warfare** down to the end of the fourth century and continued in service until the first century A.D. This was because it had many advantages, although outclassed in direct combat by the larger warships that followed soon after it. The four was a relatively fast ship, well suited to **reconnaissance** and able to operate close inshore because of its shallow draft; it could carry around 75 troops.

(5) Five (*penteres*). A fairly large warship, probably developed from the **trireme** by **Dionysius I** of **Syracuse**'s craftsmen circa 399. It was about 145 feet (45 meters) long, around 17 feet (5 meters) wide, and displaced around 98.5 tons (100 tonnes). The deck, from which 70 to 120 marines could operate, was probably about 10 feet (3 meters) above the waterline. The type was fairly slow to spread during the fourth century, often being used as a flagship rather than as a ship of the line. It was more popular in the navies of Phoenicia (although the Phoenician five may have been of a slightly different design), and this may have influenced its popularity in the fleets of **Alexander the Great** and his successors. A heavier ship than the trireme or the four, the five was better able to cope with bad weather and was significantly higher, allowing the men on its deck an advantage over those on ships closer to the waterline. Conversely, the five could be slower and was less easy to maneuver. It probably had 300 oarsmen who operated three rows of oars on each side, like a trireme, but with the oars on two of the levels on each side worked by two men instead of one (see figure 7b).

(6) Six (*hexeres*). A larger warship, probably invented under **Dionysius I** of **Syracuse** sometime in the first quarter of the fourth century. The six was a large ship, around 145 feet (45 meters) long and more than 118 tons (120 tonnes) laden. It probably had three rows of oars on each side, like a trireme or a five, but with the oars on all three of the levels on each side worked by two men. Athens apparently never used the six, but it was used by the *diadochoi* and later Hellenistic monarchs.

(7) Seven (*hepteres*). A large warship, developed from the six, probably by adding extra (standing) oarsmen at the lowest level. It had a clear advantage over smaller craft in heavy weather. Sevens (and larger ships) were probably equipped with towers, carried catapults and considerable numbers of troops, and used firepower and boarding rather than ramming. The seven was a large ship and most navies possessed only a few, so they (and the larger ships) were often used as flagships.

(8) Eight (*okteres*). A large warship of unknown origin and oar system (although it was possibly a version of the four, with four men operating each oar and the oars arranged in two rows on each side of the ship). **Philip V** of **Macedon** had at least two eights in his fleet at **Chios** in 201/0 (Polyb. 16.3). Like the six and the seven, it was a large ship and presumably able both to carry a good contingent of *epibatai* (marines) and to cope with heavier seas.

(9) Nine (*enneres*). A large warship, perhaps based on the four, with two rows of oars on each side, the oars in one row manned by four men, those in the other by five. Attested from the last quarter of the fourth century, nines were often, but not exclusively, used as flagships. They were probably not very maneuverable, although capable of mounting heavy catapults and carrying a large contingent of *epibatai* (marines).

(10) Ten (*dekeres*). A large warship, possibly invented under **Alexander the Great**. The ten was probably developed from the nine by adding an extra oarsman to one of the rows of oars on each side. **Antigonus II (Gonatas)** used tens, and the flagship of **Philip V** of **Macedon**'s fleet at **Chios** (201/0) was a ten (Polyb. 16.2-3). *See also* ACTIUM.

WEAPONS. Although in the Classical period the Greeks contrasted their use of the spear with the **Persian** use of the **bow** (cf. Aeschylus, *Persians*, 234-244), Greek warfare employed a wide variety of weapons. The spear was a consistent feature of Greek heavy infantry, common in all periods. From the mid-fourth century, though, the *sarissa* began to replace it as part of the change from the **hoplite phalanx** to the **Macedonian** phalanx. The sword was generally regarded as a secondary weapon for heavy infantry, there as a back up if the spear broke—which must have been fairly common. However, there is evidence that the sword played a more prominent role in **early Greek warfare** and several **Mycenaean** gemstones and seals show individual combat using the sword.

Psiloi (light infantry) tended to favor missile weapons (although some **peltasts** employed a thrusting spear) including the bow, **javelin**, and the **sling**. Poorer skirmishers might only be equipped with daggers or rocks. During the Classical period these types of weapons were consistently regarded as less manly than the spear or sword, because they involved long-range combat. The **cavalry** was equipped with javelins or a thrusting spear, or a combination of the two, often backed up with a sword. *See* also ARTILLERY; *GASTRAPHETES; KOPIS; MACHAIRA; XIPHOS*.

WOUNDS. The most common wounds in Greek warfare must have been puncture wounds from spears, **javelins**, arrows, catapult bolts, and swords, as well as cuts or gashes from spears and swords. Other injuries would include fractures and internal injuries caused by blows, trampling (incurred after falling in a **phalanx** or other close formation), and rocks thrown by hand or **artillery**. Puncture wounds to chest or abdomen were probably generally fatal, given the risk of infection and limited state of medical knowledge (although **Alexander the Great** survived an arrow wound that punctured his lung; Arr. *Anab.* 6.10). However, in some areas Greek medicine was relatively sophisticated and allowed for the successful treatment of a variety of wounds.

Medical theory and practice included the use of wine or vinegar as an antiseptic, bandaging, cauterization, the treatment of compressed fractures of the skull, and the reduction and setting of other fractures. The Hippocratic Corpus (written in the fifth and fourth centuries) contains at least three relevant treatises: *On Wounds*, *On Head Wounds*, and *On Fractures* and the advice in these is at times surprisingly modern. This corpus, though, is not a consistent body of theory and it is clear that practice varied considerably between doctors.

Most battlefield first aid would have been carried out by the injured man's attendants or friends but specialist doctors regularly accompanied Hellenistic armies—even if mainly to treat the leaders. Consequently, the chances of survival for senior **officers**, normally well-protected by body **armor** anyway, were quite good. **Philip II** of **Macedon**, for example, survived the loss of an eye, a broken collarbone, and serious injuries to limbs (Demosthenes, 18 [*On the Crown*], 67). Alexander the Great also survived several wounds (Arr. *Anab.* 2.17, 3.30, 4.3, 4.23, 4.26, 6.10) and **Demetrius I (Poliorcetes)** recovered from a serious facial wound from a catapult bolt at Messene (Plut., *Demetrius*, 33.2-3). **Philopoemen** apparently fully recovered from having both thighs pierced by a javelin (Polyb. 2.69; Plut., *Philopoemen*, 6.4-6) and **Pyrrhus I** of Epirus recovered from at least two wounds, including a head wound, during his career (Plut., *Pyrrhus*, 7.5, 24.2). Conversely, in the fifth century **Miltiades** died of an injury to his leg (Hdt. 6.134-136), and **Plutarch** (*Cimon*, 18) preserves an alternative account of Cimon's death in which he died of wounds received in **Cyprus**. In 145 **Ptolemy VI (Philometor)** died of wounds shortly after his victory at the battle of Oneoparas. Although the ordinary soldier would not receive such good treatment as his leaders, Alexander apparently regularly visited his wounded after battle (Arr. *Anab.* 1.16.5, 2.12.1). The wounded, however, were often an easy target for an enemy—**Darius III** killed Alexander's sick and wounded when he surprised their camp prior to the battle of **Issus** (Arr. *Anab.* 2.7).

- X -

XANTHIPPUS (XANTHIPPOS; fl. c. 255). A **Spartan mercenary** hired by **Carthage** in 255 as part of a large number of men recruited after the **Roman** defeat of the Carthaginian army near Adys. He convinced the Carthaginians that they could succeed with proper training and by engaging the Romans on the plain where the Carthaginian **elephants** and **cavalry** would be most effective. Given command, he deployed the elephants in front of the heavy infantry, drawn up in **phalanxes**, with the cavalry on each wing. Under his leadership the Carthaginians won a decisive victory over the Roman expeditionary force led by Regulus. About 13,000 Romans died, 500 were captured, and 2,000 escaped.

XENOPHON (c. 428-354). The most important contemporary source for the final years of the **Peloponnesian War** of 431-404, the exploits of the **Ten Thousand**, and the struggle for hegemony in early fourth-century Greece. Xenophon was a wealthy Athenian who almost certainly served in the Athenian **cavalry** at the end of the Peloponnesian War and may well have been associated with it in the overthrow of democracy in 404. He certainly found it advisable to leave **Athens** in 401 after the democratic restoration and accompanied **Cyrus the Younger** on his abortive attempt to oust his brother from the throne of **Persia**.

When Cyrus was killed, Xenophon played a major part in leading Cyrus' force of Greek **mercenaries** (the Ten Thousand) back to Greece—a journey of some 2,000 miles (3,200 kilometers), much of it through hostile territory. Shortly after his return, he was exiled from Athens and formed close links with **Sparta** (he lived there for a time and the Spartans later gave him an estate near **Olympia**). He returned to Athens in 366/5 and his eldest son was killed while serving in the Athenian cavalry at **Mantinea** (2) in 362.

Xenophon wrote many books, including an account of the march of the Ten Thousand (the *Anabasis*), a manual for cavalry commanders (*Hipparchikos*), technical treatises on horsemanship and state affairs, philosophical works, probably the first historical novel (the *Cyropaedia*). He also wrote political and general history. Most of his works survive and provide excellent information on military theory and practice. Xenophon was a good, thoughtful soldier with a particular interest in leadership (a vital factor in volunteer citizen armies). However, he sometimes omits important details and tends to be pro-Spartan and anti-Theban (although this has been exaggerated). *See also* RECONNAISSANCE.

Figure 6. Trireme

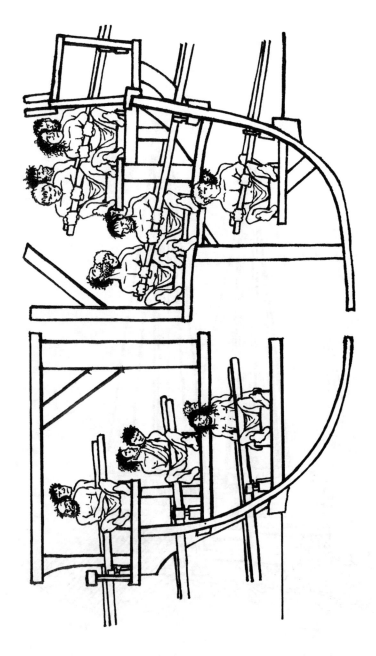

Figure 7. Rowing Systems: Trireme and Five (Quinquereme)

XERXES I (d. 465). A **Persian** king, son of **Darius I**, who reigned 486-465 and continued his father's war against Greece with a large invasion in 480. **Herodotus** portrays him as a ruler who could be cruel and capricious and who was unwilling to take advice. After the Greek victory at **Salamis** in 480, Xerxes returned home, worried that news of the defeat would lead to revolts among his subject peoples. The Greek victories at **Plataea** (1) and **Mycale** led to the final destruction of his expeditionary force and the revolt of **Ionia**. He was murdered in a palace intrigue in 465. *See also* PERSIAN WARS.

XIPHOS. A straight sword used by both infantry and **cavalry** (see figure 5a). *See* also *KOPIS*; *MACHAIRA*; WEAPONS.

- Z -

ZEUGITAI **(sing.** *ZEUGITES*)**.** The Solonic census class at **Athens** which was the minimum property qualification requiring service as a **hoplite**.

GLOSSARY

The following Greek terms, or anglicized versions of Greek terms, appear in the dictionary. The military terms listed below generally have their own entry in the dictionary, where a fuller explanation can be found if required.

agoge—the state training system at Sparta

chora—a state's territory/agricultural hinterland

eisphora—an extraordinary wartime tax levied on the rich at Athens

epibatai (sing. *epibates*)—marines

epilektoi—lit. "chosen ones"; elite troops, usually a standing force of hoplites

hegemon—leader; usually of an alliance but also of an army

helots—the serf-like class in Lacedaemon formed from the original population and the inhabitants of neighboring Messenia

hipparch (*hipparchos*; pl. *hipparchoi*)—cavalry general/cavalry commander

hippotoxotai (sing. *hippotoxotes*)—horse archers

lochos (pl. *lochoi*)—a subunit, generally of infantry, often of company size

mora (pl. *morai*)—a regiment of the Spartan army

perioeci—lit. "dwellers around"; inhabitants of Lacedaemon, of inferior status to Spartiates

phylarch (*phylarchos*; pl. *phylarchoi*)—cavalry squadron commander

phyle (pl. *phylai*)—cavalry squadrons (also tribes)

polis (pl. *poleis*)—city-state

psiloi (sing. *psilos*)—light infantry

satrapy—province of the Persian empire; governed by a satrap

Spartiate—full citizen of Sparta (who had been through the *agoge*)

strategeia—the office of general (*strategos*)

strategos (pl. *strategoi*)—general/admiral

tagus (tagos)—chief civil and military magistrate of Thessaly

taxis (pl. *taxeis*)—a group of infantry, of varying size; at Athens, an infantry regiment

taxiarch (*taxiarchos*)—commander of a *taxis*

trierarch (*trierarchos*)—ship's captain; at Athens the trierarch paid for the crew and ship's equipment

BIBLIOGRAPHY

This bibliography provides a representative selection of books and articles in English, with a few titles in other languages (principally French and German). There is no shortage of written work on ancient Greek warfare but the wide geographical and chronological scope of the subject creates problems with structuring a bibliography. The obvious choice, a chronological or geographical structure, has some advantages but makes it very difficult to place thematic works on important topics such as hoplites, cavalry, or individual Greek states.

A compromise structure, based on themes, topics, and chronology, is therefore adopted as the most useful to the largest number of readers. The sections within the bibliography cover troop types, equipment, and specific wars, periods, or geographical regions. The thematic sections are listed first, with the chronological periods following. However, even this approach has some difficulties, the main one being overlap.

For example, many of the specialist works on hoplite warfare cover engagements in both the Persian and Peloponnesian Wars and the various wars in the struggle for hegemony in the fourth century. Material relevant to the later fourth-century wars is also contained in works on Philip II of Macedon—the two subjects are inextricably linked. To help overcome this problem, works which overlap more than one section are listed in all the most relevant ones.

There is no one general work that covers the entire period dealt with in this dictionary. However, volumes 2-10 of the *Cambridge Ancient History* series provide a very detailed account of the Greek world from the Mycenaean period to its conquest by Rome. As an alternative, reasonably good coverage can be gained by using a general history of Greece—such as Nicholas Hammond's *History of Greece to 322 B.C.* or Russell Meiggs and John Bury's *History of Greece*—in combination with a similar work on Rome such as Max Cary and Howard Scullard's *History of Rome* or Arthur Boak and William Sinnigen's *History of Rome to A.D. 565*. Together these will provide the reader with the historical context for ancient Greek warfare from the early period to the Roman conquest, as well as solid coverage of the major wars and other relevant military history. For the Hellenistic monarchies, not so well covered in these books, Peter Green's *Alexander to Actium* provides fairly detailed coverage.

To follow the course of wars and campaigns, a detailed map of the Greek world is also essential. Peter Levi's *Atlas of the Greek World* is very good and also contains brief introductions to various aspects of Greek society and history. *Muir's Atlas of Ancient and Classical History*, edited by Reginald F. Treharne and Harold Fullard, also contains useful maps of all parts of the ancient Greek world and neighboring areas. The most detailed coverage, though, is in Richard Talbert's *Barrington Atlas of the Greek and Roman World*. For a close study of aspects of mainland Greek topography, see William K. Pritchett's multivolume *Studies in Greek Topography*.

For a general introduction to Greek warfare, there are several excellent large-format works that present the subject in an accurate and interesting manner, with numerous illustrations and reconstructions to aid understanding. John Warry's *Warfare in the Classical World* and Peter Connelly's *Greece and Rome at War* are very good examples of this. Despite the fact that at first glance they look like "coffee table" books, they are often better on ancient combat than apparently more weighty general works on warfare. An example of this is John Keegan's *History of Warfare*, which, for all its merits in other areas, contains basic errors about ancient Greek warfare. In contrast, Nick Sekunda's *Ancient Greeks* and *Army of Alexander the Great*, published in a series primarily intended for military modelers and wargamers, are of much more use to the reader interested in the organization, equipment, and employment of Greek armies in the Classical and early Hellenistic periods.

For more detailed coverage of the mechanics and operation of hoplite warfare, Victor Davis Hanson's *Western Way of War* has an excellent discussion. However, he focuses almost exclusively on the hoplite, presenting perhaps an oversimplified picture of Greek warfare, and his views on the influence of hoplite warfare on modern Western warfare are open to question. For the cavalry arm, see my own *Cavalry of Classical Greece* and, for the administrative aspects of Athenian cavalry, Glenn Bugh's *Horsemen of Athens*.

CONTENTS

Abbreviations of Journal Titles used in the Bibliography

ABSA	*Annual of the British School at Athens*
AHB	*Ancient History Bulletin*
AJA	*American Journal of Archeology*
AJPh	*American Journal of Philology*
AncW	*Ancient World*
C&M	*Classica et Mediævalia*
ClAnt	*Classical Antiquity*
CJ	*Classical Journal*
CPh	*Classical Philology*
CQ	*Classical Quarterly*
CQ²	*Classical Quarterly (2nd series)*
EMC	*Échos du monde classique*
G&R²	*Greece and Rome (2nd series)*
JHS	*Journal of Hellenic Studies*
MM	*Mariner's Mirror*

Reference

Fornara, Charles W. *The Athenian Board of Generals from 501 to 404.* Historia Einzelschriften no. 16. Wiesbaden: F. Steiner, 1971.

Levi, Peter. *Atlas of the Greek World.* New York: Facts on File, 1991.

Pritchett, William K. *The Greek State at War.* 5 vols. Berkeley: University of California Press, 1971-1991.

———. *Studies in Ancient Greek Topography.* 6 vols. Berkeley: University of California Press, 1965-1989.

Sage, Michael. *Warfare in Ancient Greece: A Sourcebook.* London: Routledge, 1996.

Talbert, Richard J., and Roger S. Bagnall. *The Barrington Atlas of the Greek and Roman World.* Princeton, N.J.: Princeton University Press, 2000.

Treharne, Reginald F., and Harold Fullard, eds. *Muir's Atlas of Ancient and Classical History.* 6th ed. London: George Philip, 1973.

General Greek History/Regional History

Bengtson, Hermann, Edda Bresciani, Werner Caskel, Maurice Meuleau, and Morton Smith. *The Greeks and the Persians from the Sixth to the Fourth Centuries.* London: Weidenfeld and Nicolson, 1969.

Bury, John B., and Russell Meiggs. *A History of Greece.* 4th ed. London: Macmillan, 1975.

Cambridge Ancient History. Vols. 2-10. 2nd and 3rd ed. Cambridge: Cambridge University Press, 1973-1994.

Cartledge, Paul. *Hellenistic and Roman Sparta: A Tale of Two Cities.* London: Routledge, 1989.

————. *Sparta and Lakonia: A Regional History, 1300-362 B.C.* London: Routledge and Kegan Paul, 1979.

Demand, Nancy H. *Thebes in the Fifth Century.* London: Routledge and Kegan Paul, 1982.

Dunbabin, Thomas J. *The Western Greeks: The History of Sicily and South Italy from the Foundation of the Greek Colonies to 480 B.C.* Oxford: Clarendon Press, 1968.

Errington, R. Malcolm. *A History of Macedonia.* Trans. by Catherine Errington. Berkeley: University of California Press, 1990.

Fine, John V. A. *The Ancient Greeks: A Critical History.* Cambridge, Mass.: Harvard University Press, 1983.

Finley, Moses I. *Ancient Sicily.* London: Chatto and Windus, 1979.

Fol, Alexander, and Ivan Marazov. *Thrace and the Thracians.* London: Cassell, 1977.

Green, Peter. *Alexander to Actium: The Historical Evolution of the Hellenistic Age.* Berkeley: University of California Press, 1990.

Gude, Mabel. *A History of Olynthus.* Baltimore: Johns Hopkins University Press, 1933.

Hammond, Nicholas G. L. *A History of Greece to 322 B.C.* Oxford: Clarendon, 1967.

————. *A History of Macedonia.* Vol. 1, *Historical Geography and Prehistory.* Oxford: Clarendon, 1972.

————. *The Macedonian State: The Origins, Institutions, and History.* Oxford: Clarendon, 1989.

Hammond, Nicholas G. L., and Guy T. Griffith. *A History of Macedonia.* Vol. 2, *550-336 B.C.* Oxford: Clarendon, 1979.

Hammond, Nicholas G. L., and Frank W. Walbank. *A History of Macedonia.* Vol. 3, *336-167 B.C.* Oxford: Clarendon, 1979.

Hornblower, Simon. *The Greek World, 479-323 B.C.* London: Methuen, 1983.

Kelly, Thomas. *A History of Argos to 500 B.C.* Minneapolis: University of Minnesota Press, 1976.

Legon, Ronald P. *Megara: The Political History of a Greek City-state to 336 B.C.* Ithaca, N.Y.: Cornell University Press, 1981.

Lomas, Kathryn. *Rome and the Western Greeks, 350 B.C.-A.D. 200.* London: Routledge, 1993.

McGregor, Malcolm F. *The Athenians and Their Empire.* Vancouver: University of British Columbia Press, 1987.

Osborne, Robin. *Greece in the Making, 1200-479 B.C.* London: Routledge, 1996.

Powell, Anton. *Classical Sparta: Techniques behind Her Success.* London: Routledge, 1989.

———, ed. *The Greek World.* London: Routledge, 1995.

Salmon, John B. *Wealthy Corinth. A History of the City to 338 B.C.* Oxford: Clarendon, 1984.

Sealey, Raphael, ed. *A History of the Greek City States ca. 700-338 B.C.* Berkeley: University of California Press, 1976.

Snodgrass, Anthony M. *Archaic Greece: The Age of Experiment.* London: Dent, 1980.

———. "Greek Archaeology and Greek History." *ClAnt* 4 (1985): 193-207.

Tritle, Lawrence A., ed. *The Greek World in the Fourth Century.* London: Routledge, 1997.

Webber, Christopher. *The Thracians, 700 B.C.-46 A.D.* Oxford: Osprey Military, 2001.

West, Allen B. *The History of the Chalcidic League.* 1912. Reprint. New York: Arno, 1973.

General Greek Military History

Adcock, Frank E. *The Greek and Macedonian Art of War.* Berkeley: University of California Press, 1962.

Ashworth, Lucien. "Cities, Ethnicity, and Insurgent Warfare in the Hellenic World." *War and Society* 14(2) (1996): 1-20.

Connolly, Peter. *Greece and Rome at War.* London: Greenhill Books, 1998.

Delbruck, Hans. *History of the Art of War within the Framework of Political History.* Vol. 1. Trans. by W. J. Renfroe, Jr. Westport, Conn.: Greenwood Press, 1975.

Ducrey, Pierre. *Warfare in Ancient Greece.* Trans. by J. Lloyd. New York: Schocken, 1986.

Ferrill, Arther. *The Origins of War*. Boulder, Colo.: Westview Press, 1997.

Gabriel, Richard A., and Donald W. Boose. *The Great Battles of Antiquity*. Westport, Conn.: Greenwood Press, 1994.

Garlan, Yvon. *War in the Ancient World*. Trans. by J. Lloyd. London: Chatto and Windus, 1975.

Harvey, Paul. "New Harvests Reappear: The Impact of War on Agriculture." *Athenaeum* 64 (1986): 205-218.

Keegan, John. *A History of Warfare*. London: Hutchinson, 1993.

Sekunda, Nicholas V. *The Ancient Greeks*. London: Osprey, 1986.

Shay, Jonathan. *Achilles in Vietnam: Combat Trauma and the Undoing of Character*. New York: Maxwell Macmillan, 1994.

Van Creveld, Martin L. *Technology and War from 2000 B.C. to the Present*. New York: Free Press, 1991.

Van Wees, Hans. *Status Warriors: War, Violence, and Society in Homer and History*. Amsterdam: Gieben, 1992.

Vernant, Jean-Pierre, ed. *Problèmes de la guerre en Grèce ancienne*. Paris: Mouton, 1968.

Warry, John. *Warfare in the Classical World*. London: Salamander, 1980.

Wheeler, Everett L. *Stratagem and the Vocabulary of Military Trickery*. Leiden: E.J. Brill, 1988.

Whitehead, David. "Κλοπὴ Πολέμου· 'Theft' in Ancient Greek Warfare." *C&M* 39 (1988): 43-53.

Arms and Armor

Anderson, John K. "Hoplite Weapons and Offensive Arms." In *Hoplites: The Classical Greek Battle Experience*, edited by Victor D. Hanson, 15-37. London: Routledge, 1991.

Fortenberry, Diane. "Single Greaves in the Late Helladic Period." *AJA* 95 (1991): 623-627.

Gardiner, Edward N. "Throwing the Javelin." *JHS* 27 (1907): 249-273.

Hammond, Nicholas G. L. "Training in the Use of a *Sarissa* and Its Effect in Battle, 359-333 B.C." *Antichthon* 14 (1980): 53-63.

Harris, Harold A. "Greek Javelin Throwing." *G&R*[2] 10 (1963): 26-36.

Hoffmann, Herbert. *Early Cretan Armorers*. With Antony E. Raubitschek. Cambridge, Mass.: Fogg Art Museum, 1972.

Hurwit, Jefrey M. "The Dipylon Shield Once More." *ClAnt* 4 (1985): 121-126.

Jarva, Eero. "On the Shield-apron in Ancient Greek Panoply." *Acta Archaeologica* 57 (1986): 1-25.

Manti, Peter A. "The Sarissa of the Macedonian Infantry." *AncW* 23 (1992): 30-42.

Markle, Minor M. "Macedonian Arms and Tactics under Alexander." In *Macedonia and Greece in Late Classical and Early Hellenistic Times*, edited by Beryl Barr-Sharrar and Eugene N. Borza, 87-111. Washington, D.C.: National Gallery of Art, 1982.

———. "The Macedonian Sarissa, Spear, and Related Armor." *AJA* 81 (1977): 323-339.

———. "A Shield Monument from Veria and the Chronology of Macedonian Shield Types." *Hesperia* 68 (1999): 219-254.

———. "Use of the Sarissa by Philip and Alexander of Macedon." *AJA* 82 (1978): 483-497.

McLeod, Wallace. "The Range of the Ancient Bow." *Phoenix* 19 (1965): 1-14.

Mixter, John R. "The Length of the Macedonian Sarissa during the Reigns of Philip II and Alexander the Great." *AncW* 23 (1992): 21-29.

Snodgrass, Anthony M. *Arms and Armour of the Greeks*. London: Thames and Hudson, 1967.

———. *Early Greek Armour and Weapons from the End of the Bronze Age to 600 B.C.* Edinburgh: Edinburgh University Press, 1964.

von Bothmer, Dietrich. "Armorial Adjuncts." *Metropolitan Museum Journal* 24 (1989): 65-70.

Cavalry, Mercenaries, and Light-Armed Troops

Aperghis, Georges G. "Alexander's Hipparchies." *AncW* 28 (1997): 133-148.

Best, Jan G. P. *Thracian Peltasts and Their Influence on Greek Warfare*. Groningen: Wolters-Noordhoff, 1969.

Bugh, Glenn R. *The Horsemen of Athens*. Princeton, N.J.: Princeton University Press, 1988.

Daniel, Thomas. "The Taxeis of Alexander and the Change to Chiliarch, the Companion Cavalry and the Change to Hipparchies: A Brief Assessment." *AncW* 23 (1992): 43-57.

Evans, James A. S. "Cavalry about the Time of the Persian Wars: A Speculative Essay." *CJ* 82 (1987): 97-106.

Foulon, Eric. "Μισθοφοροι et Ξενοι Hellénistiques." *Revue dés études grecques* 108 (1995): 211-218.

Griffith, Guy T. *The Mercenaries of the Hellenistic World.* Cambridge: Cambridge University Press, 1935.

Hammond, Nicholas G. L. "Cavalry Recruited in Macedonia down to 322 B.C." *Historia* 47 (1998): 404-425.

Kennett, W. Harl. "Alexander's Cavalry Battle at the Granicus." In *Polis and Polemos: Essays on Politics, War, and History in Ancient Greece in Honor of Donald Kagan*, edited by Charles D. Hamilton and Peter Krentz, 303-326. Claremont, Calif.: Regina, 1997.

Krasilnikoff, Jens A. "Aegean Mercenaries in the Fourth to Second Centuries B.C.: A Study in Payment, Plunder, and Logistics of Ancient Greek Armies." *C&M* 43 (1992): 23-36.

McKechnie, Paul. "Greek Mercenary Troops and Their Equipment." *Historia* 43 (1994): 297-305.

Miller, Harvey F. "The Practical and Economic Background to the Greek Mercenary Explosion." *G&R²* 31 (1984): 153-160.

Parke, Herbert W. *Greek Mercenary Soldiers from the Earliest Times to the Battle of Ipsus.* Chicago: Ares, 1981.

Spence, Iain G. "Athenian Cavalry Numbers in the Peloponnesian War: *IG* I³ 375 Revisited." *Zeitschrift für Papyrologie und Epigraphik* 67 (1987): 167-175.

―――. *The Cavalry of Classical Greece.* Oxford: Clarendon, 1993.

―――. "Perikles and the Defence of Attika during the Peloponnesian War." *JHS* 110 (1990): 91-109.

Stagakis George. "Homeric Warfare Practices." *Historia* 34 (1985): 129-152.

Trundle, Matthew F. "*Epikouroi, Xenoi,* and *Misthophoroi* in the Classical Greek World." *War and Society* 16 (1998): 1-12.

―――. "Identity and Community among Greek Mercenaries in the Classical World: 700-322 B.C.E." *AHB* 13 (1999): 28-38.

Vos, Maria F. *Scythian Archers in Archaic Attic Vase Painting.* Groningen: J. B. Wolters, 1963.

Webber, Christopher. *The Thracians, 700 B.C.-46 A.D.* Oxford: Osprey Military, 2001.

Whitehead, David. "Who Equipped Mercenary Troops in Classical Greece?" *Historia* 40 (1991): 105-113.

Worley, Leslie J. *Hippeis: The Cavalry of Ancient Greece.* Boulder, Colo.: Westview Press, 1994.

Xenophon. *Hiero the Tyrant and Other Treatises.* Trans. by Robin Waterfield. Harmondsworth, England: Penguin, 1997 (contains the treatises on horsemanship and the cavalry commander).

Hoplite Warfare

Anderson, John K. "Hoplites and Heresies: A Note." *JHS* 104 (1984): 152.

———. "Hoplite Weapons and Offensive Arms." In *Hoplites: The Classical Greek Battle Experience*, edited by Victor D. Hanson, 15-37. London: Routledge, 1991.

———. *Military Theory and Practice in the Age of Xenophon*. Berkeley: University of California Press, 1970.

Cartledge, Paul. "Hoplites and Heroes: Sparta's Contribution to the Technique of Ancient Warfare." *JHS* 97 (1977): 11-27.

Cawkwell, George L. "Orthodoxy and Hoplites." *CQ²* 33 (1983): 375-389.

Goldsworthy, Adrian K. "The *Othismos*, Myths, and Heresies: The Nature of Hoplite Battle." *War in History* 4 (1997): 1-26.

Hanson, Victor D., "Hoplite Technology in Phalanx Battle." In *Hoplites: The Classical Greek Battle Experience*, edited by Victor D. Hanson, 63-84. London: Routledge, 1991.

———. *Warfare and Agriculture in Classical Greece*. Rev. ed. Berkeley: University of California Press, 1998.

———. *The Western Way of War: Infantry Battle in Classical Greece*. New York: Knopf, 1989.

———. ed. *Hoplites: The Classical Greek Battle Experience*. London: Routledge, 1991.

Holladay, A. James. "Hoplites and Heresies." *JHS* 102 (1982): 94-103.

Jackson, Alastar H. "Hoplites and the Gods: The Dedication of Captured Arms and Armour." In *Hoplites: The Classical Greek Battle Experience*, edited by Victor D. Hanson, 228-249. London: Routledge, 1991.

Krentz, Peter. "Casualties in Hoplite Battles." *Greek, Roman, and Byzantine Studies* 26 (1985): 13-20.

———. "The Nature of the Hoplite Battle." *ClAnt* 4 (1985): 50-61.

Lazenby, John F. "The Killing Zone." In *Hoplites: The Classical Greek Battle Experience*, edited by Victor D. Hanson, 87-109. London: Routledge, 1991.

———. *The Spartan Army*. Warminster, England: Aris and Phillips, 1985.

Lorimer, Hild L. "The Hoplite Phalanx." *ABSA* 42 (1947): 76-138.

Luginbill, Robert D. "*Othismos:* The Importance of the Mass-Shove in Hoplite Warfare." *Phoenix* 48 (1994): 51-61.

Ober, Josiah. "Hoplites and Obstacles." In *Hoplites: The Classical Greek Battle Experience*, edited by Victor D. Hanson, 173-196. London: Routledge, 1991.

Ridley, Ronald T. "The Hoplite as Citizen: Athenian Military Institutions in Their Social Context." *Antiquité Classique* 48 (1979): 508-548.

Salmon, John. "Political Hoplites?" *JHS* 97 (1977): 84-101.

Sarikakis, Theodore C. *The Hoplite General in Ancient Athens/The Athenian Generals of the Hellenistic Age*. Chicago: Ares, 1976.

Sekunda, Nicholas V. *The Spartan Army*. London: Osprey, 1998.

Sinclair, Robert. K. "Diodorus Siculus and Fighting in Relays." *CQ²* 16 (1966): 249-255.

Snodgrass, Anthony M. "The Hoplite Reform and History." *JHS* 85 (1965): 110-122.

Storch, Rudolph H. "The Archaic Greek Phalanx, 750-650 B.C." *AHB* 12 (1998): 1-7.

Van Wees, Hans. "The Homeric Way of War: The *Iliad* and the Hoplite Phalanx." *G&R²* 41 (1994): 1-18, 131-155.

Fortifications, Siege Warfare, and Artillery

Aineias the Tactician. *How to Survive under Siege*. Trans. by David Whitehead. Oxford: Clarendon, 1990.

Fleming, Stuart. "Classical Artillery." *Archaeology* 41 (1988): 56-57.

Garlan, Yvon. *Recherches de poliorcétique grecque*. Paris: de Boccard, 1974.

Kern, Paul B. *Ancient Siege Warfare*. Bloomington: Indiana University Press, 1999.

———. "Military Technology and Ethical Values in Ancient Greek Warfare: The Siege of Plataea." *War and Society* 6(2) (1988): 1-20.

Keyser, Paul T. "The Use of Artillery by Philip II and Alexander the Great." *AncW* 25 (1994): 27-59.

Lawrence, Arnold W. *Greek Aims in Fortification*. Oxford: Clarendon, 1979.

Marsden, Eric W. *Greek and Roman Artillery: Historical Development*. Oxford: Clarendon, 1969.

———. *Greek and Roman Artillery: Technical Treatises*. Oxford: Clarendon, 1971.

McNicoll, Anthony W. "Developments in Techniques of Siegecraft and Fortification in the Greek World circa 400-100 B.C." In *La Fortification dans l'histoire du monde grec*, edited by Henri Tréziny and Pierre Leriche, 305-313. Paris: Edition du Centre national de la Recherche scientifique, 1986.

———. *Hellenistic Fortifications from the Aegean to the Euphrates*. Rev. ed. by Nicholas P. Milner. Oxford: Clarendon, 1997.

Munn, Mark. *The Dema Wall and the Boeotian War of 378-375 B.C.* Berkeley: University of California Press, 1993.

Ober, Josiah. "Early Artillery Towers: Messenia, Boiotia, Attica, Megarid." *AJA* 91 (1987): 569-604.

———. *Fortress Attica: Defense of the Athenian Land Frontier, 404-322 B.C.* Leiden: E. J. Brill, 1985.

Romane, Julian P. "Alexander's Siege of Gaza, 332 B.C." *AncW* 18 (1988): 21-30.

———. "Alexander's Siege of Tyre." *AncW* 6 (1987): 79-90.

Tréziny, Henri, and Pierre Leriche, ed. *La fortification dans l'histoire du monde grec.* Paris: Edition du Centre national de la recherche scientifique, 1986.

Van de Maele, Symphorien, and John M. Fossey, eds. *Fortifications antiquae.* Monographies en archéologie et histoire classique de l'Université McGill, No. 12. Amsterdam: Gieben, 1992.

Winter, Frederick E. *Greek Fortifications.* Toronto: University of Toronto Press, 1971.

———. "The Use of Artillery in Fourth-Century and Hellenistic Times." *EMC* 16 (1997): 247-292.

Logistics, Finance, and Organizations

Adams, W. Lindsay T. "Antipater and Cassander: Generalship on Restricted Resources in the Fourth Century." *AncW* 10 (1984): 79-88.

Austin, Michel, M. "Hellenistic Kings, War, and the Economy." *CQ²* 36 (1986): 450-466.

Bar-Kochva, Bezalel. *The Seleucid Army: Organization and Tactics in the Great Campaigns.* Cambridge: Cambridge University Press, 1976.

Burford, Alison. "Heavy Transport in Classical Antiquity." *Economic History Review*, 2nd ser., 13 (1960): 1-18.

Engels, Donald W. *Alexander the Great and the Logistics of the Macedonian Army.* Berkeley: University of California Press, 1978.

French, Alfred. "A Note on the Size of the Athenian Armed Forces in 431 B.C." *AHB* 7 (1993): 43-48.

Hamel, Debra. *Athenian Generals: Military Authority in the Classical Period.* Leiden: Brill Academic, 1998.

Hammond, Nicholas G. L. "Army Transport in the Fifth and Fourth Centuries." *Greek, Roman, and Byzantine Studies* 24 (1983): 27-31.

———. "Casualties and Reinforcements of Citizen Soldiers in Greece and Macedonia." *JHS* 109 (1989): 56-68.

Hippocrates. "Fractures, Joints, Mochlion," "In the Surgery," and "On Wounds in the Head." In *Hippocrates*. Vol. 3. Trans. by E. T. Withington. London: Heinemann, 1948.

Kallet-Marx, Lisa. *Money, Expense, and Naval Power in Thucydides' History 1-5.24*. Berkeley: University of California Press, 1993.

Krasilnikoff, Jens A. "Aegean Mercenaries in the Fourth to Second Centuries B.C.: A Study in Payment, Plunder, and Logistics of Ancient Greek Armies." *C&M* 43 (1992): 23-36.

Lazenby, John F. "Logistics in Classical Greek Warfare." *War in History* 1 (1994): 3-18.

———. *The Spartan Army*. Warminster, England: Aris and Phillips, 1985.

Pélékidis, Chrysis. *Histoire de l'éphébie attique des origines à 31 avant Jésus-Christ*. Paris: de Boccard, 1962.

Salazar, Christine F. *The Treatment of War Wounds in Graeco-Roman Antiquity*. Leiden: Brill, 2000.

Sekunda, Nicholas V. *The Spartan Army*. London: Osprey, 1998.

Siewert, Peter. *Die Trittyen Attikas und die Heeresreform des Kleisthenes. Vestigia, Beiträge zur alten Geschichte*. Vol 33. Munich: C. H. Beck, 1982.

Whitehead, David. "Who Equipped Mercenary Troops in Classical Greece?" *Historia* 40 (1991): 105-113.

Customs/Laws of War, Intelligence, and Prisoners of War

Ager, Sheila L. "Why War? Some Views on International Arbitration in Ancient Greece." *EMC* 37 (1993): 1-13.

Balcer, Jack. "The Athenian Episkopos and the Achaemenid 'King's Eye.'" *AJPh* 98 (1977): 252-263.

Bauslaugh, Robert A. *The Concept of Neutrality in Classical Greece*. Berkeley: University of California Press, 1991.

Ducrey, Pierre. *Le Traitement des prisonniers de guerre dans la Grèce antique*. Paris: de Boccard, 1968.

Engels, Donald W. "Alexander's Intelligence System." *CQ²* 30 (1980): 327-340.

Gerolymatos, André. *Espionage and Treason: A Study of the Proxenia in Political and Military Intelligence Gathering in Classical Greece*. Amsterdam: J. C. Greben, 1986.

Jackson, Alastar H. "Some Recent Work on the Treatment of Prisoners of War in Ancient Greece." *Talanta* 2 (1970): 38-53.

Lonis, Raoul. *Les Usages de la guerre entre Grecs et Barbares.* Paris: Belles Lettres, 1969.

Panagopoulos, Andreas. *Captives and Hostages in the Peloponnesian War.* Athens: Grigoris Publications, 1978.

Pownall, Frances S. "What Makes a War a Sacred War?" *EMC* 42 (n.s. 17) (1998): 35-55.

Russell, Frank S. *Information Gathering in Classical Greece.* Ann Arbor: University of Michigan Press, 1999.

Sargent, Rachel L. "The Use of Slaves by the Athenians in Warfare." *CPh* 22 (1927): 201-212, 264-279.

Shay, Jonathan. *Achilles in Vietnam: Combat Trauma and the Undoing of Character.* New York: Maxwell Macmillan, 1994.

Tritle, Lawrence A. "Hector's Body: Mutilation of the Dead in Ancient Greece and Vietnam." *AHB* 11 (1997): 123-136.

Naval Warfare

General

Casson, Lionel. *The Ancient Mariners: Seafarers and Sea Fighters of the Mediterranean in Ancient Times.* 2nd ed. Princeton, N.J.: Princeton University Press, 1991.

————. *Ships and Seafaring in Ancient Times.* Austin: University of Texas Press, 1994.

Dotson, John E. "Economics and Logistics of Galley Warfare." In *The Age of the Galley: Mediterranean Oared Vessels since Pre-Classical Times,* edited by Robert Gardiner, 217-223. London: Conway Maritime Press, 1995.

Meijer, Fik. *A History of Seafaring in the Classical World.* London: Croom Helm, 1986.

Morrison, John S., and Roderick T. Williams. *Greek Oared Ships, 900-322 B.C.* Cambridge: Cambridge University Press, 1968.

Parker, Anthony J. *Ancient Shipwrecks of the Mediterranean and the Roman Provinces.* British Archaeological Reports International Series no. 580. Oxford: Tempus Reparatum, 1992.

Pryor, John H. "The Geographical Conditions of Galley Navigation in the Mediterranean." In *The Age of the Galley: Mediterranean Oared Vessels since Pre-Classical Times,* edited by Robert Gardiner, 206-216. London: Conway Maritime Press, 1995.

Starr, Chester G. *The Influence of Sea Power on Ancient History.* Oxford: Oxford University Press, 1989.
Wallinga, Herman T. "The Trireme and History." *Mnemosyne* 43 (1990): 132-149.
Whitehead, Ian. "The περιπλους." *G&R²* 34 (1987): 178-185.

Pre-Classical Period

Johnstone, Paul. *The Sea-craft of Prehistory.* 2nd ed. London: Routledge, 1988.
Mark, Samuel E. "*Odyssey* 5.234-53 and Homeric Ship Construction: A Reappraisal." *AJA* 95 (1991): 441-445.
Morrison, John S. *Long Ships and Round Ships: Warfare and Trade in the Mediterranean, 3000 B.C.-A.D. 500.* London: Her Majesty's Stationery Office, 1980.
Wachsmann, Shelley. "Paddled and Oared Ships before the Iron Age." In *The Age of the Galley: Mediterranean Oared Vessels since Pre-Classical Times,* edited by Robert Gardiner, 10-35. London: Conway Maritime Press, 1995.
Wallinga, Herman T. "The Ancestry of the Trireme, 1200-525 B.C." In *The Age of the Galley: Mediterranean Oared Vessels since Pre-Classical Times,* edited by Robert Gardiner, 36-48. London: Conway Maritime Press, 1995.
————. *Ships and Sea-power before the Great Persian War: The Ancestry of the Ancient Trireme.* Leiden: E. J. Brill, 1993.

Classical Period/Athens

Amit, Moshe. *Athens and the Sea: A Study in Athenian Sea Power.* Collection Latomus no. 74. Brussels, 1965.
Cawkwell, George L. "Athenian Naval Power in the Fourth Century." *CQ²* 34 (1984): 334-345.
Falkner, Caroline. "The Battle of Syme, 411 B.C. (Thuc. 8.42)." *AHB* 9 (1995): 117-124.
Figuera, Thomas J. "Aigina and the Naval Strategy of the Late Fifth and Early Fourth Centuries." *Rheinisches Museum für Philologie* 133 (1990): 15-51.
Gabrielsen, Vincent. "The Athenian Navy in the Fourth Century B.C." In *The Age of the Galley: Mediterranean Oared Vessels since Pre-*

Classical Times, edited by Robert Gardiner, 234-240. London: Conway Maritime Press, 1995.

———. *Financing the Athenian Fleet: Public Taxation and Social Relations*. Baltimore: Johns Hopkins University Press, 1994.

Haas, Christopher J. "Athenian Naval Power before Themistocles." *Historia* 34 (1985): 29-46.

Hammond, Nicholas G. L. "The Manning of the Fleet in the Decree of Themistokles." *Phoenix* 40 (1986): 143-148.

Jordan, Boromir. *The Athenian Navy in the Classical Period*. Berkeley: University of California Press, 1975.

Kallet-Marx, Lisa. *Money, Expense, and Naval Power in Thucydides' History 1-5.24*. Berkeley: University of California Press, 1993.

McDougall, Iain. "The Persian Ships at Mycale." In *Owls to Athens: Essays on Classical Subjects Presented to Sir Kenneth Dover*, edited by Elizabeth M. Craik, 143-149. Oxford: Clarendon, 1990.

Morrison, John S. "The Greek Ships at Salamis and the Diekplous." *JHS* 111 (1991): 196-200.

Papalas, Anthony. "Polycrates of Samos and the First Greek Trireme Fleet." *MM* 85 (1999): 3-19.

Rosivact, Vincent J. "Manning the Athenian Fleet, 433-426 B.C." *American Journal of Ancient History* 10 (1985): 41-66.

Sealey, Raphael. "Die spartanische Nauarchie." *Klio* 58 (1976): 335-358.

Strauss, Barry S. "A Note on the Topography and Tactics of the Battle of Aegospotami." *AJPh* 108 (1987): 741-745.

Stylianou, Petros J. "How Many Naval Squadrons Did Athens Send to Evagoras?" *Historia* 37 (1988): 463-471.

Whitehead, Ian. "Xerxes' Fleet in Magnesia: The Anchorage at Sepias." *MM* 74 (1988): 283-287.

Hellenistic Period/Macedon

Berthold, Richard M. "Lade, Pergamum, and Chios: Operations of Philip V in the Aegean." *Historia* 24 (1975): 150-163.

Green, Peter. *From Alexander to Actium: The Historical Evolution of the Hellenistic Age*. Berkeley: University of California Press, 1990.

Hammond, Nicholas G. L. "The Macedonian Navies of Philip and Alexander until 330 B.C." *Antichthon* 26 (1992): 30-41.

McDonald, Alexander H., and Frank W. Walbank. "The Treaty of Apamea (188 B.C.): The Naval Clauses." *Journal of Roman Studies* 59 (1969): 30-39.

Morrison, John S. "Athenian Sea-power in 323/2 B.C.: Dream and Reality." *JHS* 107 (1987): 88-97.
———. "Hellenistic Oared Warships, 399-31 B.C." In *The Age of the Galley: Mediterranean Oared Vessels since Pre-Classical Times*, edited by
Robert Gardiner, 66-77. London: Conway Maritime Press, 1995.
Morrison, John S., and John F. Coates. *Greek and Roman Oared Warships*. Oxford: Oxbow, 1996.
Walbank, Frank W. "Sea Power and the Antigonids." In *Philip II, Alexander the Great, and the Macedonian Heritage*, edited by W. Lindsay Adams and Eugene N. Borza, 213-236. Washington, D.C.: University
Press of America, 1982.

Naval Infrastructure, Ship Architecture, Mechanics, and Construction

Basch, Lucien. "The Eleusis Museum Trireme and the Greek Trireme." *MM*
74 (1988): 163-197, 304.
———. "The Isis of Ptolemy II Philadelphus." *MM* 71 (1985): 129-151.
Blackman, David. "Naval Installations." In *The Age of the Galley: Mediter-
ranean Oared Vessels since Pre-Classical Times*, edited by Robert
Gardiner, 224-233. London: Conway Maritime Press, 1995.
Borza, Eugene N. "Timber and Politics in the Ancient World: Macedon and
the Greeks." *Transactions and Proceedings of the American Philologi-
cal Association* 131 (1987): 32-52. (Reprinted in *Makedonika: Essays
by Eugene N. Borza*, edited by Carol G. Thomas, 85-112. Claremont,
Calif.: Regina Books, 1995.)
Bouzek, J. "Some Underwater Observations in Ancient Harbours in the
Aegean." *Graecolatina Pragensia* 9 (1982): 133-141.
Casson, Lionel. "Greek and Roman Shipbuilding: New Findings." *Ameri-
can Neptune* 45 (1985): 10-19.
Casson, Lionel, and J. Richard Steffy. *The Athlit Ram*. College Station:
Texas A&M University Press, 1991.
Coates, John. "The Naval Architecture and Oar Systems of Ancient Gal-
leys." In *The Age of the Galley: Mediterranean Oared Vessels since
Pre-Classical Times*, edited by Robert Gardiner, 127-141. London:
Conway Maritime Press, 1995.
Foley, Verrard, and Werner Soedel. "Ancient Oared Warships." *Scientific
American* (April 1981): 116-129.
Mark, Samuel E. "*Odyssey* 5.234-53 and Homeric Ship Construction: A
Reappraisal." *AJA* 95 (1991): 441-445.

Meiggs, Russell. *Trees and Timber in the Ancient World*. Oxford: Clarendon, 1982.

Morrison, John S. "Hellenistic Oared Warships, 399-31 B.C." In *The Age of the Galley: Mediterranean Oared Vessels since Pre-Classical Times*, edited by Robert Gardiner, 66-77. London: Conway Maritime Press, 1995.

———. "The Trireme." In *The Age of the Galley. Mediterranean Oared Vessels since Pre-Classical Times*, edited by Robert Gardiner, 49-65. London: Conway Maritime Press, 1995.

Morrison, John S., and John F. Coates. *The Athenian Trireme: The History and Reconstruction of an Ancient Greek Warship*. Cambridge: Cambridge University Press, 1986.

Papalas, Anthony J. "The Development of the Trireme." *MM* 83 (1997): 259-271.

Parker, Anthony J. *Ancient Shipwrecks of the Mediterranean and the Roman Provinces*. Oxford: Tempus Reparatum, 1992.

Shaw, J. Timothy. "Oar Mechanics and Oar Power in Ancient Galleys." In *The Age of the Galley: Mediterranean Oared Vessels since Pre-Classical Times*, edited by Robert Gardiner, 163-171. London: Conway Maritime Press, 1995.

———, ed. *The Trireme Project: Operational Experience, 1987-1990: Lessons Learnt*. Oxbow Monographs no. 31. Oxford: Oxbow, 1993.

Sleeswyk, André W., and Fik Meijer. "The Water Supply of the *Argo* and Other Oared Ships." *MM* 84 (1998): 131-138.

Steffy, J. Richard, with additional notes by Patrice Pomey, Lucien Basch, and Honor Frost. "The Athlit Ram: A Preliminary Investigation of Its Structure." *MM* 69 (1983): 229-250.

Welsh, Frank. *Building the Trireme*. London: Constable, 1988.

Westerdahl, Christer, ed. *Crossroads in Ancient Shipbuilding: Proceedings of the Sixth International Symposium on Boat and Ship Archaeology, Roskilde 1991*. Oxford: Oxbow, 1994.

Early Greek Warfare

Ahlberg, Gudrun. *Fighting on Land and Sea in Greek Geometric Art*. Stockholm: Lund, 1971.

Connor, W. Robert. "Early Greek Land Warfare as Symbolic Expression." *Past and Present* 119 (1988): 3-29.

Crouwel, Joost H. *Chariots and Other Means of Land Transport in Bronze Age Greece*. Amsterdam: Allard Pierson Museum, 1981.

————. *Chariots and Other Wheeled Vehicles in Iron Age Greece.* Amsterdam: Allard Pierson Museum, 1992.

Drews, Robert. *The End of the Bronze Age: Changes in Warfare and the Catastrophe circa 1200 B.C.* Princeton, N.J.: Princeton University Press, 1993.

Frost, Frank J. "The Athenian Military before Cleisthenes." *Historia* 33 (1984): 283-294.

Greenhalgh, Peter A. L. *Early Greek Warfare: Horsemen and Chariots.* Cambridge: Cambridge University Press, 1973.

Homer. *The Iliad.* Trans. by Robert Fagles. Harmondsworth, England: Penguin, 1991.

Kirk, Geoffrey S. "War and the Warrior in the Homeric Poems." In *Problèmes de la guerre en Grèce ancienne,* edited by Jean-Pierre Vernant, 93-119. Paris: Mouton, 1968.

Parker, Victor. "The Dates of the Messenian Wars." *Chiron* 21 (1991): 25-47.

Sandard, Nancy. "North and South at the End of the Mycenaean Age: Aspects of an Old Problem." *Oxford Journal of Archaeology* 2 (1983): 43-68.

Singor, Hendricus W. "Nine against Troy: On Epic *Phalanges, Promakhoi,* and an Old Structure in the Story of the *Iliad.*" *Mnemosyne* 44 (1991): 17-62.

Snodgrass, Anthony M. *Early Greek Armour and Weapons from the End of the Bronze Age to 600 B.C.* Edinburgh: Edinburgh University Press, 1964.

Stagakis, George. "Homeric Warfare Practices." *Historia* 34 (1985): 129-152.

Storch, Rudolph H. "The Archaic Greek Phalanx, 750-650 B.C." *AHB* 12 (1998): 1-7.

Tritle, Lawrence A. "Hector's Body: Mutilation of the Dead in Ancient Greece and Vietnam." *AHB* 11 (1997): 123-136.

Udwin, Victor M. *Between Two Armies: The Place of the Duel in Epic Culture.* Leiden: E. J. Brill, 1999.

Van Wees, Hans. "The Homeric Way of War: The *Iliad* and the Hoplite Phalanx." *G&R*² 41 (1994): 1-18, 131-155.

————. "Kings in Combat: Battles and Heroes in the *Iliad.*" *CQ*² 36 (1986): 285-303.

————. "Leaders of Men? Military Organisation in the *Iliad.*" *CQ*² 38 (1988): 1-24.

————. *Status Warriors: War, Violence, and Society in Homer and History.* Amsterdam: Gieben, 1992.

The Persian Wars

Balcer, Jack M. *The Persian Conquest of the Greeks, 545-450 B.C.* Konstanz: Universitätsverlag Konstanz, 1995.

———. "The Persian Wars against Greece: A Reassessment." *Historia* 38 (1989): 127-143.

Bengtson, Hermann, Edda Bresciani, Werner Caskel, Maurice Meuleau, and Morton Smith. *The Greeks and the Persians from the Sixth to the Fourth Centuries.* London: Weidenfeld and Nicolson, 1969.

Burn, Andrew R. *Persia and the Greeks: The Defence of the West, c. 546-478.* 2nd ed. London: Duckworth, 1984.

Diodorus of Sicily. Vol. 4. Trans. by Charles H. Oldfather. London: Heinemann (Loeb Classical Library), 1956.

Doenges, Norman A. "The Campaign and Battle of Marathon." *Historia* 47 (1998): 1-17.

Donlan, Walter D., and James Thompson. "The Charge at Marathon: Herodotus 6.112." *CJ* 71 (1976): 339-343.

———. "The Charge at Marathon Again." *Classical World* 72 (1978/9): 419-420.

Evans, James A. S. "Cavalry about the Time of the Persian Wars: A Speculative Essay." *CJ* 82 (1987): 97-106.

———. "Herodotos and the Battle of Marathon." *Historia* 42 (1993): 279-307.

Hammond, Nicholas G. L. "Sparta at Thermopylae." *Historia* 45 (1996): 1-20.

Herodotus. *The Histories.* Trans. by Aubrey de Sélincourt (rev. John Marincola). Harmondsworth, England: Penguin, 1996.

Hignett, Charles. *Xerxes' Invasion of Greece.* Oxford: Clarendon, 1963.

Kagan, Donald. *Problems in Ancient History.* Vol. 1. 2nd ed. New York: Macmillan, 1975.

Karavites, Peter. "Macedonian Pragmatism and the Persian Wars." *AncW* 28 (1997): 119-126.

Keaveney, Arthur. "The Medisers of Thessaly." *Eranos* 93 (1995): 30-38.

Lazenby, John F. *The Defence of Greece, 490-479 B.C.* Warminster, England: Aris and Phillips, 1993.

Lenardon, Robert J. *The Saga of Themistocles.* London: Thames and Hudson, 1978.

Plutarch, *The Rise and Fall of Athens.* Trans. by Ian Scott-Kilvert. Harmondsworth, England: Penguin, 1960.

Sekunda, Nicholas, V. *The Persian Army, 560-330 B.C.* London: Osprey, 1992.

Shrimpton, Gordon. "The Persian Cavalry at Marathon." *Phoenix* 34 (1980): 20-37.
Wardman, Alan E. "Tactics and the Tradition of the Persian Wars." *Historia* 8 (1959): 49-60.
West, William C. "The Trophies of the Persian Wars." *CPh* 64 (1969): 7-19.
Whitehead, Ian. "Xerxes' Fleet in Magnesia: The Anchorage at Sepias." *MM* 74 (1988): 283-287.

The Peloponnesian Wars

Badian, Ernst. *From Plataea to Potidaea*. Baltimore: Johns Hopkins University Press, 1993.
Bloedow, Edmund F. "Hermocrates' Strategy against the Athenians in 415 B.C." *AHB* 7 (1993): 115-124.
Brunt, Peter A. "Spartan Policy and Strategy in the Archidamian War." *Phoenix* 19 (1965): 255-280.
Buck, Robert J. "The Sicilian Expedition." *AHB* 2 (1988): 73-79.
Cawkwell, George. *Thucydides and the Peloponnesian War*. London: Routledge, 1997.
———. "Thucydides' Judgment of Periclean Strategy." In *Studies in the Greek Historians in Memory of Adam Parry*, edited by Donald Kagan, 53-70. Cambridge: Cambridge University Press, 1975.
de Ste. Croix, Geoffrey E. M. *The Origins of the Peloponnesian War*. London: Duckworth, 1972.
Diodorus of Sicily. Vols. 4, 5, and 6. Trans. by Charles H. Oldfather. London: Heinemann (Loeb Classical Library), 1956, 1950, 1954.
Falkner, Caroline. "Sparta's Colony at Herakleia Trachinia and Spartan Strategy in 426." *EMC* 43 (n.s. 18) (1999): 45-58.
French, Alfred. "A Note on the Size of the Athenian Armed Forces in 431 B.C." *AHB* 7 (1993): 43-48.
Hale, John. "General Phormio's Art of War: A Greek Commentary on a Chinese Classic." In *Polis and Polemos: Essays on Politics, War, and History in Ancient Greece in Honor of Donald Kagan*, edited by Charles D. Hamilton and Peter Krentz, 85-103. Claremont, Calif.: Regina, 1997.
Hanson, Victor D. "Thucydides and the Desertion of Attic Slaves during the Decelean War." *ClAnt* 11 (1992): 210-228.
Holladay, A. James. "Sparta's Role in the First Peloponnesian War." *JHS* 97 (1977): 54-63.

Kagan, Donald. *The Archidamian War.* Ithaca, N.Y.: Cornell University Press, 1974.
———. *The Fall of the Athenian Empire.* Ithaca, N.Y.: Cornell University Press, 1987.
———. *The Outbreak of the Peloponnesian War.* Ithaca, N.Y.: Cornell University Press, 1969.
———. *The Peace of Nicias and the Sicilian Expedition.* Ithaca, N.Y.: Cornell University Press, 1981.
———. *Problems in Ancient History.* Vol. 1. 2nd ed. New York: Macmillan, 1966.
Kallet-Marx, Lisa. *Money, Expense, and Naval Power in Thucydides' History 1-5.24.* Berkeley: University of California Press, 1993.
Kelly, Thomas. "Thucydides and Spartan Strategy in the Archidamian War." *American Historical Review* 87 (1982): 25-54.
Kern, Paul B. "Military Technology and Ethical Values in Ancient Greek Warfare: The Siege of Plataea." *War and Society* 6(2) (1988): 1-20.
Krentz, Peter. "The Strategic Culture of Periclean Athens." In *Polis and Polemos: Essays on Politics, War, and History in Ancient Greece in Honor of Donald Kagan,* edited by Charles D. Hamilton and Peter Krentz, 55-72. Claremont, Calif.: Regina, 1997.
Lewis, David M. "The Origins of the First Peloponnesian War." In *Selected Papers in Greek and Near Eastern History,* edited by Peter J. Rhodes, 9-21. Cambridge: Cambridge University Press, 1997.
Losada, Luis A. *The Fifth Column in the Peloponnesian War.* Leiden: E. J. Brill, 1972.
Luginbill, Robert D. "Thucydides' Evaluation of the Sicilian Expedition: 2.65.11." *AncW* 28 (1997): 127-132.
Panagopoulos, Andreas. *Captives and Hostages in the Peloponnesian War.* Athens: Grigoris Publications, 1978.
Plant, Ian. "The Battle of Tanagra: A Spartan Initiative?" *Historia* 43 (1994): 259-274.
Plutarch. *The Rise and Fall of Athens.* Trans. by Ian Scott-Kilvert. Harmondsworth, England: Penguin, 1960.
Rhodes, Peter J. "Thucydides on the Causes of the Peloponnesian War." *Hermes* 115 (1987): 154-165.
Roisman, Joseph. *The General Demosthenes and His Use of Military Surprise.* Historia Einzelschriften no. 78. Stuttgart: F. Steiner, 1993.
Sealey, Raphael. "The Causes of the Peloponnesian War." *CPh* 70 (1975): 89-109.
Spence, Iain G. "Athenian Cavalry Numbers in the Peloponnesian War: *IG* I^3 375 Revisited." *Zeitschrift für Papyrologie und Epigraphik* 67 (1987): 167-175.

————. "Perikles and the Defence of Attika during the Peloponnesian War." *JHS* 110 (1990): 91-109.

Strassler, Robert B. "The Opening of the Pylos Campaign." *JHS* 110 (1990): 110-125.

Thucydides. *History of the Peloponnesian War*. Trans. by Rex Warner. Harmondsworth, England: Penguin, 1972.

Westlake, Henry D. "Seaborne Raids in Periclean Strategy." *CQ* 39 (1945): 75-84.

Wilson, John B. *Athens and Corcyra: Strategy and Tactics in the Peloponnesian War*. Bristol, England: Bristol Classical Press, 1987.

Xenophon. *A History of My Times*. Trans. by Rex Warner. Harmondsworth, England: Penguin, 1979.

Fourth-Century Warfare

Anderson, John K. *Military Theory and Practice in the Age of Xenophon*. Berkeley: University of California Press, 1970.

Buck, Robert J. "The Athenians at Thebes in 379/8 B.C." *AHB* 6 (1992): 103-109.

Buckler, John. *The Theban Hegemony, 371-362 B.C.* Cambridge, Mass.: Harvard University Press, 1980.

Cargill, Jack. *The Second Athenian League: Empire or Free Alliance?* Berkeley: University of California Press, 1981.

Cartledge, Paul. *Agesilaos and the Crisis of Sparta*. London: Duckworth, 1987.

Caven, Brian. *Dionysius I, Warlord of Sicily*. New Haven, Conn.: Yale University Press, 1990.

Cawkwell, George L. "The Common Peace of 366/5." CQ^2 11 (1961): 80-86.

————. "The Foundation of the Second Athenian Confederacy." CQ^2 23 (1973): 47-59.

Cook, Margaret L. "Ismenias' Goals in the Corinthian War." In *Essays in the Topography, History, and Culture of Boiotia*, edited by Albert Schachter, 57-63. Montreal: McGill University Press, 1990.

Demosthenes and Aeschines. Trans. by Arnold N. W. Saunders. Harmondsworth, England: Penguin, 1975.

Diodorus of Sicily. Vols. 6 (trans. by Charles H. Oldfather), 7 (trans. by Charles L. Sherman), 8 (trans. by C. Bradford Welles), 9 and 10 (trans. by Russel M. Geer), 1954, 1952, 1963, 1947, and 1954.

Greek Political Oratory. Trans. by Arnold N. W. Saunders. Harmondsworth, England: Penguin, 1985.

Hamilton, Charles D. *Agesilaus and the Failure of Spartan Hegemony*. Ithaca, N.Y.: Cornell University Press, 1991.

———. "Sparta." In *The Greek World in the Fourth Century*, edited by Lawrence A. Tritle, 41-65. London: Routledge, 1997.

———. *Sparta's Bitter Victories: Politics and Diplomacy in the Corinthian War*. Ithaca, N.Y.: Cornell University Press, 1979.

Heskel, Julia. "Macedonia and the North, 400-336." In *The Greek World in the Fourth Century*, edited by Lawrence A. Tritle, 167-188. London: Routledge, 1997.

Munn, Mark. *The Dema Wall and the Boeotian War of 378-375 B.C.* Berkeley: University of California Press, 1993.

———. "Thebes and Central Greece." In *The Greek World in the Fourth Century*, edited by Lawrence A. Tritle, 66-106. London: Routledge, 1997.

Plutarch. *The Age of Alexander*. Trans. by Ian Scott-Kilvert. Harmondsworth, England: Penguin, 1973.

Rahe, Paul A. "The Military Situation in Western Asia on the Eve of Cunaxa." *AJPh* 101 (1980): 79-96.

Ruzicka, Stephen. "Athens and the Politics of the Eastern Mediterranean in the Fourth Century B.C." *AncW* 23 (1992): 63-70.

———. "The Eastern Greek World." In *The Greek World in the Fourth Century*, edited by Lawrence A. Tritle, 107-136. London: Routledge, 1997.

Schwenck, Cynthia. "Athens." In *The Greek World in the Fourth Century*, edited by Lawrence A. Tritle, 8-40. London: Routledge, 1997.

Seager, Robin. "The King's Peace and the Balance of Power in Greece, 386-62." *Athenaeum* 52 (1974): 36-63.

Sealey, Raphael. *Demosthenes and His Time: A Study in Defeat*. New York: Oxford University Press, 1993.

Talbert, Richard J. A. "The Greeks in Sicily and South Italy." In *The Greek World in the Fourth Century*, edited by Lawrence A. Tritle, 137-165. London: Routledge, 1997.

Tritle, Lawrence A. *Phocion the Good*. London: Croom Helm, 1988.

———. ed. *The Greek World in the Fourth Century*. London: Routledge, 1997.

Westlake, Henry D. "Spartan Intervention in Asia, 400-397 B.C." *Historia* 35 (1986): 405-426.

———. *Thessaly in the Fourth Century B.C.* London: Methuen, 1935.

Winter, Frederick E. "The Use of Artillery in Fourth-Century and Hellenistic Times." *EMC* 16 (1997): 247-292.

Xenophon. *A History of My Times*. Trans. by Rex Warner. Harmondsworth, England: Penguin, 1979.
———. *The Persian Expedition*. Trans. by Rex Warner. Harmondsworth, England: Penguin, 1972.

Philip II and Alexander of Macedon

Adams, W. Lindsay, and Eugene N. Borza, eds. *Philip II, Alexander the Great, and the Macedonian Heritage*. Washington, D.C.: University Press of America, 1982.
Aperghis, Georges G. "Alexander's Hipparchies." *AncW* 28 (1997): 133-148.
Arrian. *The Campaigns of Alexander*. Trans. by Aubrey de Sélincourt. Harmondsworth, England: Penguin, 1971.
Badian, Ernst. "Agis III: Revisions and Reflections." In *Ventures into Greek History*, edited by Ian Worthington, 258-292. Oxford: Clarendon, 1994.
Baynham, Elizabeth, J. "Antipater: Manager of Kings." In *Ventures into Greek History*, edited by Ian Worthington, 331-356. Oxford: Clarendon, 1994.
Bosworth, A. Brian. *Conquest and Empire: The Reign of Alexander the Great*. Cambridge: Cambridge University Press, 1988.
———. "A Cut Too Many? Occam's Razor and Alexander's Footguard." *AHB* 11 (1997): 47-56.
Buckler, John. *Philip II and the Sacred War*. New York: E. J. Brill, 1989.
Burn, Andrew R. "The Generalship of Alexander." *G&R* 12 (1965): 140-154.
Cawkwell, George L. "Philip and Athens." In *Philip of Macedon*, edited by Miltiades B. Hatzopoulos and Louisa D. Hatzopoulos, 100-110. Athens: Ekdotike Athenon, 1980.
———. "Philip and the Amphictyonic League." In *Philip of Macedon*, edited by Miltiades B. Hatzopoulos and Louisa D. Hatzopoulos, 78-89. Athens: Ekdotike Athenon, 1980.
———. *Philip of Macedon*. London: Faber, 1978.
Curtius Rufus, Quintus. *The History of Alexander*. Trans. by John Yardley. Harmondsworth, England: Penguin, 1984.
Daniel, Thomas. "The Taxeis of Alexander and the Change to Chiliarch, the Companion Cavalry and the Change to Hipparchies: A Brief Assessment." *AncW* 23 (1992): 43-57.

Demosthenes and Aeschines. Trans. by Arnold N. W. Saunders. Harmondsworth, England: Penguin, 1975.

Devine, Albert M. "The Battle of Gaugamela: A Tactical and Source-Critical Study." *AncW* 13 (1986): 87-116.

————. "The Battle of the Hydaspes: A Tactical and Source-Critical Study." *AncW* 16 (1987): 91-113.

————. "The Strategies of Alexander the Great and Darius III in the Issus Campaign (333 B.C.)." *AncW* 12 (1985): 25-38.

Diodorus of Sicily. Vols. 7 (trans. by Charles L. Sherman), 8 (trans. by C. Bradford Welles). London: Heinemann (Loeb Classical Library), 1952 and 1963.

Ellis, Jack. R. *Philip II and Macedonian Imperialism*. London: Thames and Hudson, 1976.

Engels, Donald W. "Alexander's Intelligence System." *CQ²* 30 (1980): 327-340.

————. *Alexander the Great and the Logistics of the Macedonian Army*. Berkeley: University of California Press, 1978.

Fuller, John F. C. *The Generalship of Alexander the Great*. London: Eyre and Spottiswoode, 1958.

Greek Political Oratory. Trans. by Arnold N. W. Saunders. Harmondsworth, England: Penguin, 1985.

Green, Peter. *Alexander of Macedon, 356-323 B.C.: A Historical Biography*. Berkeley: University of California Press, 1991.

————. *Alexander the Great*. London: Weidenfeld and Nicolson, 1970.

Griffith, Guy T. "Philip as a General and the Macedonian Army." In *Philip of Macedon*, edited by Miltiades B. Hatzopoulos and Louisa D. Hatzopoulos, 58-77. Athens: Ekdotike Athenon, 1980.

Hamilton, James R. *Alexander the Great*. Pittsburgh: University of Pittsburgh Press, 1982.

Hammond, Nicholas G. L. *Alexander the Great: King, Commander, and Statesman*. 3rd ed. Bristol: Bristol Classical Press, 1996.

————. "Cavalry Recruited in Macedonia down to 322 B.C." *Historia* 47 (1998): 404-425.

————. "The Macedonian Navies of Philip and Alexander until 330 B.C." *Antichthon* 26 (1992): 30-41.

————. *Philip II of Macedon*. Baltimore: Johns Hopkins University Press, 1994.

————. "Training in the Use of a *Sarissa* and Its Effect in Battle, 359-333 B.C." *Antichthon* 14 (1980): 53-63.

Hatzopoulos, Miltiades B., and Hatzopoulos Louisa D., eds. *Philip of Macedon*. Athens: Ekdotike Athenon, 1980.

Heckel, Waldemar. "Alexander at the Persian Gates." *Athenaeum* 58 (1980): 168-174.

———. *The Marshals of Alexander's Empire.* London: Routledge, 1992.

———. "Resistance to Alexander the Great." In *The Greek World in the Fourth Century,* edited by Lawrence A. Tritle, 189-227. London: Routledge, 1997.

Keegan, John. *The Mask of Command.* Harmondsworth, England: Penguin, 1988.

Kennett, W. Harl. "Alexander's Cavalry Battle at the Granicus." In *Polis and Polemos: Essays on Politics, War, and History in Ancient Greece in Honor of Donald Kagan,* edited by Charles D. Hamilton and Peter Krentz, 303-326. Claremont, Calif.: Regina, 1997.

Keyscr, Paul T. "The Use of Artillery by Philip II and Alexander the Great." *AncW* 25 (1994): 27-59.

Manti, Peter A. "The Sarissa of the Macedonian Infantry." *AncW* 23 (1992): 31-42.

Markle, Minor M. "Macedonian Arms and Tactics under Alexander." In *Macedonia and Greece in Late Classical and Early Hellenistic Times,* edited by Beryl Barr-Sharrar and Eugene N. Borza, 87-111. Washington, D.C.: National Gallery of Art, 1982.

———. "The Macedonian Sarissa, Spear, and Related Armor." *AJA* 81 (1977): 323-339.

———. "The Strategy of Philip in 346 B.C." *CQ²* 24 (1974): 253-268.

———. "Use of the Sarissa by Philip and Alexander of Macedon." *AJA* 82 (1978): 483-497.

Marsden, Eric W. *The Campaign of Gaugamela.* Liverpool, England: Liverpool University Press, 1964.

Milns, Robert D. *Alexander the Great.* London: Robert Hale, 1968.

Mixter, John R. "The Length of the Macedonian Sarissa during the Reigns of Philip II and Alexander the Great." *AncW* 23 (1992): 21-29.

Plutarch. *The Age of Alexander.* Trans. by Ian Scott-Kilvert. Harmondsworth, England: Penguin, 1973.

Pritchett, William K. "Observations on Chaironeia." *AJA* 62 (1958): 307-311.

Rahe, Paul A. "The Annihilation of the Sacred Band at Chaeronea." *AJA* 85 (1981): 84-87.

Romane, Julian P. "Alexander's Siege of Gaza, 332 B.C." *AncW* 18 (1988): 21-30.

———. "Alexander's Siege of Tyre." *AncW* 16 (1987): 79-90.

Sekunda, Nicholas, V. *The Persian Army, 560-330 B.C.* London: Osprey, 1992.

Sekunda, Nicholas V., and John Warry. *Alexander the Great: The Armies and Campaigns, 334-323 B.C.* London: Osprey, 1998.

Wilcken, Ulrich. *Alexander the Great.* New York: W. W. Norton, 1967.

Hellenistic Warfare

Adams, W. Lindsay T. "Antipater and Cassander: Generalship on Restricted Resources in the Fourth Century." *AncW* 10 (1984): 79-88.

————. "The Successors of Alexander." In *The Greek World in the Fourth Century*, edited by Lawrence A. Tritle, 228-248. London: Routledge, 1997.

Austin, Michel, M. "Hellenistic Kings, War, and the Economy." *CQ* 36 (1986): 450-466.

Bar-Kochva, Bezalel. *The Seleucid Army: Organization and Tactics in the Great Campaigns.* Cambridge: Cambridge University Press, 1976.

Billows, Richard A. *Antigonos the One-Eyed and the Creation of the Hellenistic State.* Berkeley: University of California Press, 1990.

Devine, Albert M. "Diodorus' Account of the Battle of Gabiene." *AncW* 12 (1985): 87-96.

————. "Diodorus' Account of the Battle of Gaza." *Acta Classica* 27 (1984): 31-40.

————. "Diodorus' Account of the Battle of Paraitacene." *AncW* 12 (1985): 75-86.

Diodorus of Sicily. Vols. 11 and 12. Trans. by Francis R. Walton. London: Heinemann (Loeb Classical Library), 1957.

Errington, R. Malcolm. *Philopoemen.* Oxford: Oxford Press, 1969.

Gabbert, Janice J. *Antigonus II Gonatas: A Political Biography.* London: Routledge, 1997.

Green, Peter. *Alexander to Actium: The Historical Evolution of the Hellenistic Age.* Berkeley: University of California Press, 1990.

Griffith, Guy T. *The Mercenaries of the Hellenistic World.* Cambridge: Cambridge University Press, 1935.

Heckel, Waldemar. *The Marshals of Alexander's Empire.* London: Routledge, 1992.

Lévêque, Pierre. "La Guerre à l'époque hellénistique." In *Problèmes de la guerre en Grèce ancienne, à la mémoire d'André Aymard*, edited by Jean-Pierre Vernant, 261-287. Paris: Mouton, 1968.

Lund, Helen S. *Lysimachus: A Study in Early Hellenistic Kingship.* London: Routledge, 1992.

Nachtergael, Georges. *Les Galates en Grèce et les Sôtéria de Delphes.* Brussels: Académie Royale de Belgique, 1975.

Plutarch. *The Age of Alexander.* Trans. by Ian Scott-Kilvert. Harmondsworth, England: Penguin, 1973.

Polybius. *The Rise of the Roman Empire.* Trans. by Ian Scott-Kilvert. Harmondsworth, England: Penguin, 1979.

Tarn, William W. *Antigonus Gonatas.* Oxford: Clarendon, 1913.

———. *Hellenistic Military and Naval Developments.* 2nd ed. Chicago: Ares, 1984.

Urban, Ralf. "Das Heer des Kleomenes bei Sellasia." *Chiron* 3 (1973): 95-102.

Wheatley, Pat. "Young Demetrius Poliorcetes." *AHB* 13 (1999): 1-13.

Winter, Frederick E. "The Use of Artillery in Fourth-Century and Hellenistic Times." *EMC* 16 (1997): 247-292.

Roman Conquest of Greece and the Greek East

Badian, Ernst. "Rome and Antiochus the Great: A Study in Cold War." In *Studies in Greek and Roman History,* edited by Ernst Badian, 112-139. Oxford: Blackwell, 1968. (Reprinted from *Classical Philology* 54 (1959): 81-99.)

Diodorus of Sicily. Vols. 11 and 12. Trans. by Francis R. Walton. London: Heinemann (Loeb Classical Library), 1957.

Errington, R. Malcolm. *The Dawn of Empire: Rome's Rise to World Power.* Ithaca, N.Y.: Cornell University Press, 1972.

Green, Peter. *Alexander to Actium: The Historical Evolution of the Hellenistic Age.* Berkeley: University of California Press, 1990.

Kallet-Marx, Robert M. *Hegemony to Empire: The Development of Roman Imperium in the East from 148-62 B.C.* Berkeley: University of California Press, 1995.

Livy. *Rome and the Mediterranean.* Trans. by Henry Bettenson. Harmondsworth, England: Penguin, 1976.

———. *The War with Hannibal.* Trans. by Aubrey de Sélincourt. Harmondsworth, England: Penguin, 1972.

Lomas, Kathryn. *Rome and the Western Greeks, 350 B.C.-A.D. 200.* London: Routledge, 1993.

Polybius. *The Rise of the Roman Empire.* Trans. by Ian Scott-Kilvert. Harmondsworth, England: Penguin, 1979.

Sherwin-White, Adrian N. *Roman Foreign Policy in the East 168 B.C. to A.D. 1.* London: Duckworth, 1984.

ABOUT THE AUTHOR

Iain Spence (RFD; B.A. Hons, University of New England; PhD, University College, London; Diploma of Education, University of New England; Graduate Diploma of Strategic Studies, Joint Services Staff College; jssc; Fellow of the Royal Historical Society) is a senior lecturer in the School of Classics, History and Religion at the University of New England, Armidale, N.S.W., Australia. His main area of expertise is in ancient Greek military and social history, concentrating particularly on the place of cavalry in the warfare of the Classical period. He is a member of the editorial boards of *Antichthon*, the journal of the Australian Society of Classical Studies, and *Ancient History: Resources for Teachers*. He was elected a fellow of the Royal Historical Society in 1998. In addition to articles in scholarly journals he is the author of *The Cavalry of Classical Greece* (1993). He also holds the rank of Colonel in the Australian Army Reserve. Following postings as Commanding Officer of the 12th/16th Hunter River Lancers (APC) and as an instructor at the Australian Army Command and Staff College, he is currently the Commandant of the Regional Training Centre (N.S.W.).